D1641661

QUANTUM DYNAMICAL SYSTEMS

Quantum Dynamical Systems

ROBERT ALICKI

Institute of Theoretical Physics and Astrophysics,
University of Gdańsk, Poland

and

MARK FANNES

Instituut voor Theoretische Fysica, K.U. Leuven,
Belgium

OXFORD

UNIVERSITY PRESS

This book has been printed digitally and produced in a standard specification
in order to ensure its continuing availability

OXFORD
UNIVERSITY PRESS

Great Clarendon Street, Oxford OX2 6DP

Oxford University Press is a department of the University of Oxford.
It furthers the University's objective of excellence in research, scholarship,
and education by publishing worldwide in

Oxford New York

Auckland Cape Town Dar es Salaam Hong Kong Karachi
Kuala Lumpur Madrid Melbourne Mexico City Nairobi
New Delhi Shanghai Taipei Toronto
With offices in
Argentina Austria Brazil Chile Czech Republic France Greece
Guatemala Hungary Italy Japan South Korea Poland Portugal
Singapore Switzerland Thailand Turkey Ukraine Vietnam

Oxford is a registered trade mark of Oxford University Press
in the UK and in certain other countries

Published in the United States
by Oxford University Press Inc., New York

ISBN 978-0-19-850400-9

To our families
Majka and Monika
Rita, Sofie, Geert, Katrien and Thomas

PREFACE

In the 1930s, J. von Neumann (1932) developed in his book *Mathematische Grundlagen der Quantenmechanik* a precise and self-contained description of the quantum mechanics of a finite collection of particles. This work synthesized the contributions of the founding fathers such as E. Schrödinger, W. Heisenberg and P. A. M. Dirac and it provided a rigorous mathematical framework for the theory. Quantum systems are described in terms of observables that act as self-adjoint operators on the Hilbert space of the system and the expectation functionals are given in terms of density matrices. The dynamics is determined by a self-adjoint Hamiltonian. He also introduced the 'von Neumann' entropy of a density matrix as a measure of the indeterminacy of a quantum state. In explicit models, the time evolution of quantum systems was very hard to investigate. One was restricted to perturbative or semiclassical approaches around integrable systems. Nowadays, the computer offers completely new possibilities for numerical investigations of complex dynamical behaviour.

In the 1960s and 1970s, new insights in the behaviour of large quantum systems led to a rigorous description of such systems in algebraic terms. The observables of the system form an abstract C*-algebra. The positive, normalized functionals on the algebra are the expectation functionals and the dynamics is given in terms of a group of automorphisms. Each stationary state determines uniquely a representation of the abstract system in terms of the quantum mechanical Hilbert space picture. A new phenomenon now appears: often the physically relevant states turn out to be disjoint when some overall parameters, such as temperature or order parameters in equilibrium states, are changed. This means that there is no transformation linking the representations of different states. This is the basis for understanding divergences in perturbative computations and many phenomena in phase transitions. A lot of ideas and techniques were borrowed from quantum field theory and from the rigorous probabilistic approach in classical statistical mechanics. The ergodic theory of infinite quantum systems started along similar lines as that of classical systems. Some notions, such as stochastic processes, differentiable dynamics on manifolds, Lyapunov exponents and dynamical entropy had no obvious quantum counterparts.

Around the same time, the quantum theory of open systems and measurements was established. The main source of inspiration was the theory of lasers in quantum optics. Models describing the interaction of a small system weakly coupled to a heat bath were studied. This led to many new interesting objects such as quantum master equations, quantum stochastic processes, quantum noises, quantum stochastic differential equations and quantum communication channels. On the mathematical level, this initiated the study of semigroups of completely pos-

itive maps and their generators. The same mathematical techniques are involved
in the theory of general measurements and continuous observations of quantum
dynamical systems.

In the 1980s, numerical experiments provided evidence for interesting complex phenomena in quantum model systems with classical chaotic counterparts
such as the peculiar statistics of the eigenvalue distribution and the relation
with the canonical ensembles of random matrices. One also observed an ergodic
behaviour of the energy eigenfunctions and the so-called quantum diffusion of
observables. Experimentally, chaotic effects have been observed in non-linear microscopic systems, e.g. the erratic response of some complex molecules to periodic pulses, the Rydberg atoms in strong fields, and the dynamics of periodically
pulsed spins.

Deeper insight into non-commutative algebraic structures is a main goal of
the non-commutative geometry of A. Connes (1994) developed in the 1990s. One
can expect a strong influence of these ideas on the theory of quantum dynamical
systems in the same way as differential geometry proved very useful in classical
mechanics.

We feel that the time has come to unify a number of these aspects and
techniques. We present a general framework for describing quantum dynamical
systems, both finite and infinite, conservative and dissipative and illustrate this
description with many explicit examples. Classical systems are included in the
scheme as the special class of commutative dynamical systems. Along the road,
we provide the necessary mathematical tools which come mainly from functional
analysis and operator algebras. Next, we study dynamical systems from a statistical point of view by constructing symbolic models of the system. We argue that
the non-commutative shifts on spin chains are appropriate models. They are the
natural counterpart of the classical shift dynamics on symbol sequences and are
the key to the introduction of dynamical entropy. The actual construction can
be interpreted as repeated measurements of the system during its evolution. It
involves completely positive maps and leads to a quantum stochastic process.
This formalism is then used to study the ergodic properties of infinite quantum dynamical systems and chaos in finite systems. Although some numerical
examples are presented, we do not aim at overviewing this vast subject in any
way. For particular models, the ideas from non-commutative geometry provide
characterizations of ergodic properties of quantum dynamical systems such as
the relation between quantum Lyapunov exponents and dynamical entropy.

We try to present the topics we deal with in a mathematically rigorous but
friendly way, whenever possible. Special care is taken to motivate the mathematical machinery by introductory examples and demands from the underlying
physics. The arguments that we present in numerous examples are often sketchy
and these examples can therefore be regarded as exercises.

This book addresses graduate students in physics and mathematics, theoretical and mathematical physicists with interest in the mathematical aspects of
quantum physics and mathematicians interested in applications of ergodic the-

ory, operator algebras and statistics to physics. We assume as prior knowledge a standard undergraduate course in quantum mechanics and basic knowledge on probability theory, measure theory, and linear operators on Hilbert spaces. We include a number of references and brief comments at the end of most chapters in the form of notes. Our only aim is to provide some help for the interested reader and not at all to give a comprehensive overview of the literature.

ACKNOWLEDGEMENTS

It is a pleasure to thank W. Miklaszewski and P. Spincemaille for their kind assistance with the computation and realization of the figures. The writing of this book would have been impossible without a good deal of back and forth travelling of the authors between their home institutes. We wish to thank the University of Gdańsk and the K. U. Leuven for funding and hospitality. More thanks for funding are also due to the *Komitet Badań Naukowych*, grant KBN 2PO3B 042 16, and to the *Fonds voor Wetenschappelijk Onderzoek—Vlaanderen*.

CONTENTS

1

INTRODUCTION

Ask a physicist to write down a single formula which embraces the very essence of quantum mechanics, then with a large probability—*probability* is an inevitable word in quantum mechanics—you will see

$$QP - PQ = i\hbar. \tag{1.1}$$

The physicist will tell you that Q and P are mathematical objects representing position and momentum of the physical system and can be seen as infinite Hermitian matrices. The product in formula (1.1) is a matrix product and the non-commutation of Q and P reflects a fundamental feature of the microworld most strikingly expressed by Heisenberg's uncertainty relations.

In all theories of physical systems, one has to know the dynamical behaviour, e.g. in the form of an evolution equation for the observables of the system. In our case, an observable X is represented as a formal function, say a polynomial, in the non-commuting variables Q and P. The evolution equation is another icon of quantum mechanics:

$$\frac{dX(t)}{dt} = \frac{i}{\hbar} \left(HX(t) - X(t)H \right), \tag{1.2}$$

where H is the very distinguished *energy observable called Hamiltonian*. Using only formal manipulations on matrices, we can easily guess the solution of eqn (1.2):

$$X(t) = U(t)XU(t)^{-1}, \quad t \in \mathbb{R}$$

with $U(t)$ a unitary matrix, formally given by

$$U(t) := \exp\left(\frac{it}{\hbar}H\right) = \sum_{n=0}^{\infty} \frac{(itH)^n}{n!}.$$

The final step in the construction of quantum theory is to produce an expression for the mean value of an arbitrary observable X. Such a prescription defines a *state of the quantum system*. We do not give here the mathematical details, but we assume for the moment that states are linear, positive and normalized functionals on the algebra of observables.

The simple choice

$$H = \frac{1}{2}(P^2 + Q^2)$$

together with (1.1) leads to the time evolution

$$Q(t) = Q \cos t - P \sin t \quad \text{and} \quad P(t) = Q \sin t + P \cos t \tag{1.3}$$

of Q and P. This justifies the name *quantum harmonic oscillator* for this system.

Although quite a good deal of theoretical physics can be done using systems of harmonic oscillators, we need more for our purposes. As we want to build the theory of quantum dynamical systems, we need models with more sophisticated dynamical behaviour than that given by (1.3).

This 'more' can already be achieved by imposing less restrictions on the commutation relations (1.1), relaxing them to a discrete version—'Be wise, discretise!' M. Kac—

$$UV = e^{i\theta} VU, \tag{1.4}$$

with $\theta \in [0, 2\pi[$. To see the relation with (1.1), calculate formally the commutation relation between $U(\alpha) := \exp(i\alpha Q)$ and $V(\beta) := \exp(-i\beta P)$ and choose $\theta = \hbar\alpha\beta \mod 2\pi$. An algebra of observables $\mathfrak{A}_\theta$ can now be constructed by taking functions, e.g. polynomials, of the non-commuting variables U and V. In particular $\mathfrak{A}_\theta$ is spanned by elements of the form

$$W(k_1, k_2) := e^{-i\theta k_1 k_2} U^{k_1} V^{k_2}.$$

In order to define the dynamics of our system, we follow again M. Kac's wisdom and introduce discrete time steps $n = \ldots, -1, 0, 1, \ldots$ instead of a continuous parameter $t \in \mathbb{R}$. The proposed dynamics is not only a linear homomorphism Θ of the algebra $\mathfrak{A}_\theta$, but, similarly to that given by (1.3), it is linear in the sense that

$$\begin{aligned} \Theta\big(W(k_1, k_2)\big) &= W(k_1', k_2') \quad \text{with} \\ k_1' &= t_{11}k_1 + t_{21}k_2 \quad \text{and} \quad k_2' = t_{12}k_1 + t_{22}k_2. \end{aligned} \tag{1.5}$$

The matrix $T := [t_{\mu,\nu}]$ has integer entries and $\det T = 1$ in order to preserve the commutation relation (1.4) and hence the products of observables in $\mathfrak{A}_\theta$.

The structure of $\mathfrak{A}_\theta$ depends heavily on the parameter θ. If $\theta = 0$, then all observables commute and $\mathfrak{A}_0$ is isomorphic to the algebra of functions on the two-dimensional torus $\mathbb{T}^2$. In this case, the dynamical system given by (1.5) is the famous Arnold cat—a beloved animal of classical dynamical systems theory, which possesses all nice ergodic properties. For a rational multiple p/q of π, $\mathfrak{A}_\theta$ is a finite-dimensional matrix algebra with functions on $\mathbb{T}^2$ as entries. Together with the periodicity condition $W(k_1, k_2)^q = \mathbf{1}$, it corresponds to the *canonical quantization of the Arnold cat*. These models are useful in the theory of *quantum chaos* where one wishes to trace remnants of classical ergodic behaviour surviving quantization. Mathematically, the most interesting case arises for irrational multiples of π for which one speaks about *quantum deformation of the torus*. The infinite-dimensional algebra $\mathfrak{A}_\theta$ is now called *the irrational rotation algebra* and it provides a canonical example of quantum geometry. The corresponding dynamical system given by (1.5) is the θ-deformed Arnold cat and will be extensively used throughout this book as a paradigm for ideas ranging from non-commutative ergodicity through quantum dynamical entropy to quantum Lyapunov exponents.

This simple model introduces us into the exciting and challenging new world of non-commutative mathematics, part of which is the theory of quantum dynamical systems.

2

BASIC TOOLS FOR QUANTUM MECHANICS

There are different ways of introducing the mathematical formalism of quantum mechanics, perhaps the most elegant one is the axiomatic road. Nevertheless, for practical reasons we follow a shortcut often preferred by physicists which leads us from heuristic ideas supported by firm experimental results to a rigorous formalism of operators acting on Hilbert spaces. An enormous body of experimental facts concerning microscopic systems can be summarized in a few postulates.

Superposition principle   The pure states of quantum mechanics, which may be identified with elementary events, satisfy the *superposition principle*: if ψ and φ are pure states, then their linear combination with complex coefficients $\alpha\psi + \beta\varphi$ defines a new pure state of the system. This feature justifies another name for quantum mechanics: *wave mechanics*. It is supported by numerous experiments on diffraction and interference of matter waves. Mathematically, this postulate introduces a *complex linear space* $\mathfrak{H}$ as a fundamental object.

Probability amplitude   Having two states, an initial one ψ prepared by a certain device and a final one φ selected by a measuring apparatus, one can calculate a complex *transition probability amplitude* $\langle \psi, \varphi \rangle$ which is bilinear, strictly speaking sesquilinear, in its arguments to comply with the superposition principle. The probability amplitude gives the probability of finding a system in a final state φ if it is prepared in an initial state ψ:

$$\text{Prob}\{\psi \mapsto \varphi\} := |\langle \psi, \varphi \rangle|^2.$$

As $\text{Prob}\{\psi \mapsto \psi\}$ must be equal to 1, we have to use normalized vectors ψ as states, i.e., we must always have $\langle \psi, \psi \rangle = 1$. This postulate equips our linear space $\mathfrak{H}$ with an inner product $\langle \cdot, \cdot \rangle$ and therefore endows it with a *Hilbert space structure*.

Observables   An *observable* A of the system can be constructed in an operational sense as a sequence of normalized states $\{\varphi_k \mid k = 1, 2, \dots\}$ which can be selected by a measuring apparatus and corresponding values $\{\lambda_k \mid k = 1, 2, \dots\}$ of the observable. In agreement with the previous postulate, the average value of the observable A in a state ψ is given by

$$\langle A \rangle_\psi = \sum_k \lambda_k |\langle \psi, \varphi_k \rangle|^2 = \langle \psi, A\psi \rangle,$$

where we identify our observable with the linear operator A

$$A\,\psi := \sum_k \lambda_k \,\langle \varphi_k, \psi \rangle \, \varphi_k$$

acting on $\mathfrak{H}$. Slightly generalizing this construction, we may postulate that quantum observables are represented by *linear self-adjoint operators on $\mathfrak{H}$*.

Symmetries and dynamics   A special role is played by linear operators U which preserve transition amplitudes, i.e. $\langle U\,\psi, U\,\varphi \rangle = \langle \psi, \varphi \rangle$ and therefore describe symmetries of the theory in terms of unitary operators on the Hilbert space. Continuous symmetry groups, including time translations, i.e. the time evolution, provide especially important observables A corresponding to infinitesimal transformations $U(\epsilon) = \mathbf{1} - i\epsilon A + \cdots$.

This roughly outlined mathematical formalism of quantum mechanics will be refined in the next sections which offer a concise introduction to the Hilbert space theory. We do not reproduce proofs of fundamental results because of the existence of numerous excellent textbooks dealing with the Hilbert space theory.

In a first section, we review basic notions in the theory of linear operators on Hilbert spaces, without entering in the mathematical details. This material furnishes the basic vocabulary that we need in the book. We illustrate the various notions by simple examples more or less directly connected to physics. The second section contains some notions from measure theory and introduces spectral measures. Section 2.3 introduces the probabilistic scheme that is adapted to the description of simple quantum systems. We obtain hereby a basic example of a quantum probability space, which is one of the fundamental notions under consideration. Section 2.4 deals with observables in quantum systems with few degrees of freedom. This algebraic approach will prove useful later on when we consider more abstract dynamical systems for which a treatment in terms of wave functions is less appropriate. We expand on this in later chapters. The last section shows how one constructs composed systems in order to study many-particle systems.

2.1   Hilbert spaces and operators

2.1.1   *Vector spaces*

Because of the superposition principle, complex linear spaces, also called complex vector spaces or simply vector spaces, are the fundamental structures needed to describe quantum systems. A *complex vector space* $\mathfrak{V}$ is a set equipped with two operations: multiplication of vectors by complex scalars $(\alpha, x) \mapsto \alpha x$ and addition of vectors $(x, y) \mapsto x + y$. These maps satisfy standard requirements such as

$$(x + y) + z = x + (y + z) =: x + y + z \quad \text{and} \quad \alpha(x + y) = \alpha x + \alpha y.$$

We denote complex scalars by the first letters of the Greek alphabet and elements of general vector spaces by lowercase letters as x, y and z.

Example 2.1 Basic examples of complex linear spaces are the spaces $\mathbb{C}^d$ of column vectors of height d with complex entries

$$\mathbb{C}^d = \left\{ \mathbf{x} = \begin{pmatrix} x_1 \\ x_2 \\ \vdots \\ x_d \end{pmatrix} \,\middle|\, x_j \in \mathbb{C} \right\}. \tag{2.1}$$

Scalar multiplication of $\mathbf{x}$ by α consists in multiplying each entry of $\mathbf{x}$ by α and two column vectors $\mathbf{x}$ and $\mathbf{y}$ are added by adding their respective entries: $(\mathbf{x} + \mathbf{y})_j = x_j + y_j$. We use the notation $\mathbf{e}_j$ to denote the particular column vectors with all entries zero except for the jth one which is equal to 1:

$$\mathbf{e}_j := \begin{pmatrix} 0 \\ \vdots \\ 1 \\ \vdots \\ 0 \end{pmatrix} \quad \leftarrow \quad j\text{th row.} \tag{2.2}$$

A general column vector $\mathbf{x}$ as in (2.1) can now be written as

$$\mathbf{x} = x_1 \mathbf{e}_1 + x_2 \mathbf{e}_2 + \cdots + x_d \mathbf{e}_d.$$

A subset S of a complex vector space $\mathfrak{V}$ is *linearly independent* if for any choice of x_1, x_2, ... , $x_n \in S$ and $n = 1, 2, \ldots$ the only way to have $\alpha_1 x_1 + \alpha_2 x_2 + \cdots + \alpha_n x_n = 0$ is to choose $\alpha_1 = \alpha_2 = \cdots = \alpha_n = 0$. The space $\mathfrak{V}$ is said to be *finite-dimensional* if each linearly independent set is finite. It is straightforward to show that any maximally independent subset S of a finite-dimensional vector space has the same number of elements, d. This number is the *dimension of the vector space* and each vector in $\mathfrak{V}$ can be written in a unique way as a linear combination of elements of S. Such a set is therefore called a *basis*. The space $\mathbb{C}^d$ of Example 2.1 has dimension d and its *standard basis* is given by the vectors $\{\mathbf{e}_1, \mathbf{e}_2, \ldots, \mathbf{e}_d\}$. If one can find within $\mathfrak{V}$ linearly independent sets of arbitrary size, then $\mathfrak{V}$ is said to be *infinite-dimensional*. A subset $\mathfrak{V}_0$ of $\mathfrak{V}$ that is itself a vector space is called a *linear subspace of $\mathfrak{V}$* or simply a *subspace*. The *span of a subset S of $\mathfrak{V}$* is the linear subspace generated by the elements of S, i.e. the set of all finite complex linear combinations of elements of S. It will be denoted by $\mathrm{Span}(S)$.

In the d-dimensional case, there is no convergence problem. Indeed, fixing a basis, each vector is determined by its d scalar coefficients with respect to the given basis. A sequence of vectors converges iff the sequence of coefficients converges. This notion of convergence does not depend on the choice of basis. In the infinite-dimensional case, however, the topological properties of the vector space are quite important. In this book, we mostly consider convergence defined in terms of a *norm* on the space, which attributes a length to each vector.

Example 2.2 Some common examples of norms on $\mathbb{C}^d$ are

$$\|\mathbf{x}\|_1 := |x_1| + |x_2| + \cdots + |x_d|,$$
$$\|\mathbf{x}\|_2 := \left(|x_1|^2 + |x_2|^2 + \cdots + |x_d|^2\right)^{1/2},$$
$$\|\mathbf{x}\|_\infty := \max\{|x_1|, |x_2|, \cdots, |x_d|\}.$$

The norm $\|\cdot\|_2$ has the special feature that it is connected to an *inner product*

$$(\mathbf{x}, \mathbf{y}) \mapsto \langle \mathbf{x}, \mathbf{y} \rangle := \overline{x_1} y_1 + \overline{x_2} y_2 + \cdots + \overline{x_d} y_d \qquad (2.3)$$

on $\mathbb{C}^d$ by the relation $\|\mathbf{x}\|^2 = \langle \mathbf{x}, \mathbf{x} \rangle$. The space $\mathbb{C}^d$, equipped with the inner product (2.3) is the basic example of a *Hilbert space*. The notion of inner product is quite useful as it introduces a rich and manageable geometrical structure on the space and on its linear transformations.

2.1.2 *Banach and Hilbert spaces*

Although quantum mechanics lives on a Hilbert space, more general structures are also relevant. We shall, for instance, consider spaces of linear operators on Hilbert spaces and these often do not possess a natural Hilbert space structure.

A *normed space* $\mathfrak{V}$ is a complex linear space equipped with a norm $\|\cdot\|$, i.e. a functional that satisfies

$$\|x\| \geq 0 \text{ and if } \|x\| = 0, \text{ then } x = 0,$$
$$\|\alpha x\| = |\alpha| \|x\|,$$
$$\|x + y\| \leq \|x\| + \|y\|$$

for all $x, y \in \mathfrak{V}$ and $\alpha \in \mathbb{C}$. Such a norm allows us to introduce the *open balls* of radius $r \geq 0$ and centre x as the set of vectors y that satisfy $\|x - y\| < r$. The *open sets* in $\mathfrak{V}$ are arbitrary unions of open balls. A sequence $\{x_1, x_2, \dots\}$ *converges to* $x \in \mathfrak{V}$ iff $\|x - x_n\| \to 0$. Every converging sequence is a *Cauchy sequence*, i.e. for any choice of $\epsilon > 0$, $\|x_m - x_n\| < \epsilon$ whenever both m and n are large enough. The converse is not true as some Cauchy sequences might fail to have a limit in $\mathfrak{V}$. A normed space $(\mathfrak{V}, \|\ \|)$ is a *Banach space* if it is *complete*, which means that every Cauchy sequence in $\mathfrak{V}$ converges (in $\mathfrak{V}$). We reserve the notation $\mathfrak{B}$ for general Banach spaces. Two norms $\|\cdot\|$ and $\|\cdot\|'$ are *equivalent* if there exists a $\gamma > 0$ such that for all vectors x

$$\frac{1}{\gamma} \|x\|' < \|x\| < \gamma \|x\|'.$$

Obviously, two equivalent norms determine the same topology. In the case of finite-dimensional spaces, there is only one notion of convergence and all norms are equivalent.

Example 2.3 A simple example of an infinite-dimensional space is the space $\ell^0(\mathbb{N}_0)$ of sequences of complex numbers indexed by $\mathbb{N}_0$ with only a finite number of entries different from 0. Several norms can be considered on $\ell^0(\mathbb{N}_0)$ and each norm yields a corresponding completion. In the standard examples that we present below, the completion is again a space of sequences but now with infinitely many entries different from zero. Let $1 \leq p$, then the Banach space $\ell^p(\mathbb{N}_0)$ consists of all sequences $\mathbf{x} = (x_1, x_2, \ldots)$ of complex numbers x_j such that

$$\|\mathbf{x}\|_p := \left(\sum_{j=1}^{\infty} |x_j|^p \right)^{1/p} < \infty.$$

The notation $\ell^\infty(\mathbb{N}_0)$ is used for the Banach space of all uniformly bounded sequences $\mathbf{x}$ indexed by $\mathbb{N}_0$ and

$$\|\mathbf{x}\|_\infty := \sup_{j \in \mathbb{N}_0} |x_j|.$$

A *Hilbert space* is a complex linear space $\mathfrak{H}$ that comes with a complex *inner product* $\langle \cdot, \cdot \rangle$ satisfying the requirements

$$\langle \psi, \alpha\varphi + \beta\eta \rangle = \alpha \langle \psi, \varphi \rangle + \beta \langle \psi, \eta \rangle,$$
$$\langle \psi, \varphi \rangle = \overline{\langle \varphi, \psi \rangle},$$
$$\langle \psi, \psi \rangle \geq 0 \text{ and if } \langle \psi, \psi \rangle = 0, \text{ then } \psi = 0$$

for all $\psi, \varphi, \eta \in \mathfrak{H}$ and $\alpha, \beta \in \mathbb{C}$. Moreover, the space $\mathfrak{H}$ is a Banach space with respect to the norm

$$\|\psi\| := \sqrt{\langle \psi, \psi \rangle}, \quad \psi \in \mathfrak{H}.$$

A Hilbert space is a special instance of a Banach space and we mostly use the symbols φ, ψ and η for vectors in a general Hilbert space.

Example 2.4 The algebra $\mathcal{M}_d$ of complex $d \times d$ matrices $A = [a_{jk}]$ with the inner product

$$\langle A, B \rangle := \operatorname{Tr} A^* B = \sum_{j,\,k=1}^{d} \overline{a_{jk}}\, b_{kj} \tag{2.4}$$

is a complex Hilbert space of dimension d^2. In (2.4), A^* denotes the Hermitian conjugate of the matrix A,

$$A^* := [a^*_{jk}] \quad \text{with } a^*_{jk} := \overline{a_{kj}}.$$

Example 2.5 The Banach space of Example 2.3 with $p = 2$ is a Hilbert space. The inner product between two square summable sequences $\mathbf{x}$ and $\mathbf{y}$ is given by

$$\langle \mathbf{x}, \mathbf{y} \rangle := \sum_{j=1}^{\infty} \overline{x_j}\, y_j.$$

Example 2.6 There is a continuous version of Example 2.5. We can start with the vector space of complex continuous functions f on $\mathbb{R}$ that are square integrable, i.e.

$$\int_{-\infty}^{\infty} dx\, |f(x)|^2 < \infty.$$

For two such functions f and g we can consider the inner product

$$\langle f, g \rangle := \int_{-\infty}^{\infty} dx\, \overline{f(x)}\, g(x).$$

The integral makes sense because $|\overline{f}g| \leq |f|^2 + |g^2|$. This space is, however, not complete. The completion, denoted by $\mathcal{L}^2(\mathbb{R})$, will be discussed in Section 2.2. It is not essential in this example to have complex functions on $\mathbb{R}$; in fact we could replace $\mathbb{R}$ by any reasonable subset of $\mathbb{R}^d$ or even by any so-called measurable space. This more general setting covers in fact both the spaces of square summable sequences and square integrable functions.

2.1.3 *Geometrical properties of Hilbert spaces*

The following properties characterize special geometric features of Hilbert spaces. Let ψ and φ denote arbitrary elements of a Hilbert space $\mathfrak{H}$.

Schwarz inequality

$$|\langle \psi, \varphi \rangle| \leq \|\psi\|\, \|\varphi\|. \tag{2.5}$$

Moreover, the equality holds iff φ and ψ are linearly dependent, i.e. iff there exist an $\alpha \in \mathbb{C}$ such that $\varphi = \alpha\psi$ or $\psi = 0$.

Parallelogram law

$$\|\psi + \varphi\|^2 + \|\psi - \varphi\|^2 = 2(\|\psi\|^2 + \|\varphi\|^2). \tag{2.6}$$

Polarization identity

$$\langle \psi, \varphi \rangle = \tfrac{1}{4}\Big(\|\psi + \varphi\|^2 - \|\psi - \varphi\|^2 + i\|\psi + i\varphi\|^2 - i\|\psi - i\varphi\|^2 \Big). \tag{2.7}$$

Norm If a norm $\|\cdot\|$ on a Banach space $\mathfrak{B}$ satisfies the parallelogram law (2.6), then $\mathfrak{B}$ is in fact a Hilbert space with inner product given by (2.7).

Elements ψ and φ of a Hilbert space $\mathfrak{H}$ are *orthogonal* if $\langle \psi, \varphi \rangle = 0$. A sequence $\{\varphi_j \mid j = 1, 2, \ldots\}$ is *orthonormal* if $\langle \varphi_k, \varphi_l \rangle = \delta_{kl}$. For orthonormal sequences $\{\varphi_j \mid j = 1, 2, \ldots\}$, we have *Bessel's inequality*

$$\|\psi\|^2 \geq \sum_{j=1}^{\infty} |\langle \varphi_j, \psi \rangle|^2. \tag{2.8}$$

The volume $\text{Vol}(\{\varphi_1, \varphi_2, \ldots, \varphi_n\})$ of a parallelotope spanned by n vectors $\{\varphi_1, \varphi_2, \ldots, \varphi_n\}$ can be expressed in terms of the inner products between these vectors

$$\text{Vol}(\{\varphi_1, \varphi_2, \ldots, \varphi_n\})^2 = \det([\langle \varphi_j, \varphi_k \rangle]).$$

The $n \times n$ matrix $[\langle \varphi_j, \varphi_k \rangle]$ is also known as the *Gram matrix* corresponding to $\{\varphi_1, \varphi_2, \ldots, \varphi_n\}$. One can easily see that the n vectors φ_j are linearly independent iff their Gram matrix is invertible, i.e. iff its determinant, and hence the volume of the parallelotope spanned by the vectors, is non-zero.

2.1.4   *Orthonormal bases*

For Hilbert space computations, we often need to consider *orthonormal bases*. These are maximal orthonormal sets in the sense that there is no normalized vector in $\mathfrak{H}$ orthogonal to each of the elements of the basis. An orthonormal set $\{e_j \mid j = 1, 2, \ldots\}$ is a basis iff it satisfies one of the following equivalent requirements

Fourier expansion   For any $\psi \in \mathfrak{H}$, the following expansion holds

$$\psi = \sum_{j=1}^{\infty} \langle e_j, \psi \rangle \, e_j. \tag{2.9}$$

Equation (2.9) states that every vector $\varphi \in \mathfrak{H}$ can be developed in the basis $\{e_1, e_2, \ldots\}$. In particular, it implies that the linear span of $\{e_1, e_2, \ldots\}$ is dense in $\mathfrak{H}$. The $\langle e_j, \psi \rangle$ are the *generalized Fourier coefficients of* ψ.

Parseval equality   For every $\psi, \eta \in \mathfrak{H}$

$$\langle \psi, \eta \rangle = \sum_{j=1}^{\infty} \langle \psi, e_j \rangle \, \langle e_j, \eta \rangle. \tag{2.10}$$

Parseval equality   For any $\psi \in \mathfrak{H}$

$$\|\psi\|^2 = \sum_{j=1}^{\infty} |\langle \psi, e_j \rangle|^2. \tag{2.11}$$

This equality is a particular case of (2.10), putting $\eta = \psi$. Furthermore Bessel's inequality (2.8) now becomes an equality.

In the presentation of above, we have restricted ourselves to *separable* Hilbert spaces, i.e. to Hilbert spaces that contain a countable dense subset such as the complex rational combinations of a basis $\{e_1, e_2, \ldots\}$. The notions of orthonormal set and orthonormal basis can be extended to arbitrary Hilbert spaces. One can show that every Hilbert space $\mathfrak{H}$ has an orthonormal basis $\{e_\iota \mid \iota \in I\}$ and that any two orthonormal bases have the same cardinality, which is called the *dimension of* $\mathfrak{H}$. As a consequence, each infinite-dimensional separable Hilbert space $\mathfrak{H}$ is isomorphic to $\ell^2(\mathbb{N})$, i.e. there exists a one-to-one linear and norm-preserving map from $\mathfrak{H}$ onto $\ell^2(\mathbb{N})$ that identifies each vector with its sequence of generalized Fourier coefficients.

Example 2.7 For the n-dimensional Hilbert space $\mathbb{C}^n$ of Examples 2.1 and 2.2, $\{\mathbf{e}_j \mid j = 1, 2, \ldots, n\}$ of (2.2) is an orthonormal basis referred to as the *standard basis* of $\mathbb{C}^n$. Similarly, $\{\mathbf{e}_\iota \mid \iota \in I\}$ is the orthonormal basis of $\ell^2(I)$ with $\mathbf{e}_\iota$ the function that takes the value 1 at ι and zero elsewhere.

Example 2.8 Parseval's identity (2.10) is sometimes expressed in terms of distributions, see Section 2.2.2 for details. For an orthonormal basis $\{e_1, e_2, \ldots\}$ of $\mathcal{L}^2(\mathbb{R})$

$$\sum_{j=1}^{\infty} \overline{e_j}(x)\, e_j(y) = \delta(x - y). \tag{2.12}$$

The equality (2.12) as such is rather formal but it can be given a rigorous meaning if one imposes suitable regularity conditions on the orthonormal basis.

Example 2.9 The one-dimensional torus $\mathbb{T}$ is just the unit circle in $\mathbb{R}^2$. Its points are conveniently parametrized by an angle θ ranging between 0 and 2π. We denote the normalized rotation-invariant measure on $\mathbb{T}$ by $d\theta/2\pi$. A convenient basis of the space $\mathcal{L}^2(\mathbb{T}, d\theta/2\pi)$ is given by the family of functions $\{\epsilon_j \mid j \in \mathbb{Z}\}$ with

$$\epsilon_j(\theta) := \exp ij\theta, \quad 0 \le \theta < 2\pi.$$

Example 2.10 A natural basis of $\mathcal{L}^2(\mathbb{R})$ is associated with the energy of the harmonic oscillator. It consists of the set $\{x \mapsto c_n\, \mathrm{H}_n(x)\, e^{-x^2/2} \mid n \in \mathbb{N}\}$. H_n is the nth Hermite polynomial given by Rodrigues's formula

$$\mathrm{H}_n(x) = (-1)^n e^{x^2} \left(\frac{d}{dx}\right)^n e^{-x^2}.$$

The normalization of the Hermite polynomials is traditional. In order to normalize them in the Hilbert space sense, the c_n have to be chosen as $\left(1/\pi^{1/2} 2^n n!\right)^{1/2}$.

2.1.5 *Subspaces and projectors*

For any subset P of a Hilbert space $\mathfrak{H}$, one can define its *orthogonal complement*

$$\mathsf{P}^\perp := \{\varphi \in \mathfrak{H} \mid \langle \varphi, \psi \rangle = 0 \text{ for all } \psi \in \mathsf{P}\}.$$

$\mathsf{P}^\perp$ is a closed linear subspace of $\mathfrak{H}$.

Given a closed linear subspace P of a Hilbert space $\mathfrak{H}$, one of the important geometrical properties of $\mathfrak{H}$ is the unique *orthogonal decomposition* of an arbitrary element $\psi \in \mathfrak{H}$:

$$\psi = \varphi + \eta, \quad \varphi \in \mathsf{P} \quad \text{and} \quad \eta \in \mathsf{P}^{\perp}. \tag{2.13}$$

Due to the uniqueness of the decomposition (2.13), one can introduce the map $\psi \mapsto P\psi := \varphi$. P is called the *orthogonal projector* on P or simply the *projector* on P. The projector P satisfies for all $\psi, \varphi \in \mathfrak{H}$ and $\alpha, \beta \in \mathbb{C}$ the relations

$$\begin{aligned} P(\alpha\psi + \beta\varphi) &= \alpha P\psi + \beta P\varphi, \quad \|P\psi\| \le \|\psi\|, \\ P^2 &= P \quad \text{and} \quad \langle P\psi, \varphi \rangle = \langle \psi, P\varphi \rangle. \end{aligned} \tag{2.14}$$

These four conditions completely characterize orthogonal projectors, i.e. to any map satisfying (2.14), there corresponds a closed subspace $\mathsf{P} = P\mathfrak{H}$ on which P projects.

Example 2.11 Let $\{\varphi_j \mid j = 1, 2, \dots, n\}$ be an orthonormal set in $\mathfrak{H}$, then the operator

$$P\psi := \sum_{j=1}^{n} \langle \varphi_j, \psi \rangle \, \varphi_j, \quad \psi \in \mathfrak{H} \tag{2.15}$$

is the projector on $\mathrm{Span}(\{\varphi_j \mid j = 1, 2, \dots, n\})$.

Example 2.12 Let $[a, b]$ be an interval in $\mathbb{R}$. The *multiplication operator* on the Hilbert space $\mathcal{L}^2(\mathbb{R})$ given by

$$(P\psi)(x) := I_{[a,b]}(x)\,\psi(x), \quad \psi \in \mathcal{L}^2(\mathbb{R}),$$

where $I_{[a,b]}$ is the indicator function of $[a, b]$, is the projector onto the functions in $\mathcal{L}^2(\mathbb{R})$ with support in $[a, b]$. The range of P can be naturally identified with $\mathcal{L}^2([a, b])$.

The sum of two projectors P_1 and P_2 is again a projector iff $P_1 P_2 = 0$. In terms of the subspaces on which P_1 and P_2 project, this condition means that these subspaces have to be orthogonal. If this is the case $P_1 + P_2$ projects precisely on the sum of the subspaces. The product of two projectors P_1 and P_2 is again a projector iff P_1 and P_2 commute in which case it projects on the intersection of the ranges of P_1 and P_2. A projector P_1 is less or equal to P_2 if $P_1 \mathfrak{H} \subset P_2 \mathfrak{H}$, which we denote by $P_1 \le P_2$. This is equivalent to $P_1 = P_1 P_2$ or $P_1 = P_2 P_1$. Apart from taking sums and products of projectors, there exist the operations of *union* $\vee$ and *intersection* $\wedge$ which strongly resemble the analogous operations on subsets of a set. The union $P_1 \vee P_2$ of two projectors is the projector on the subspace spanned by $P_1 \mathfrak{H}$ and $P_2 \mathfrak{H}$. The intersection $P_1 \wedge P_2$ is the projector on $P_1 \mathfrak{H} \cap P_2 \mathfrak{H}$. The projectors on $\mathfrak{H}$ have the interesting structure of an *ordered lattice*:

$\exists$ a smallest element: $0 \leq P$,

$\exists$ a largest element: $P \leq \mathbf{1}$,

$\exists$ suprema: $P_k \leq \bigvee_j P_j$,

$\exists$ infima: $\bigwedge_j P_j \leq P_k$,

weak distributivity: if $P_3 \leq P_1$ then $P_1 \wedge (P_2 \vee P_3) = (P_1 \wedge P_2) \vee P_3$.

2.1.6   *Linear maps between Banach spaces*

Example 2.13 Let A be a linear map from the n-dimensional complex vector space $\mathbb{C}^n$ to $\mathbb{C}^m$:

$$\mathbf{y} = A\mathbf{x}.$$

Representing $\mathbf{x}$ and $\mathbf{y}$ as column vectors with entries x_k and y_j as in (2.1), we can find uniquely determined complex numbers a_{jk} such that

$$y_j = \sum_{k=1}^{n} a_{jk}\, x_k, \quad j = 1, 2, \dots, m,$$

i.e. A is the $m \times n$ matrix $[a_{jk}]$ with the usual action of a rectangular matrix on a column vector. The set $\mathcal{M}_{mn}$ of rectangular complex matrices with m rows and n columns carries a natural vector space structure which corresponds to the structure for the maps. If $m = n$, then the matrices are square and we denote them by $\mathcal{M}_n$ rather than $\mathcal{M}_{nn}$. Moreover, a linear map A from $\mathbb{C}^m$ to $\mathbb{C}^n$ can be composed with a linear map B from $\mathbb{C}^\ell$ to $\mathbb{C}^m$ yielding a linear map $C := A \circ B$ from $\mathbb{C}^\ell$ to $\mathbb{C}^n$. Such a composition amounts, in terms of the corresponding matrices, to the usual matrix multiplication,

$$c_{jk} = \sum_{r=1}^{m} a_{jr}\, b_{rk}, \quad j = 1, 2, \dots, n \text{ and } k = 1, 2, \dots, \ell.$$

The most general *linear maps or operators* used in this book are transformations A between two Banach spaces $\mathfrak{B}_1$ and $\mathfrak{B}_2$. Let $\mathrm{Dom}(A)$ be a linear subspace of $\mathfrak{B}_1$. We call a map A *linear with domain* $\mathrm{Dom}(A)$ if it satisfies, for all $x, y \in \mathrm{Dom}(A)$ and $\alpha, \beta \in \mathbb{C}$,

$$A(\alpha x + \beta y) = \alpha Ax + \beta Ay.$$

The map A is *bounded* if there exist a $K \geq 0$ such that

$$\|Ax\| \leq K \|x\|, \quad x \in \mathrm{Dom}(A). \tag{2.16}$$

In fact, A is continuous iff it is continuous at 0 and this is the case iff A is bounded. The smallest constant K which satisfies (2.16) is the *norm of* A and is given by

$$\|A\| := \sup_{\substack{x \in \mathrm{Dom}(A) \\ \|x\|=1}} \|A\,x\|.$$

If an operator $A : \mathfrak{B}_1 \to \mathfrak{B}_2$ is bounded on its domain and if this domain is dense in $\mathfrak{B}_1$, then there exists a unique bounded extension of A defined on the whole of $\mathfrak{B}_1$. This extension can still be denoted by A. It is straightforward to verify that $A \mapsto \|A\|$ is a norm on the linear space of all bounded linear operators from $\mathfrak{B}_1$ to $\mathfrak{B}_2$ which itself becomes a Banach space denoted by $\mathcal{B}(\mathfrak{B}_1, \mathfrak{B}_2)$. If $\mathfrak{B}_1 = \mathfrak{B}_2 = \mathfrak{B}$, we use the notation $\mathcal{B}(\mathfrak{B})$ for $\mathcal{B}(\mathfrak{B}, \mathfrak{B})$.

2.1.7 *Linear functionals and Dirac notation*

Referring to Example 2.13, linear functionals are linear maps from $\mathbb{C}^n$ to $\mathbb{C}$. If we identify complex numbers with 1×1 matrices, then linear functionals on $\mathbb{C}^n$ are row vectors. The notion of a *linear and bounded functional* is useful in the general context of Banach spaces $\mathfrak{B}$. It is a bounded, linear map F from $\mathfrak{B} \to \mathbb{C}$. The space of all linear and bounded functionals on $\mathfrak{B}$ is obviously a linear complex space and, equipped with the norm

$$\|F\| := \inf\{K \mid |F(x)| \leq K\|x\| \text{ for all } x \in \mathfrak{B}\}$$
$$= \sup\{|F(x)| \mid \|x\| = 1\},$$

it becomes a Banach space. This space is denoted by $\mathfrak{B}^*$ and is called the *dual space* of $\mathfrak{B}$.

A number of celebrated results in functional analysis concern the structure of spaces dual to important concrete classes of Banach spaces. We give two instances of such a characterization, others will appear later on.

Example 2.14 Let $1 \leq p < \infty$ and let q be such that $1/p + 1/q = 1$. With a sequence $\mathbf{f} = (f_1, f_2, \ldots,)$ in $\ell^q(\mathbb{N}_0)$, we construct the bounded linear functional

$$\mathbf{c} \mapsto \sum_{j=1}^{\infty} f_j\, c_j \tag{2.17}$$

on $\ell^p(\mathbb{N}_0)$. The boundedness is a consequence of *Hölder's inequality* which states that for $p, q, \mathbf{f}$ and $\mathbf{c}$ as above

$$\sum_{j=1}^{\infty} |f_j\, c_j| \leq \left(\sum_{j=1}^{\infty} |f_j|^q\right)^{1/q} \left(\sum_{k=1}^{\infty} |c_k|^p\right)^{1/p}.$$

The converse also holds: every continuous linear functional on $\ell^p(\mathbb{N}_0)$ is of the form (2.17). Remark that in the Hilbert space case ($p = 2$), the dual space is exactly isomorphic with the space itself. This is true in general.

Theorem 2.15 (Riesz) *Let $\mathfrak{H}$ be a Hilbert space. For any $F \in \mathfrak{H}^*$, there exists a unique vector $\varphi \in \mathfrak{H}$ such that*

$$F(\psi) = \langle \varphi, \psi \rangle, \quad \psi \in \mathfrak{H} \tag{2.18}$$

and $\|F\| = \|\varphi\|$. As a consequence, $\mathfrak{H}^*$ is isometrically anti-isomorphic to $\mathfrak{H}$.

The antilinearity of the isomorphism in Riesz's theorem arises because we express the functionals with the inner product, using the elements of the original Hilbert space as parameter. Dirac invented a special notation to deal with vectors and functionals. He proposed to denote the element ψ of the Hilbert space by a *ket* symbol $|\psi\rangle$ and the functional (2.18) by a *bra* symbol $\langle\varphi|$. The value of the functional F is then given by the *bracket*

$$F(\psi) = \langle \varphi, \psi \rangle \equiv \langle \varphi | \psi \rangle = \langle \varphi \| \psi \rangle.$$

This *Dirac notation* is quite convenient for many computations with operators on Hilbert spaces.

Example 2.16 The projector of eqn (2.15) can now be written as

$$P = \sum_j |\varphi_j\rangle\langle\varphi_j|.$$

Example 2.17 The operator identity

$$\mathbb{1} = \sum_j |e_j\rangle\langle e_j|. \tag{2.19}$$

expresses that a sequence $\{e_j \mid j = 1, 2, \dots\}$ is an orthonormal basis in $\mathfrak{H}$. Many calculations in quantum mechanics are performed inserting or cancelling the identity operator in the form (2.19). An instance of this is Parseval's equality (2.10).

It is useful to consider less severe notions of convergence than that induced by the norm. We list a number of them. For simplicity, we state the convergence for sequences. To be precise, we need to consider nets in all but the first.

Uniform convergence or norm convergence A sequence $\{x_n\}$ of vectors in a Banach space $\mathfrak{B}$ converges *uniformly* to 0 if

$$\lim_{n\to\infty} \|x_n\| = 0.$$

Weak convergence   A sequence $\{x_n\}$ of vectors in a Banach space $\mathfrak{B}$ converges *weakly* to 0 if for any continuous linear functional F

$$\lim_{n\to\infty} F(x_n) = 0.$$

Weak convergence*   A sequence of functionals $\{F_n\}$ in the dual $\mathfrak{B}^*$ of a Banach space converges *-weakly* to 0 if for any vector $x \in \mathfrak{B}$

$$\lim_{n\to\infty} F_n(x) = 0.$$

Strong operator convergence   A sequence of bounded operators $\{A_n\}$ from $\mathfrak{B}_1$ to $\mathfrak{B}_2$ converges to 0 in the *strong operator sense* if, for any vector $x \in \mathfrak{B}_1$, $\{A_n x\}$ converges uniformly to 0 in $\mathfrak{B}_2$, i.e. if

$$\lim_{n\to\infty} \|A_n x\| = 0. \tag{2.20}$$

Weak operator convergence   A sequence of bounded operators $\{A_n\}$ from $\mathfrak{B}_1$ to $\mathfrak{B}_2$ converges to 0 in the *weak operator sense* if, for any vector $x \in \mathfrak{B}_1$, $\{A_n x\}$ converges weakly to 0 in $\mathfrak{B}_2$, i.e. if for any bounded linear functional F on $\mathfrak{B}_2$

$$\lim_{n\to\infty} F(A_n x) = 0. \tag{2.21}$$

Obviously,

norm convergence $\implies$ weak convergence,

norm convergence $\implies$ weak* convergence,

uniform convergence of operators $\implies$ strong operator convergence, and

strong operator convergence $\implies$ weak operator convergence.

None of these arrows can be reversed.

Example 2.18 In $\ell^p(\mathbb{N}_0)$, the canonical basis tends weakly but not uniformly to 0.

If $\{e_1, e_2, \dots\}$ is an orthonormal basis in a Hilbert space $\mathfrak{H}$, then

$$P_n := \mathbf{1} - \sum_{j=1}^{n} |e_j\rangle\langle e_j|$$

tends to 0 in the strong operator sense, but not in operator norm.

Let S be the right-shift on $\ell^p(\mathbb{N}_0)$, then S^n tends in weak operator sense to 0 but not in the strong operator sense.

Example 2.19 The *trace* of an $n \times n$ matrix $A = [a_{jk}]$ is the number

$$\operatorname{Tr} A := \sum_j a_{jj}.$$

It has the obvious property $\operatorname{Tr} A + B = \operatorname{Tr} A + \operatorname{Tr} B$ and it satisfies the remarkable property $\operatorname{Tr} AB = \operatorname{Tr} BA$. It is the only linear functional on $\mathcal{M}_n$ which enjoys this property and which takes the value n on $\mathbf{1}$. Let F be a linear functional on $\mathcal{M}_n$ and $\{e_1, e_2, \ldots, e_n\}$ the standard orthonormal basis in $\mathbb{C}^n$. If we put

$$R := \sum_{jk} F(|e_j\rangle\langle e_k|) \, |e_k\rangle\langle e_j|,$$

then we easily verify that

$$F(A) = \operatorname{Tr} RA.$$

We shall consider a generalization of this formula later on.

An operator T on a separable Hilbert space $\mathfrak{H}$ is a *trace-class operator* if there exist two orthonormal bases $\{e_1, e_2, \ldots\}$ and $\{f_1, f_2, \ldots\}$ such that

$$\sum_{jk} |\langle e_j, T \, f_k\rangle| < \infty. \tag{2.22}$$

If T is trace-class, then it can be shown that

$$\operatorname{Tr} T := \sum_j \langle g_j, T \, g_j\rangle$$

is absolutely convergent for any orthonormal basis $\{g_1, g_2, \ldots\}$ of $\mathfrak{H}$ and that the sum is independent of the chosen basis.

Let $\{\psi_1, \psi_2, \ldots\}$ be a family of vectors in $\mathfrak{H}$ such that $\sum_j \|\psi_j\|^2 < \infty$. The operator

$$T := \sum_j |\psi_j\rangle\langle\psi_j| \tag{2.23}$$

is a trace-class operator by Parseval's equality. Similarly, AT and TA are trace-class whenever A is bounded and

$$\operatorname{Tr} TA = \operatorname{Tr} AT = \sum_j \langle\psi_j, A \, \psi_j\rangle. \tag{2.24}$$

2.1.8 *Adjoints of bounded operators*

Let A be a bounded linear operator from the Banach space $\mathfrak{B}_1$ into $\mathfrak{B}_2$. Consider now a functional $F \in \mathfrak{B}_2^*$, then the formula

$$x \in \mathfrak{B}_1 \mapsto F(Ax) =: \tilde{F}(x)$$

defines a linear and bounded functional $\tilde{F}$ on $\mathfrak{B}_1$. The map $F \mapsto \tilde{F}$ is linear. Varying the functional F we obtain the linear *dual transformation* A^T, also called *transposed*, which is bounded by $\|A\|$,

$$A^\mathsf{T} : \mathfrak{B}_2^* \to \mathfrak{B}_1^* : F \mapsto \tilde{F}.$$

Furthermore,

$$\|A^\mathsf{T}\| = \|A\|.$$

In the case of a Hilbert space $\mathfrak{H}$, it is more natural to use the antilinear identification of $\mathfrak{H}^*$ with $\mathfrak{H}$ given by Riesz's theorem, Theorem 2.15. Consider a bounded operator $A \in \mathcal{B}(\mathfrak{H}, \mathfrak{K})$ and apply essentially the same construction as above. The *Hermitian conjugate* or *adjoint* of A is the linear operator A^* from $\mathfrak{K}$ to $\mathfrak{H}$ satisfying

$$\langle \varphi, A\psi \rangle = \langle A^*\varphi, \psi \rangle, \quad \psi \in \mathfrak{H} \text{ and } \varphi \in \mathfrak{K}.$$

The operation of taking adjoints possesses the following properties:

$$
\begin{aligned}
&(\alpha A + \beta B)^* = \bar{\alpha} A^* + \bar{\beta} B^*, \\
&(AB)^* = B^* A^*, \quad B \in \mathcal{B}(\mathfrak{H}, \mathfrak{K}) \text{ and } A \in \mathcal{B}(\mathfrak{K}, \mathfrak{L}), \\
&(A^*)^* = A, \\
&\|AA^*\| = \|A\|^2 = \|A^*\|^2.
\end{aligned}
\tag{2.25}
$$

In the physics literature, the Hermitian adjoint of A is often denoted by $A^\dagger$. The essential difference between transposed and adjoint lies in the antilinearity of the latter with respect to multiplication by a complex scalar.

Example 2.20 It is a simple exercise to compute the transpose of a linear map A from $\mathbb{C}^n$ to $\mathbb{C}^m$ given in terms of an $m \times n$ matrix $[a_{jk}]$ as in Example 2.13. A^T is now an $n \times m$ matrix $[(a^\mathsf{T})_{jk}]$ with matrix elements

$$(a^\mathsf{T})_{jk} = a_{kj}.$$

If we compute the Hermitian adjoint of A, equipping $\mathbb{C}^m$ and $\mathbb{C}^n$ with their natural Hilbert space structures, we obtain $A^* = [(a^*)_{jk}]$ with

$$(a^*)_{jk} = \overline{a_{kj}}.$$

Example 2.21 Two finite sets of vectors $\{\varphi_j \mid j = 1, 2, \ldots, k\}$ and $\{\psi_j \mid j = 1, 2, \ldots, k\}$ in the Hilbert space $\mathfrak{H}$ define two operators that are adjoints of each other,

$$A = \sum_{j=1}^{k} |\varphi_j\rangle\langle\psi_j| \quad \text{and} \quad A^* = \sum_{j=1}^{k} |\psi_j\rangle\langle\varphi_j|.$$

Example 2.22 *(Standard matrix units)* Let $\{e_1, e_2, \ldots, e_n\}$ be the canonical orthonormal basis in $\mathbb{C}^n$. The rank-one operators

$$e_{jk} := |e_j\rangle\langle e_k|, \quad j, k \in \{1, 2, \ldots, n\}$$

are called the standard matrix units in $\mathcal{M}_n$. They satisfy the relations

$$e_{jk}\, e_{\ell r} = \delta_{k\ell}\, e_{jr} \quad \text{and} \quad \left(e_{jk}\right)^* = e_{kj}.$$

Example 2.23 Let a be a continuous bounded complex function on $\mathbb{R}$. It defines a multiplication operator A on $\mathcal{L}^2(\mathbb{R})$ by

$$(A f)(x) := a(x)\, f(x). \tag{2.26}$$

A is bounded with norm equal to $\|a\|_\infty := \sup_{x \in \mathbb{R}} |a(x)|$. The adjoint of A is the multiplication operator by the complex conjugated function

$$(A^* f)(x) = \overline{a(x)} f(x), \quad f \in \mathcal{L}^2(\mathbb{R}).$$

2.1.9 *Hermitian, unitary and normal operators*

A number of types of operators can be discerned according to their relation with respect to their adjoints. We list some of the most common.

- A bounded linear operator $A \in \mathcal{B}(\mathfrak{H})$ on a Hilbert space $\mathfrak{H}$ is *self-adjoint* or *Hermitian* if

$$A = A^*$$

or, equivalently, if

$$\langle A\varphi, \psi \rangle = \langle \varphi, A\psi \rangle, \quad \psi, \varphi \in \mathfrak{H}.$$

- A bounded operator $A \in \mathcal{B}(\mathfrak{H})$ is *positive* if

$$\langle \psi, A\psi \rangle \geq 0, \quad \psi \in \mathfrak{H}.$$

Any positive operator is self-adjoint and for any positive operator A there exists a unique positive operator B, called the *square root* of A and denoted by $\sqrt{A}$, such that

$$A = B^2.$$

In particular, as A^*A is always positive for any $A \in B(\mathfrak{H})$, the operator

$$|A| := \sqrt{A^*A}$$

is called the *absolute value* of A.

- A bounded operator $V \in \mathcal{B}(\mathfrak{H}, \mathfrak{K})$ is an *isometry* if it preserves the inner product

$$\langle V\varphi, V\psi \rangle = \langle \varphi, \psi \rangle, \quad \varphi, \psi \in \mathfrak{H}.$$

This is equivalent to $V^*V = \mathbf{1}_{\mathfrak{H}}$.

- A bounded operator $U \in \mathcal{B}(\mathfrak{H}, \mathfrak{K})$ is *unitary* if U possesses an inverse U^{-1} in $\mathcal{B}(\mathfrak{K}, \mathfrak{H})$ and if for all $\psi, \varphi \in \mathfrak{H}$

$$\langle U\varphi, U\psi \rangle = \langle \varphi, \psi \rangle$$

or, equivalently,

$$U^*U = UU^* = \mathbf{1}, \quad \text{i.e. } U^{-1} = U^*.$$

The unitary transformations of a Hilbert space $\mathfrak{H}$ with their composition form a group, the *unitary group of* $\mathfrak{H}$. It is denoted by $\mathcal{U}(\mathfrak{H})$.

- A bounded operator A is *normal* if

$$AA^* = A^*A.$$

The norm of a normal operator can be computed by considering its diagonal matrix elements in an arbitrary basis:

$$\|A\| = \sup_{\|\varphi\|=1} |\langle \varphi, A\varphi \rangle|. \tag{2.27}$$

Example 2.24 Let us consider a square $n \times n$ matrix $A = [a_{jk}]$ and denote by r_j and c_j the rows and columns of A considered as vectors in $\mathbb{C}^n$.

- A is normal iff $\langle r_j, r_k \rangle = \langle c_k, c_j \rangle$ for any choice of $j, k = 1, 2, \dots, n$.
- A is Hermitian if $\overline{r_j} = c_j$, where $\overline{r_j}$ is the vector obtained from r_j by taking componentwise complex conjugates.
- A is unitary iff it is isometric and this is the case iff its columns (or rows) form an orthonormal basis in $\mathbb{C}^n$, i.e. $\langle c_j, c_k \rangle = \langle r_j, r_k \rangle = \delta_{jk}$.
- Finally, A is positive iff all determinants of sub matrices $A_d := [a_{jk}]_{j,k=1,\dots,d}$ are non-negative.

Example 2.25 Any multiplication operator by a function a, see (2.26), is normal. It is self-adjoint iff

$$a(x) = \overline{a(x)}$$

and unitary iff

$$|a(x)| = 1.$$

Example 2.26 For any Hermitian operator $A \in \mathcal{B}(\mathfrak{H})$ and any analytic function

$$F(z) = \sum_{n=0}^{\infty} a_n z^n$$

with radius of convergence greater than $\|A\|$, the operator B defined by the norm convergent series

$$B := F(A) := \sum_{n=0}^{\infty} a_n A^n$$

is normal. In particular, the operator

$$e^{iA} := \sum_{n=0}^{\infty} \frac{i^n}{n!} A^n$$

is unitary for $A = A^*$.

Example 2.27 Any unitary operator U can be written in Dirac notation as

$$U = \sum_j |\psi_j\rangle\langle\varphi_j|$$

where $\{\psi_j \mid j = 1, 2, \dots\}$ and $\{\varphi_j \mid j = 1, 2, \dots\}$ are two bases in $\mathfrak{H}$. For any j we have then $U\varphi_j = \psi_j$, meaning that U is the linear map that rotates the basis $\{\varphi_j \mid j = 1, 2, \dots\}$ into $\{\psi_j \mid j = 1, 2, \dots\}$.

Example 2.28 By Plancherel's theorem, the *Fourier transform*

$$(U_F f)(x) := f^\wedge(x) = \frac{1}{\sqrt{2\pi}} \int_{-\infty}^{\infty} dy \, f(y) \exp(-ixy)$$

is a unitary operator on $\mathcal{L}^2(\mathbb{R})$.

2.1.10 *Partial isometries and polar decomposition*

An operator $V \in \mathcal{B}(\mathfrak{H})$ is a *partial isometry* if $P := V^*V$ and $Q := VV^*$ are projectors. We call P the *initial projector* and Q the *final projector* of V. It follows from this definition that V maps $P\mathfrak{H}$ isometrically onto $Q\mathfrak{H}$, i.e. for any vector $\psi \in P\mathfrak{H}$, $\|V\psi\| = \|\psi\|$ and $V = 0$ on $(P\mathfrak{H})^{\perp}$. In particular, if $P = \mathbf{1}$, then V in an isometry.

Every operator $T \in \mathcal{B}(\mathfrak{H})$ has a unique *polar decomposition*

$$T = V\,|T|,$$

where V is a partial isometry uniquely determined by the requirements

$$V(T\varphi) = |T|\varphi, \quad \varphi \in \mathfrak{H},$$
$$V\varphi = 0, \quad \varphi \in \mathrm{Ker}(T^*).$$

The *kernel*, $\mathrm{Ker}(A)$, of a bounded linear operator A is the set of all vectors that are mapped onto 0 by A. It is a closed subspace of $\mathfrak{H}$. If T is invertible, then V is unitary.

Example 2.29 Every partial isometry is of the form

$$V = \sum_j |\varphi_j\rangle\langle\psi_j|$$

with $\{\varphi_j\}$ and $\{\psi_j\}$ orthonormal sequences.

Example 2.30 The right-shift operator on $\ell^2(\mathbb{N}_0)$, defined in terms of the canonical basis $\{e_j \mid j = 1, 2, \ldots\}$,

$$V e_j := e_{j+1}, \quad j = 0, 1, \ldots$$

is an isometry with final projector projecting on $\ell^2(\mathbb{N}_0 \setminus \{1\})$.

2.1.11 Spectra of operators

We begin with

Example 2.31 For any $d \times d$ matrix M, considered as a linear operator acting on $\mathbb{C}^d$, the *spectral problem* consists in finding the non-trivial solutions of

$$M\mathbf{x} = \lambda\mathbf{x}, \quad \lambda \in \mathbb{C}, \ \mathbf{x} \in \mathbb{C}^d. \tag{2.28}$$

If we fix λ and rewrite eqn (2.28) as

$$(M - \lambda\mathbf{1})\mathbf{x} = 0, \tag{2.29}$$

we obtain for the components of $\mathbf{x}$ a set of d linear equations in d unknowns which admits in general only the zero solution. Non-trivial solutions exist only for these special values of λ for which $\det(M - \lambda\mathbf{1}) = 0$ or, equivalently, for which $M - \lambda\mathbf{1}$ is not invertible. We are therefore interested in finding the roots of the *characteristic polynomial* $\det(M - \lambda\mathbf{1})$ of M. As this is a polynomial equation of degree d, we know that it has exactly d roots $(\lambda_1, \lambda_2, \ldots, \lambda_d)$, some of them possibly coinciding. The λ_j are the *eigenvalues of M* and the set

$$\mathrm{Sp}(M) := \{\lambda_j \mid j = 1, 2, \ldots, \nu \leq d\}$$

is called the *spectrum of M*. The complement of the spectrum is the *resolvent set of M* and it consists of all λ such that $M - \lambda\mathbf{1}$ is invertible.

For any $\lambda \in \mathrm{Sp}(M)$, there exists a non-zero solution $\mathbf{x}$ of eqn (2.29). Such an $\mathbf{x}$ is called an *eigenvector of M* corresponding to the eigenvalue λ. The eigenvectors of M corresponding to a $\lambda \in \mathrm{Sp}(M)$ form a linear subspace, but in general the collection of these subspaces might fail to generate the whole of $\mathbb{C}^d$. If we call the subspace of all $\mathbf{x}$ such that $(M - \lambda\mathbf{1})^k\mathbf{x} = 0$ for k sufficiently large the *spectral subspace Q_λ*, then the dimension of Q_λ is precisely the *multiplicity of the eigenvalue λ* as a root of the characteristic equation. Two different subspaces Q_λ and Q_μ have only the zero vector in common and the Q_λ span the whole of $\mathbb{C}^d$, i.e. any vector $\mathbf{x} \in \mathbb{C}^d$ can uniquely be written as

$$\mathbf{x} = \sum_{\lambda \in \mathrm{Sp}(M)} \mathbf{x}_\lambda$$

with $\mathbf{x}_\lambda \in Q_\lambda$.

If M is a normal matrix, then the situation becomes simpler: for each eigen-value λ of M there are exactly as many linearly independent eigenvectors corresponding to λ as the multiplicity of the eigenvalue, and eigenvectors belonging to different eigenvalues are mutually orthogonal.

Example 2.32 Consider on the Hilbert space $\mathcal{L}^2([0,1])$ a multiplication operator A by a strictly increasing continuous function $a : [0,1] \mapsto \mathbb{C}$ as in (2.26). One can easily see that the eigenvalue problem analogous to (2.28)

$$(A f)(x) = a(x) f(x) = \lambda f(x),$$

has no non-trivial solution. Nevertheless, for any $\lambda = a(x_0)$, we can find an approximative solution of (2.28) with an approximative eigenfunction f_λ different from zero in the vicinity of x_0 only. Hence, the range of the function a should be the spectrum of A.

For a general bounded operator T acting on a Banach space $\mathfrak{B}$, it is easier to define those points which do not belong to the spectrum of T. The *resolvent set* of T denoted by $\mathrm{Res}(T)$ consists of all complex numbers λ for which $T - \lambda\mathbf{1}$ has a bounded inverse $R_\lambda(T) = (T - \lambda\mathbf{1})^{-1}$ called the *resolvent* of T. The *spectrum* of T, denoted by $\mathrm{Sp}(T)$, is the complement of the resolvent set of T. The spectrum contains all eigenvalues of T which form the *point spectrum* of T, the rest can be classified in different ways.

As already mentioned, Hermitian and unitary operators on Hilbert spaces play a particularly important role in quantum mechanics describing observables and symmetries respectively. The spectrum of a Hermitian operator lies on the real axis while the spectrum of a unitary operator lies on the unit circle,

$$A = A^* \;\Rightarrow\; \mathrm{Sp}(A) \subset \mathbb{R},$$
$$UU^* = U^*U = \mathbf{1} \;\Rightarrow\; \mathrm{Sp}(U) \subset \{\lambda \mid |\lambda| = 1\}.$$

Any $d \times d$ Hermitian matrix M can be diagonalized, which means that it can be written as a linear combination of mutually orthogonal projection operators. In Dirac notation

$$M = \sum_{j=1}^{d} \lambda_j |\varphi_j\rangle\langle\varphi_j|. \tag{2.30}$$

On $\mathcal{L}^2(\mathbb{R})$, the multiplication operator by a real function $a : \mathbb{R} \to \mathbb{R}$ can be approximated by a combination of projectors given by multiplication operators by suitable indicator functions

$$(A f)(x) \approx \sum_{k} \lambda_k I_{\Delta_k}(x) f(x), \tag{2.31}$$

where $\Delta_k = a^{-1}(]\lambda_{k-1}, \lambda_k])$ and $\cdots < \lambda_{k-1} < \lambda_k < \lambda_{k+1} < \cdots$ cover the range of a. An important feature of the representation (2.30) of the matrix M, and also of the multiplication operator (2.31), is that one can again easily express any reasonable function of M or A in terms of the same projectors, e.g.

$$f(M) = \sum_{j=1}^{d} f(\lambda_j) \, |\varphi_j\rangle\langle\varphi_j|.$$

The special representations (2.30) and (2.31) will find their abstract generalization in the *Spectral Theorem*.

2.1.12   *Unbounded operators*

In most cases, the interesting physical observables of a quantum system are described by unbounded self-adjoint operators. We return to the general situation of operators T between Banach spaces $\mathfrak{B}_1$ and $\mathfrak{B}_2$. An operator T' is an *extension* of T, denoted as $T \subset T'$, if

$$\mathrm{Dom}(T) \subset \mathrm{Dom}(T') \quad \text{and} \quad T x = T' x, \ x \in \mathrm{Dom}(T).$$

An unbounded operator T is *closed* if for every sequence $\{x_n\} \subset \mathrm{Dom}(T)$ the conditions

$$\lim_{n\to\infty} x_n = x \quad \text{and} \quad \lim_{n\to\infty} T x_n = y$$

imply

$$x \in \mathrm{Dom}(T) \quad \text{and} \quad Tx = y.$$

An unbounded operator is *closable* if it possesses a closed extension and the minimal closed extension is the *closure* of T.

Consider an operator A (generally unbounded) acting on a Hilbert space $\mathfrak{H}$ with domain $\mathrm{Dom}(A)$ dense in $\mathfrak{H}$. The adjoint operator A^* is defined on $\mathrm{Dom}(A^*)$ which consists of all $\varphi \in \mathfrak{H}$ such that the formula $F(\psi) := \langle \varphi, A\psi \rangle$ defines a bounded linear functional on $\mathrm{Dom}(A)$ and can therefore be extended to the whole of $\mathfrak{H}$. Due to Riesz's theorem, Theorem 2.15, there exists a unique element of $\mathfrak{H}$, denoted by $A^*\varphi$, such that

$$F(\psi) = \langle A^*\varphi, \psi \rangle = \langle \varphi, A\psi \rangle, \quad \psi \in \mathrm{Dom}(A), \ \varphi \in \mathrm{Dom}(A^*).$$

The adjoint of an operator is always closed and if $A \subset B$, then $B^* \subset A^*$. An operator A is *self-adjoint* if $A = A^*$.

In quantum mechanics, a relevant unbounded observable is typically given by a differential operator A_0 defined on a certain standard domain $\mathrm{Dom}(A_0)$ of infinitely differentiable wave functions and satisfies the following condition:

$$\langle \varphi, A_0\psi \rangle = \langle A_0\varphi, \psi \rangle, \quad \psi, \varphi \in \mathrm{Dom}(A_0). \tag{2.32}$$

Condition (2.32) expresses that A_0^* is densely defined and that it extends A_0. Such an operator is called *symmetric*. In order to correspond to a proper physical

observable with a clear probabilistic interpretation, one needs self-adjoint operators. One of the important issues in quantum mechanics is to find out whether A_0
admits a self-adjoint extension and to classify all such extensions. A symmetric
operator which possesses a unique self-adjoint extension is called *essentially self-
adjoint*. A useful criterion for essential self-adjointness of a symmetric operator
A is to check whether the equation

$$A\psi \pm i\psi = \varphi$$

has a unique solution ψ for every φ in a dense set of $\mathfrak{H}$.

Example 2.33 The *Schwartz space* $\mathcal{S}(\mathbb{R})$ consists of the infinitely differentiable
complex functions whose derivatives vanish rapidly at ∞. The multiplication
operator Q by the indeterminate x on $\mathcal{S}(\mathbb{R})$ is symmetric. Moreover, as the
equation

$$Q\psi \pm i\psi = \varphi$$

admits for any $\varphi \in \mathcal{S}(\mathbb{R})$ the unique solution

$$\psi(x) = \frac{\varphi(x)}{x \pm i},$$

Q is essentially self-adjoint. We denote its closure, which is the *position operator*,
again by Q. Also, the momentum operator

$$(P\varphi)(x) := -i\hbar\frac{d\varphi(x)}{dx}$$

is essentially self-adjoint on $\mathcal{S}(\mathbb{R})$. Integration by parts shows already that P is
symmetric:

$$\int_{\mathbb{R}} dx\, \overline{\varphi(x)}\left(-i\hbar\frac{d\psi(x)}{dx}\right) = \int_{\mathbb{R}} dx\, \overline{\left(-i\hbar\frac{d\varphi(x)}{dx}\right)}\psi(x). \qquad (2.33)$$

By using Fourier transformation, it can be shown that P is essentially self-
adjoint. The closure of P, which is self-adjoint, can again be denoted by P and
it is the *momentum operator* in standard quantum mechanics. The space $\mathcal{S}(\mathbb{R})$ is
a common domain for position and momentum and they satisfy the commutation
relation

$$[Q, P] = i\hbar\mathbf{1}$$

on this domain.

2.2 Measures

2.2.1 *Measures and integration*

A natural notion of size of a subset of the real line, say a union of disjoint
intervals, is the total length of the subset. The surface of a subset would be the

natural extension in two dimensions. In the sequel we need quite an extension of this notion to that of a *measure*. It is useful to imagine that we have a certain space $\mathfrak{X}$ at hand, called a *measurable space* and that a measure describes a mass distribution in this space. The measure of a given subset $S \subset \mathfrak{X}$ corresponds to the total mass in S. In order to develop a mathematically consistent theory, one cannot allow for general subsets S of $\mathfrak{X}$ but has to restrict attention to a collection of sufficiently regular sets called the *σ-algebra of measurable subsets*.

More precisely, a *σ-algebra* is a collection Σ of subsets of $\mathfrak{X}$ such that

- $\emptyset \in \Sigma$,
- $\mathfrak{X} \setminus S \in \Sigma$ whenever $S \in \Sigma$,
- finite or countably infinite unions of elements in Σ belong again to Σ.

A *measure* μ is a mapping from Σ to $\mathbb{R}^+$ that satisfies

$$\mu(\emptyset) = 0,$$

$$\mu\left(\bigcup_j S_j\right) = \sum_j \mu(S_j) \quad \text{whenever } \{S_1, S_2, \dots\} \text{ is an at most} \tag{2.34}$$
$$\text{countable collection of mutually disjoint sets in } \Sigma.$$

A set $S \in \Sigma$ is of *zero measure* if $\mu(S) = 0$. The σ-algebra Σ is *complete* w.r.t. μ if it contains all subsets of sets of measure zero. It is always possible to complete a σ-algebra. A *property holds μ-almost everywhere* or, in short, almost everywhere if it holds on the complement of a set of zero measure. We write that the property holds μ-a.e. or simply a.e. A measure μ is *bounded* if $\mu(\mathfrak{X}) < \infty$.

Next, one passes from the level of subsets of $\mathfrak{X}$ to that of complex functions on $\mathfrak{X}$. Such a function is said to be *measurable* if both its real and imaginary parts are measurable and a real function $f : \mathfrak{X} \to \mathbb{R}$ is measurable if

$$f^{-1}(]-\infty, t]) := \{x \in \mathfrak{X} \mid f(x) \leq t\}$$

belongs to Σ. Integrals of measurable functions are constructed in several steps. The *indicator function* I_S of a subset S of $\mathfrak{X}$ takes the value 1 on S and vanishes outside S. Let S be a measurable subset of $\mathfrak{X}$, then I_S is measurable and we put

$$\int_{\mathfrak{X}} \mu(dx)\, I_S(x) := \mu(S). \tag{2.35}$$

A non-negative measurable *simple function* s is a finite positive combination of indicator functions of measurable subsets $g = \sum_j \lambda_j I_{S_j}$, $\lambda_j \geq 0$ and we put

$$\int_{\mathfrak{X}} \mu(dx)\, g(x) := \sum_j \lambda_j \mu(S_j). \tag{2.36}$$

The consistency of formula (2.36) follows from the additivity of a measure on finite unions of disjoint measurable sets. Formula (2.35) is a particular case of (2.36) and it encodes the linearity of the integral in its most basic form. The integral of a non-negative measurable function f is obtained by approximating f from below with simple functions, i.e. f is said to be an *integrable function* if

$$\int_{\mathfrak{X}} \mu(dx)\, f(s) := \sup\left\{ \int_{\mathfrak{X}} \mu(dx) g(x) \ \Big| \ g \text{ measurable simple}\right.$$
$$\left.\text{function and } 0 \le g \le f \right\} < \infty.$$

A complex measurable function f is integrable if $|f|$ is integrable and

$$\int_{\mathfrak{X}} \mu(dx)\, f(x) := \int_{\mathfrak{X}} \mu(dx)\, (\Re e f)^+(x) - \int_{\mathfrak{X}} \mu(dx)\, (\Re e f)^-(x)$$
$$+ i \int_{\mathfrak{X}} \mu(dx)\, (\Im m f)^+(x) - i \int_{\mathfrak{X}} \mu(dx)\, (\Im m f)^-(x).$$

For a real function f, $f^\pm$ denote the positive and negative parts

$$f^\pm(x) := \max\{0, \pm f(x)\}.$$

A particular case of the construction of above is obtained by considering for $(\mathfrak{X}, \Sigma)$ a discrete space, that is a countable set I with σ-algebra consisting of all subsets of I, and for μ the *counting measure* which assigns to each subset of I its cardinality. In this case, integrability is equivalent to summability.

Often, we consider a space $\mathfrak{X}$ that carries already a natural notion of convergence determined by specifying its *open sets*. The smallest σ-algebra which contains all open sets is called the algebra of *Borel subsets* of $\mathfrak{X}$. The prototype of such a situation is $]0, 1[$, where every open set is a countable union of open intervals. Sets such as $]a, b]$ with $0 \le a$ and $b < 1$ are, for example, Borel sets that are neither closed nor open. The measure of any Borel set is obtained as the infimum of the measures of open sets containing the given set. It is therefore sufficient to specify a Borel measure on open sets.

Example 2.34 There exists a unique measure on $[0, 1]$, called the *Lebesgue measure*, which assigns the value $b - a$ to an open interval $]a, b[$. This measure is uniquely characterized by the property that it is translation invariant, i.e. that it assigns the same measure to $]a + c, b + c[$ as to $]a, b[$, a, b and c chosen in such a way that all intervals lie in $[0, 1]$ and that $[0, 1]$ has measure 1. The usual Riemann notation is kept for the Lebesgue measure:

$$b - a = \int_{[0,1]} dx\, I_{]a,b[}(x) =: \int_a^b dx.$$

A larger class of measures is obtained by modulating the Lebesgue measure with a density ρ which is a non-negative integrable function on $[0, 1]$. The measure of an open interval $]a, b[$ is now

$$\mu_\rho(]a, b[) := \int dx\, \rho(x)\, I_{]a,b[}(x) =: \int_a^b dx\, \rho(x).$$

Measures of the type μ_ρ are called *absolutely continuous with respect to the Lebesgue measure*. In general a *continuous measure* μ is characterized by the property that for any $a \in [0, 1]$

$$\lim_{\epsilon \downarrow 0} \mu(]a - \epsilon, a + \epsilon[) = 0.$$

There exist many measures that are continuous but not absolutely continuous (with respect to Lebesgue). Finally, there are the *atomic measures* on $[0, 1]$ that live on countable unions of points. The prototype of such a measure is the *Dirac measure* δ_c that assigns the value 1 to $]a, b[$ if $a < c < b$ and zero otherwise. In particular, $\delta_c(]a, c]) = 1$ whenever $a < c$.

The *product construction* allows to construct from given measure spaces more complicated spaces. In its simplest version, one starts with two measure spaces $(\mathfrak{X}_1, \Sigma_1, \mu_1)$ and $(\mathfrak{X}_2, \Sigma_2, \mu_2)$. The Cartesian product $\mathfrak{X}_{12} := \mathfrak{X}_1 \times \mathfrak{X}_2$ is equipped with the σ-algebra Σ_{12} generated by the *measurable rectangles* $S_1 \times S_2$ with $S_j \in \Sigma_j$. One shows that there is a unique measure μ_{12} on Σ_{12} that is fully determined by the requirement

$$\mu_{12}(S_1 \times S_2) := \mu_1(S_1) \mu_2(S_2).$$

$(\mathfrak{X}_{12}, \Sigma_{12}, \mu_{12})$ is the *product of the measure spaces* $(\mathfrak{X}_1, \Sigma_1, \mu_1)$ and $(\mathfrak{X}_2, \Sigma_2, \mu_2)$.

We now sketch an alternative functional approach to measures. By $\mathcal{C}(\mathfrak{X})$ we denote the complex Banach space of all continuous complex functions on a compact Hausdorff space $\mathfrak{X}$ equipped with the supremum norm

$$\|f\|_\infty := \sup_{x \in \mathfrak{X}} |f(x)|, \quad f \in \mathcal{C}(\mathfrak{X}).$$

(A space $\mathfrak{X}$ is a *Hausdorff space* if different points are well-separated in the sense that we can find closed non-intersecting neighbourhoods of these points.) The dual space $\mathcal{C}(\mathfrak{X})^*$ of $\mathcal{C}(\mathfrak{X})$ carries a natural notion of positivity: a continuous linear functional F on $\mathcal{C}(\mathfrak{X})$ is positive if $F(f) \geq 0$ for every non-negative f.

Example 2.35 Let μ be a bounded measure on $\mathfrak{X}$, then μ defines a bounded linear functional F on $\mathcal{C}(\mathfrak{X})$ by integrating out the variable $x \in \mathfrak{X}$ with respect to the measure, i.e.

$$F(f) := \int_{\mathfrak{X}} \mu(dx) f(x).$$

F is obviously linear and is bounded as

$$|F(f)| = \left| \int_{\mathfrak{X}} \mu(dx) f(x) \right| \leq \|f\|_\infty \mu(\mathfrak{X}), \tag{2.37}$$

and in fact equality is attained in (2.37) for constant functions. Moreover, F is manifestly positive.

The following theorem states that Example 2.35 is general and it therefore provides us with a functional analytic approach to measure theory.

Theorem 2.36 (Banach–Saks–Kakutani) *There is, for a compact Hausdorff space $\mathfrak{X}$, a one-to-one correspondence between the positive elements in the dual of $\mathcal{C}(\mathfrak{X})$ and the bounded measures on $\mathfrak{X}$ given by*

$$F(f) = \int_{\mathfrak{X}} \mu(dx)\, f(x), \quad f \in \mathcal{C}(\mathfrak{X}).$$

A general continuous linear functional F on $\mathcal{C}(\mathfrak{X})$ can be decomposed in a linear combination

$$F = F_1 - F_2 + iF_3 - iF_4$$

of at most four positive functionals F_j, which leads us to the notion of *signed measure* as the corresponding linear combination of the measures μ_j. Then Theorem 2.36 holds for general functionals, replacing measure by signed measure. This whole setting has a natural extension to locally compact Hausdorff spaces, replacing continuous functions on $\mathfrak{X}$ by continuous functions on $\mathfrak{X}$ which tend to zero at infinity and signed measures by *Radon measures*.

2.2.2 *Distributions*

It is often useful to differentiate objects that are not regular enough to allow for a derivative in the usual sense of real analysis. These could be functions with discontinuities but also measures or even wilder objects. The theory of distributions is able to cope with such situations. In order to keep the presentation minimal, we consider here only distributions that live on $\mathbb{R}$, but a complete theory can be set up for smooth manifolds, possibly with boundaries. The basic idea is that distributions are continuous functionals on spaces of very regular functions, called test functions, and then to transpose operations on test functions by duality to distributions.

A complex function f on $\mathbb{R}$ is a *test function* if it is smooth, i.e. infinitely many times differentiable, and if it has *compact support*, which means that f vanishes outside an interval. We denote the space of all test functions by $\mathcal{D}(\mathbb{R})$. This is a complete linear space with respect to the following notion of convergence: a sequence $\{f_n\}$ of test functions converges to 0 if

$$\lim_{n \to \infty} \sup_{x \in \mathbb{R}} |f_n^{(k)}(x)| = 0$$

for all $k = 0, 1, 2, \ldots$ and if there exists a *fixed finite interval* $[a, b]$ such that

$$f(x) = 0, \quad x \notin [a, b].$$

A *distribution* is a continuous linear functional on $\mathcal{D}(\mathbb{R})$. Convergence of distributions is to be understood pointwise: a sequence $\{T_n\}$ of distributions converges to the distribution T if

$$\lim_{n \to \infty} T_n(g) = T(g)$$

for any test function.

Example 2.37 Let ℓ be a *locally integrable* complex function on $\mathbb{R}$, meaning that the restriction of ℓ to any finite interval is integrable, then

$$L(g) := \int_{\mathbb{R}} dx\, \ell(x)\, g(x), \quad g \in \mathcal{D}(\mathbb{R}),$$

is a distribution, such L are called *regular*.

Let $c \in \mathbb{R}$, then

$$\delta_c(g) := g(c), \quad g \in \mathcal{D}(\mathbb{R}),$$

is a distribution that is often written in pseudo-regular way:

$$g(c) = \int_{\mathbb{R}} dx\, \delta(x - c)\, g(x),$$

where the *Dirac δ function* is supposed to satisfy

$$\delta(x) = \begin{cases} 0 & \text{whenever } x \neq 0, \\ \int_{\mathbb{R}} dx\, \delta(x) = 1. \end{cases}$$

An important property of distributions is that they can be differentiated any number of times and that taking derivatives is a continuous operation. The derivative T' of a distribution T is defined through the relation

$$T'(g) := -T(g'), \quad g \in \mathcal{D}(\mathbb{R}).$$

The continuity of T' is a direct consequence of the strong convergence requirements on test functions, so is the continuity of taking derivatives with respect to T: for any $g \in \mathcal{D}(\mathbb{R})$

$$\lim_{n \to \infty} (T_n)'(g) = -\lim_{n \to \infty} T_n(g') = -T(g') = T'(g).$$

Example 2.38

$$(\delta_c)'(g) = -\delta_c(g') = -g'(c).$$

2.2.3 *Hilbert spaces of functions*

Many Hilbert spaces arise as spaces of square integrable complex functions on measurable spaces. More precisely, let $(\mathfrak{X}, \Sigma, \mu)$ be a triple which consists of a measurable space $\mathfrak{X}$, a σ-algebra Σ of measurable subsets of $\mathfrak{X}$ and a σ-additive measure μ. A complex function f on $\mathfrak{X}$ is *square integrable* if f^2 or, equivalently,

$|f|^2$ is integrable. A deep result in measure theory states that the space $\mathcal{L}^2(\mathfrak{X}, \mu)$ of square integrable functions with inner product given by

$$\langle f, g \rangle := \int_{\mathfrak{X}} \mu(dx)\, \overline{f(x)}\, g(x)$$

is a Hilbert space. In fact, one should be somewhat more precise and rather consider the space of equivalence classes of square integrable complex functions on $\mathfrak{X}$, where two functions are equivalent if they coincide μ-a.e.

A particular case of this construction arises when $\mathfrak{X}$ is a countable set I and μ is the counting measure on I. The notion of equivalence becomes trivial in this situation as there are no sets of measure zero, except for the empty set, and a complex function on I is usually called a sequence. Doing so, we obtain the Hilbert space $\ell^2(I)$ of square summable complex sequences indexed by I.

The notion of a $d \times d$ matrix as an operator on $\mathbb{C}^d$ extends to a certain degree to $\mathcal{L}^2$-spaces in terms of operators defined by kernels. A function $k : \mathfrak{X} \times \mathfrak{X} \to \mathbb{C}$ is a kernel

    if it is measurable in both its arguments;

    if $(K\,f)(x) := \int_{\mathfrak{X}} \mu(dy)\, k(x, y)\, f(y)$ makes sense μ-a.e. in x; and   (2.38)

    if $f \mapsto K\,f$ defines a bounded linear map.

In general, one cannot give the necessary and sufficient conditions to guarantee the boundedness of K. A sufficient condition, however, is that k be a *Hilbert–Schmidt integral kernel*, namely

$$\int_{\mathfrak{X} \times \mathfrak{X}} \mu(dx)\mu(dy)\, |k(x, y)|^2 \leq \infty.$$

That k defines a bounded operator follows from Schwarz's inequality

$$\int_{\mathfrak{X}} \mu(dx) \left| \int_{\mathfrak{X}} \mu(dy)\, k(x, y) f(y) \right|^2$$
$$\leq \int_{\mathfrak{X}} \mu(dx) \int_{\mathfrak{X}} \mu(dy)\, |k(x, y)|^2 \int_{\mathfrak{X}} \mu(dz)\, |f(z)|^2$$
$$\leq \int_{\mathfrak{X} \times \mathfrak{X}} \mu(dx)\mu(dy)\, |k(x, y)|^2 \|f\|^2,$$

and therefore

$$\|K\|^2 \leq \int_{\mathfrak{X} \times \mathfrak{X}} \mu(dx)\mu(dy)\, |k(x, y)|^2.$$

2.2.4   *Spectral measures*

In Section 2.2.1, measures and integrals were complex, we now extend these notions to operator-valued objects. The spectral measures will more precisely be projection-valued and we shall integrate real (complex) functions with respect to such a spectral measure in order to obtain a general self-adjoint (normal) operator, i.e. we shall describe how to approximate in a suitable sense any self-adjoint (normal) operator by a limit of linear combinations of projection operators.

Definition 2.39 *A spectral measure is a map E from a σ-algebra Σ of subsets of a measurable space $\mathfrak{X}$ to the projectors on a Hilbert space $\mathfrak{H}$ that satisfies*

(a) $E(\emptyset) = 0$ *and* $E(\mathfrak{X}) = \mathbf{1}$

(b) *if $\{S_j \mid j \in J\}$ is a countable family of mutually disjoint measurable subsets of $\mathfrak{X}$, then $E(\cup_j S_j) = \sum_j E(S_j)$.*

Definition 2.39 implies that spectral projectors belonging to disjoint sets are orthogonal. The sum of mutually orthogonal projectors in (b) converges in strong operator sense, i.e. there exists a projector P such that

$$\lim_K \left\| P\varphi - \sum_{j \in K} E(S_j)\varphi \right\| = 0, \quad \varphi \in \mathfrak{H},$$

where K are finite sets tending to J. This is a consequence of Bessel's inequality.

Example 2.40 Consider a family $\{P_j \mid j = 1, 2, \ldots, n\}$ of mutually orthogonal projectors on a Hilbert space $\mathfrak{H}$ adding up to $\mathbf{1}$. For any subset S of $\{1, 2, \ldots, n\}$ put

$$E(S) := \sum_{j \in S} P_j,$$

then E is a spectral measure on $\{1, 2, \ldots, n\}$.

Example 2.41 Let a be a real measurable function on $\mathbb{R}$ and put for any Borel subset S of $\mathbb{R}$

$$E(S) := P_{a^{-1}(S)},$$

where, for any Borel subset M of $\mathbb{R}$, P_M is the orthogonal projector on the subspace $\mathcal{L}^2(M)$, i.e. the multiplication operator by the indicator function I_M of M. E is then a spectral measure (on $\mathbb{R}$).

The spectral family of Example 2.41 can be used to reconstruct any multiplication operator. On $\mathcal{L}^2(\mathbb{R})$, the multiplication operator by a real function $a : \mathbb{R} \to \mathbb{R}$ can be approximated, in the sense of strong convergence, by finite linear combinations of spectral projectors

$$(A f)(x) \approx \sum_k \lambda_k I_{\Delta_k}(x) f(x)$$

where $\Delta_k = a^{-1}(]\lambda_{k-1}, \lambda_k])$ and $\cdots < \lambda_{k-1} < \lambda_k < \lambda_{k+1} < \cdots$ cover the range of a. This representation is the germ of the *Spectral Theorem*.

Theorem 2.42 (Spectral Theorem) *Any self-adjoint operator A on a Hilbert space $\mathfrak{H}$ can be written on its domain as an integral in the sense of strong convergence*

$$A\varphi = \int_{\mathbb{R}} \lambda\, E^A(d\lambda)\, \varphi, \quad \varphi \in \mathrm{Dom}(A), \tag{2.39}$$

where E^A is a uniquely determined spectral measure on the Borel subsets of $\mathbb{R}$. E^A is the spectral measure of A. Moreover,

$$\mathrm{Dom}(A) = \Big\{ \varphi \,\Big|\, \int_{\mathbb{R}} \lambda^2 \, \|E^A(d\lambda)\,\varphi\|^2 < \infty \Big\}.$$

The Spectral Theorem contains more information than given by (2.39). First of all, the non-trivial contribution to the integral in (2.39) comes from the spectrum of A and hence one can write

$$A = \int_{\mathrm{Sp}(A)} \lambda\, E^A(d\lambda).$$

Moreover, for any measurable function $f : \mathrm{Sp}(A) \mapsto \mathbb{R}$, the expression $f(A)$ makes sense and defines a self-adjoint operator with spectral representation

$$f(A) = \int_{\mathrm{Sp}(A)} f(\lambda)\, E^A(d\lambda). \tag{2.40}$$

If φ is an eigenvector of A with eigenvalue α, then $\|E^A(d\lambda)\,\varphi\|^2 = \delta(\lambda-\alpha)\, d\lambda$. We can decompose $\mathfrak{H}$ as $\mathfrak{H}_d \oplus \mathfrak{H}_c$, where $\mathfrak{H}_d$ is the closed subspace spanned by the eigenvectors of A. Both subspaces $\mathfrak{H}_d$ and $\mathfrak{H}_c$ are globally invariant under A. Writing a vector $\varphi \in \mathrm{Dom}(A)$ as $\varphi_d + \varphi_c$, we see that

$$\|E^A(d\lambda)\,\varphi\|^2 = \|E^A(d\lambda)\,\varphi_d\|^2 + \|E^A(d\lambda)\,\varphi_c\|^2$$

precisely provides the decomposition of $\|E^A(d\lambda)\,\varphi\|^2$ in an atomic and a continuous part. It is sometimes useful to refine the procedure, writing $\mathfrak{H}_c$ as $\mathfrak{H}_{\mathrm{a.c.}} \oplus \mathfrak{H}_{\mathrm{s.c.}}$, where $\mathfrak{H}_{\mathrm{a.c.}}$ is spanned by the vectors $\varphi \in \mathrm{Dom}(A)$ such that $\|E^A(d\lambda)\,\varphi\|^2$ is absolutely continuous w.r.t. the Lebesgue measure. A has a *pure point spectrum* if $\mathfrak{H}_c = \{0\}$, it has a *purely continuous spectrum* if $\mathfrak{H}_d = \{0\}$. Finally, A has an *absolutely continuous spectrum* if $\mathfrak{H}_d = \mathfrak{H}_{\mathrm{s.c.}} = \{0\}$.

A spectral representation of the type (2.40) is valid for any normal operator which has in general a complex spectrum. One has therefore to consider a spectral measure on $\mathbb{C}$ instead of on $\mathbb{R}$. In particular, for a unitary operator U we have

$$U = \int_{|z|=1} z\, E^U(dz), \tag{2.41}$$

where the spectral measure E^U lives on the unit circle in $\mathbb{C}$.

2.3 Probability in quantum mechanics

Probability lies at the heart of any observation or experiment in physics. Repeated observations of identically prepared systems produce relative frequencies of outcomes and one would like to recover these frequencies in terms of a model of the system under consideration. Essentially two models are in current use in physics: classical probability theory that deals with larger systems at more or less standard conditions of energy, temperature, density ... and quantum probability that applies to either systems at atomic or subatomic level or at very low energies.

Classical probability is now universally described by *Kolmogorov's model*, i.e. there is a basic space of elementary events, equipped with a structure of meaningful events and a functional that assigns a weight to such events. This is precisely the mathematical structure of measure theory $(\mathfrak{X}, \Sigma, \mu)$ as sketched in Section 2.1 with a *probability measure* μ which is normalized in the sense that $\mu(\mathfrak{X}) = 1$. The elementary outcomes form the measurable space $\mathfrak{X}$, the meaningful events are the questions that can be given a yes/no answer and they correspond to the measurable subsets of $\mathfrak{X}$. Finally, the probability that the event $S \in \Sigma$ occurs is given by

$$\mathrm{Prob}(S) = \mu(S). \tag{2.42}$$

The probability measures on a given space $\mathfrak{X}$ have a convex structure. This allows us to look at extreme probability measures, namely those that cannot be written as a non-trivial convex combination of probability measures. These extreme measures, also called *pure states*, are in fact in one-to-one correspondence with the elementary outcomes of the system; they are the Dirac measures living at a single point of the measurable space, see Example 2.34 for the case $\mathfrak{X} = [0, 1]$. A system can be described by classical probability if we are able to construct a Kolmogorovian model that reproduces the observed relative frequencies in terms of probabilities given by (2.42).

The probabilistic model of quantum theories is quite different from the classical. It can be introduced at various degrees of generality. We present here a basic version that is well-suited for the description of systems with few degrees of freedom. In *von Neumann's model*, which preceded in fact Kolmogorov's model, the role of the events is played by the *lattice of projectors* on a Hilbert space $\mathfrak{H}$. The elementary outcomes are identified with the one-dimensional projectors. Probabilities are assigned to events P as in (2.42)

$$\mathrm{Prob}(P) = \mu(P), \tag{2.43}$$

where μ satisfies the following requirements, similar to those in (2.34):

- $\mu(0) = 0$,
- $\mu(\mathbf{I}) = 1$,
- $\mu\left(\sum_j P_j\right) = \sum_j \mu(P_j)$ whenever $\{P_1, P_2, \dots\}$ is an at most countable collection of mutually orthogonal projectors.

The last requirement is the analogue of the σ-additivity of measures, orthogonal projectors replacing the notion of disjoint subsets. Again, the functionals μ in (2.43) have a convex structure and the extreme ones, called pure states, play the role of Dirac measures on the quantum probability space. *Gleason's theorem* settles the structure of the states on the lattice of projections.

Theorem 2.43 (Gleason) *Let $\mathfrak{H}$ be a separable Hilbert space of dimension greater or equal to three. Any state on the lattice of projectors on $\mathfrak{H}$ is a countable convex combination of pure states and any pure state is of the form*

$$\text{Prob}(P) = \langle \varphi, P\,\varphi \rangle,$$

with $\varphi \in \mathfrak{H}$ and $\|\varphi\| = 1$.

2.3.1 *Pure states*

Gleason's theorem provides a firm ground to the idea that pure states are associated with Hilbert space vectors and that they give rise to an interpretation of self-adjoint operators as observables. The connection goes through the Spectral Theorem, which allows us to associate probability distributions to observables.

Let E^A be the spectral measure of a given self-adjoint operator A as in Theorem 2.42. For any Borel subset S of $\mathbb{R}$ we may consider the event that the observed value of A belongs to S. By Gleason's theorem, we can write

$$\text{Prob}(\text{Observed value of } A \text{ belongs to } S) = p_\psi^A(S)$$
$$:= \langle \psi, E^A(S)\psi \rangle = \|E^A(S)\,\psi\|^2. \tag{2.44}$$

It is easy to check, using the properties of the spectral measure E^A, that p_ψ^A uniquely defines a probability measure on $\mathbb{R}$ which is the probability distribution of the measured values of the observable A in the state ψ. From the spectral representation (2.39), it follows that the probability distribution p_ψ^A is concentrated on the spectrum of A and hence the possible measured values of the observable belongs to its spectrum. The average value of the observable $f(A)$ can be obtained from the formula

$$< f(A) >_\psi = \int_{\text{Sp}(A)} f(\lambda)\,\langle \psi, E^A(d\lambda)\,\psi \rangle = \langle \psi, f(A)\,\psi \rangle \tag{2.45}$$

whenever the integral in (2.45) makes sense. In particular, the *mean-square deviation* of A is given by

$$\Delta_\psi(A)^2 := \langle \psi, A^2\,\psi \rangle - \langle \psi, A\,\psi \rangle^2.$$

From the condition under which Schwarz inequality (2.5) becomes equality it follows that $\Delta_\psi(A) = 0$ iff $A\psi = \lambda\psi$, i.e. only the eigenvalues of A can be measured with absolute certainty and then only in states that are eigenvectors of A.

Consider two observables

$$A = \int_{\mathbb{R}} \lambda \, E^A(d\lambda) \quad \text{and} \quad B = \int_{\mathbb{R}} \lambda \, E^B(d\lambda).$$

A and B *commute* if

$$E^A(S_1) \, E^B(S_2) = E^B(S_2) \, E^A(S_1)$$

for any two Borel subsets S_1 and S_2 of $\mathbb{R}$. For bounded observables it is equivalent to $AB = BA$, for unbounded domains have to be taken into account. For commuting observables and a state $\psi \in \mathfrak{H}$ with $\|\psi\| = 1$ we have the *joint probability measure* on $\mathbb{R}^2$,

$$E^{AB}(S_1 \times S_2) := \langle \psi, E^A(S_1) E^B(S_2) \, \psi \rangle = \|E^A(S_1) E^B(S_2) \, \psi\|^2,$$

which gives the probability that the observable A takes values in S_1 and that simultaneously the observable B takes values in S_2. For any real measurable function f on $\mathbb{R}^2$ the operator expression $f(A, B)$ makes sense and its average in the state ψ is given by the integral

$$< f(A, B) >_\psi = \langle \psi, f(A, B) \, \psi \rangle$$
$$:= \int_{\mathbb{R}^2} f(\lambda_1, \lambda_2) \, \|E^{AB}(d\lambda_1, d\lambda_2) \, \psi\|^2.$$

While commuting observables are jointly measurable, non-commuting satisfy *uncertainty relations* in the form

$$\Delta_\psi(A)\Delta_\psi(B) \geq \tfrac{1}{2} \left| \langle \psi, [A, B] \, \psi \rangle \right| \tag{2.46}$$

for ψ in a joint domain of A, A^2, B, B^2, AB and BA.

Example 2.44 The inequality (2.46), applied to the position and momentum operators Q and P of Example 2.33, is known as the *Heisenberg uncertainty relation* and reads

$$\Delta_\psi(Q)\Delta_\psi(P) \geq \frac{\hbar}{2}. \tag{2.47}$$

A quantum state $\langle \cdot \rangle_\psi$ can be viewed as the collection of all classical probability measures p_ψ^A that it defines by letting A vary through the self-adjoint operators on the Hilbert space. There are of course restrictions and compatibility conditions on the family of probability measures p_ψ^A when A varies. For example, for non-commuting observables there exists no canonical joint probability distribution in any state. In Chapter 8, we shall extend the definition of an observable to describe unsharp or fuzzy measurements. Such an operational approach allows us to define joint probability distributions for non-commuting observables which are, however, non-unique and depend on the measurement procedure.

2.3.2   Mixed states, density matrices

Quantum mechanics deals with isolated physical systems and pure states that are identified with normalized vectors in a Hilbert space. In quantum statistical mechanics, we consider open quantum systems and their mixed states which reflect an additional degree of randomness related to the incomplete information about our system.

Writing expression (2.44) in a slightly different form, we obtain, using formulas (2.23) and (2.24),

$$p_\psi^A(S) = \mathrm{Tr}\,|\psi\rangle\langle\psi|\,E^A(S).$$

We can then produce mixtures of probability distributions

$$\sum_j \lambda_j\, p_{\psi_j}^A(S) = \mathrm{Tr}\sum_j \lambda_j\,|\psi_j\rangle\langle\psi_j|\,E^A(S) = \mathrm{Tr}\,\rho\,E^A(S) \qquad (2.48)$$

with $\lambda_j \geq 0$ and $\sum_j \lambda_j = 1$. The positive operator

$$\rho := \sum_j \lambda_j\,|\psi_j\rangle\langle\psi_j|$$

which appears in (2.48) has trace equal to 1 and is called *a density operator* or *density matrix*. It will be identified with a mixed state of a quantum system. The mean value of an observable A now reads

$$<A>_\rho = \mathrm{Tr}\,\rho\,A = \int_{\mathbb{R}} \lambda\,\mathrm{Tr}\,\rho\,E^A(d\lambda),$$

whenever the integral exists.

Example 2.45 In some applications, one deals with a density matrix ρ on $\mathcal{L}^2(\mathbb{R})$ given by an integral kernel that is a continuous function ρ on $\mathbb{R}^2$ which fulfils, for $\psi \in \mathcal{L}^2(\mathbb{R})$, the following conditions:

$$(\rho\,\psi)(x) := \int_{\mathbb{R}} dy\, \rho(x,y)\,\psi(y),$$
$$\int_{\mathbb{R}^2} dx\,dy\, \rho(x,y)\,\overline{\psi(x)}\,\psi(y) \geq 0,$$
$$\int_{\mathbb{R}} dx\, \rho(x,x) = 1.$$

If A is a bounded integral kernel operator defined by the kernel a, see (2.38), then we can explicitly express $\mathrm{Tr}\,\rho\,A$ in terms of the kernels ρ and a:

$$\mathrm{Tr}\,\rho\,A = \int_{\mathbb{R}^2} dx\,dy\, \rho(x,y)\,a(y,x). \qquad (2.49)$$

Often, the expression (2.49) makes sense for a much wider class of ρ and A than we have considered here.

2.4   Observables in quantum mechanics

We shall, in the remainder of this book, not use the quantum logic approach that was sketched at the beginning of Section 2.3 but rather turn to the collection of observables of the system as the basic object. This will also be useful in describing more abstract dynamical systems. Using unbounded observables, even if formally self-adjoint (in fact symmetric), can quickly lead to mathematically ill-defined problems and it may sometimes be quite difficult to make the physically correct choices in such situations. One should be very careful with domain questions when handling unbounded observables and this poses severe mathematical difficulties. For example a sum or a product of two unbounded observables is generally not a well-defined operator with a dense domain in the Hilbert space. The basic unbounded observables that describe the system usually satisfy some commutation relations. Fortunately, as we shall see in this section, such physically relevant unbounded operators, even non-commuting families, may have a common domain of analytic vectors which is large enough to determine them uniquely, a so-called common core. It is then convenient to use as basic observables some bounded functions, such as exponentials, of the unbounded operators. Doing so, one often obtains certain Lie groups of unitary and hence bounded operators. Therefore, it is sufficient to consider as basic mathematical structure the algebra $\mathcal{B}(\mathfrak{H})$ of all bounded operators on a Hilbert space $\mathfrak{H}$ or some suitably chosen subalgebra.

2.4.1   Compact operators

The first example of a norm-closed subalgebra contains observables with discrete spectrum corresponding to the measurement procedures which involve filtering of a discrete set of quantum states.

An operator $G \in \mathcal{B}(\mathfrak{H})$ is a *finite-rank operator* if it can be written in Dirac notation as

$$G = \sum_{j=1}^{k} |\psi_j\rangle\langle\phi_j|, \quad \psi_j, \phi_j \in \mathfrak{H}.$$

Finite-rank operators form a subalgebra $\mathcal{C}_0(\mathfrak{H}) \subset \mathcal{B}(\mathfrak{H})$ which is closed under the adjoint operation. It is, however, not norm-closed. The norm closure of $\mathcal{C}_0(\mathfrak{H})$, denoted by $\mathcal{C}(\mathfrak{H})$, is the *algebra of compact operators*. The general form of a compact operator is

$$C = \sum_{j=1}^{\infty} \gamma_j \, |\varphi_j\rangle\langle\psi_j|,$$

where $\{\varphi_1, \varphi_2, \dots\}$ and $\{\psi_1, \psi_2, \dots\}$ are two orthonormal bases of $\mathfrak{H}$, $\gamma_j \in \mathbb{C}$ and $\lim_{n\to\infty} \gamma_j = 0$. $\mathcal{C}(\mathfrak{H})$ is closed under taking adjoints and AC and CA are compact whenever C is compact and A is bounded.

Some subclasses of compact operators play an important role. Among them the Hilbert–Schmidt and trace-class operators. A linear operator K on $\mathfrak{H}$ is a

Hilbert–Schmidt operator if there exists an orthonormal basis $\{\varphi_1, \varphi_2, \dots\}$ of $\mathfrak{H}$ such that

$$\sum_{j=1}^{\infty} \|K\,\varphi_j\|^2 < \infty. \tag{2.50}$$

It follows from (2.50) that K is compact, therefore *a fortiori* bounded, and that the sum in (2.50) is independent of the chosen basis. The product of two Hilbert–Schmidt operators is trace-class and this allows us to introduce a canonical inner product between Hilbert–Schmidt operators:

$$\langle K, L \rangle := \operatorname{Tr} K^* L = \sum_{j=1}^{\infty} \langle K\,\varphi_j, L\,\varphi_j \rangle. \tag{2.51}$$

The space $\mathcal{T}_2(\mathfrak{H})$ of Hilbert–Schmidt operators is a linear space which is complete with respect to the norm induced by the inner product (2.51). It is therefore a Hilbert space.

A compact operator T is a *trace-class operator* if there exists in $\mathfrak{H}$ a basis $\{\varphi_1, \varphi_2, \dots\}$ such that the infinite series $\sum_j \langle \varphi_j, |T|\,\varphi_j \rangle$ converges where the absolute value $|T|$ of T is the square root of T^*T. The sum of this series and of $\sum_j \langle \varphi_j, T\,\varphi_j \rangle$ is independent of the choice of the basis and

$$\operatorname{Tr} T := \sum_j \langle \varphi_j, T\,\varphi_j \rangle$$

is the *trace of* T. This definition of trace-class operators is equivalent to the one given in (2.22).

The space of all trace-class operators equipped with the *trace norm*

$$\|T\|_1 := \operatorname{Tr} |T|$$

is a Banach space and is denoted by $\mathcal{T}(\mathfrak{H})$. Let $C \in \mathcal{C}(\mathfrak{H})$, $T \in \mathcal{T}(\mathfrak{H})$ and $A \in \mathcal{B}(\mathfrak{H})$, then $CA, AC \in \mathcal{C}(\mathfrak{H})$, $TA, AT \in \mathcal{T}(\mathfrak{H})$,

$$\begin{aligned} \left|\operatorname{Tr} CT\right| &\le \|CT\|_1 \le \|C\|\,\|T\|_1, \\ \left|\operatorname{Tr} TA\right| &\le \|TA\|_1 \le \|A\|\,\|T\|_1. \end{aligned} \tag{2.52}$$

These inequalities are optimal in the sense that

$$\sup_{\|C\|\le 1} \left|\operatorname{Tr} TC\right| = \|T\|_1 \quad \text{and} \quad \sup_{\|T\|_1 \le 1} \left|\operatorname{Tr} TA\right| = \|A\|.$$

The inequalities (2.52) show that the linear functionals

$$\begin{aligned} F(C) &:= \operatorname{Tr} TC, \quad T \in \mathcal{T}(\mathfrak{H}), \ C \in \mathcal{C}(\mathfrak{H}) \\ F(T) &:= \operatorname{Tr} TA, \quad T \in \mathcal{T}(\mathfrak{H}), \ A \in \mathcal{B}(\mathfrak{H}) \end{aligned} \tag{2.53}$$

are continuous, and the following theorem states that these forms of the functionals are general.

Theorem 2.46 *Let $\mathfrak{H}$ be a separable Hilbert space, then*

$$\mathcal{C}(\mathfrak{H})^* = \mathcal{T}(\mathfrak{H}) \quad and \quad \mathcal{T}(\mathfrak{H})^* = \mathcal{B}(\mathfrak{H})$$

in the sense of isometric isomorphisms of Banach spaces, with the pairing between dual elements given as in (2.53).

A functional F on $\mathcal{C}(\mathfrak{H})$ is *positive* if it assigns a non-negative value to any positive operator, it is *normalized* if $F(\mathbf{1}) = 1$. Theorem 2.46 characterizes the density matrices as the positive, normalized, continuous, linear functionals on the algebra of the compact operators. Indeed, plugging the projector on a vector ψ in (2.53), positivity requires that $\langle \psi, T \psi \rangle$ be positive. Moreover, normalization means $\operatorname{Tr} T = 1$, hence T has to be a density matrix. A density matrix also determines a functional on the algebra of all bounded linear operators on the Hilbert space, but there exist many more. Functionals of the form

$$F(A) = \operatorname{Tr} TA, \quad T \in \mathcal{T}(\mathfrak{H}), \quad A \in \mathcal{B}(\mathfrak{H})$$

are *normal* and they play an important role in the theory of von Neumann algebras.

The algebra $\mathcal{B}(\mathfrak{H})$ of bounded operators is not always the best choice for algebra of observables. This is because all infinite-dimensional separable Hilbert spaces are isomorphic, see Section 2.1.4, which implies the isomorphism of the corresponding $\mathcal{B}(\mathfrak{H})$'s, $\mathcal{C}(\mathfrak{H})$'s, ... Such an isomorphism is, however, irrelevant from the physical point of view. In the classical theory it would correspond to the statement that all phase spaces have the same cardinality.

The relevant structure for a classical phase space Γ is that of a differentiable manifold. This determines, for example, the degrees of freedom of the system. In order to compare this with the quantum situation, we have to express the structure on the level of functions on phase space which leads us to consider the Abelian algebra $\mathcal{S}(\Gamma)$ of infinitely differentiable complex functions on Γ (with rapid decrease at infinity in the non-compact case) equipped with its natural derivations. In classical mechanics, this structure is more relevant than that of the continuous complex functions on Γ (vanishing at infinity) which would only characterize Γ up to homeomorphisms.

Similarly, the quantum topology, which encodes important features of the physical system, can be prescribed by choosing a distinguished norm-closed sub-algebra $\mathfrak{A} \subset \mathcal{B}(\mathfrak{H})$, but the truly relevant structure is determined by specifying a subalgebra $\mathfrak{A}_0$ of smooth operators, equipped with its corresponding derivations. We now present a few examples; later on more will follow.

2.4.2 *Weyl quantization*

In this section, we present an algebraic version of standard quantum mechanics with d degrees of freedom, the counterpart of classical systems with phase space $\mathbb{R}^d \times \mathbb{R}^d$.

On the Schwartz space $\mathcal{S}(\mathbb{R}^d)$, the *position operators* $\mathbf{Q} = (Q_1, Q_2, \dots, Q_d)$,

$$(Q_j\, \varphi)(\mathbf{x}) := x_j \varphi(\mathbf{x}),$$

and the *momentum operators* $\mathbf{P} = (P_1, P_2, \dots, P_d)$,

$$P_j := -i\hbar \frac{\partial}{\partial x_j},$$

are essentially self-adjoint, see Example 2.33. The space $\mathcal{S}(\mathbb{R}^d)$ is a common core for $\mathbf{Q}$ and $\mathbf{P}$ and they satisfy the commutation relations

$$(Q_j\, P_k - P_k\, Q_j)\varphi = i\hbar\delta_{jk}\, \varphi, \quad \varphi \in \mathcal{S}(\mathbb{R}^d)$$

on this domain.

As the Q_j are commuting multiplication operators, we can easily write down the joint spectral measure $E^{\mathbf{Q}}$ of $\mathbf{Q}$. For any Borel subset S of $\mathbb{R}^d$, $E^{\mathbf{Q}}(S)$ is the multiplication operator by the indicator function of S

$$\left(E^{\mathbf{Q}}(S)\, \varphi\right)(\mathbf{x}) = \begin{cases} \varphi(\mathbf{x}) & \text{for } \mathbf{x} \in S, \\ 0 & \text{otherwise.} \end{cases}$$

The Fourier transform

$$\varphi^\wedge(\mathbf{p}) := \frac{1}{(2\pi\hbar)^{d/2}} \int_{\mathbb{R}^d} d\mathbf{x}\, e^{-i\mathbf{p}\cdot\mathbf{x}/\hbar}\, \varphi(\mathbf{x})$$

is a unitary transformation from the position representation $\mathcal{L}^2(\mathbb{R}^d)$ to the momentum representation $\mathcal{L}^2(\mathbb{R}^d)$ that transforms position into momentum,

$$(\mathbf{Q}\, \varphi)^\wedge = \mathbf{P}\, \varphi^\wedge,$$

and the same relation relates the joint spectral measure of $\mathbf{P}$ to that of $\mathbf{Q}$.

Next, we want to use smooth functions of both position and momentum but, even for polynomials in two indeterminates, there are several possibilities of associating an operator with such a function. This is due to the lack of commutativity between position and momentum operators and it is therefore important to agree on the ordering of these operators in terms that contain several non-commuting factors.

In order to avoid domain problems, it is useful to pass to an exponentiated version of the canonical commutation relations, known as the *Weyl relations*. Using the spectral measures of $\mathbf{P}$ and $\mathbf{Q}$ we easily obtain the explicit actions of $\{\exp(i\mathbf{p} \cdot \mathbf{Q}/\hbar) \mid \mathbf{p} \in \mathbb{R}^d\}$ and $\{\exp(-i\mathbf{q} \cdot \mathbf{P}/\hbar) \mid \mathbf{q} \in \mathbb{R}^d\}$:

$$\left(e^{i\mathbf{p}\cdot\mathbf{Q}/\hbar}\, \varphi\right)(\mathbf{x}) = e^{i\mathbf{p}\cdot\mathbf{x}/\hbar}\, \varphi(\mathbf{x}),$$
$$\left(e^{-i\mathbf{q}\cdot\mathbf{P}/\hbar}\, \varphi\right)(\mathbf{x}) = \varphi(\mathbf{x} - \mathbf{q}).$$

We can then compute the adjoints and commutators of the unitary operators $\exp(i\mathbf{p} \cdot \mathbf{Q}/\hbar)$ and $\exp(i\mathbf{q} \cdot \mathbf{P}/\hbar)$:

$$\left(e^{i\mathbf{p}\cdot\mathbf{Q}/\hbar}\right)^* = e^{-i\mathbf{p}\cdot\mathbf{Q}/\hbar}, \qquad \left(e^{-i\mathbf{q}\cdot\mathbf{P}/\hbar}\right)^* = e^{i\mathbf{q}\cdot\mathbf{P}/\hbar},$$

$$e^{-i\mathbf{q}\cdot\mathbf{P}/\hbar}\, e^{i\mathbf{p}\cdot\mathbf{Q}/\hbar} = e^{-i\mathbf{q}\cdot\mathbf{p}/\hbar}\, e^{i\mathbf{p}\cdot\mathbf{Q}/\hbar}\, e^{-i\mathbf{q}\cdot\mathbf{P}/\hbar}.$$

The *Weyl operators* are defined by

$$W(\mathbf{q},\mathbf{p}) := e^{i\mathbf{q}\cdot\mathbf{p}/2\hbar}\, e^{-i\mathbf{q}\cdot\mathbf{P}/\hbar}\, e^{i\mathbf{p}\cdot\mathbf{Q}/\hbar}$$

$$= e^{-i\mathbf{q}\cdot\mathbf{p}/2\hbar}\, e^{i\mathbf{p}\cdot\mathbf{Q}/\hbar}\, e^{-i\mathbf{q}\cdot\mathbf{P}/\hbar} = e^{i(\mathbf{p}\cdot\mathbf{Q}-\mathbf{q}\cdot\mathbf{P})/\hbar}, \tag{2.54}$$

where the last equality can be checked by differentiation. The explicit action of $W(\mathbf{q},\mathbf{p})$ reads

$$\big(W(\mathbf{q},\mathbf{p})\,\varphi\big)(\mathbf{x}) = e^{i\mathbf{p}\cdot(\mathbf{x}-\mathbf{q}/2)/\hbar}\,\varphi(\mathbf{x}-\mathbf{q}).$$

The Weyl operators satisfy the basic algebraic relations that are equivalent to the *canonical commutation relations*

$$\begin{aligned} W(\mathbf{q},\mathbf{p})^* &= W(-\mathbf{q},-\mathbf{p}), \\ W(\mathbf{q}_1,\mathbf{p}_1)\,W(\mathbf{q}_2,\mathbf{p}_2) &= e^{-i(\mathbf{q}_1\cdot\mathbf{p}_2-\mathbf{q}_2\cdot\mathbf{p}_1)/2\hbar}\,W(\mathbf{q}_1+\mathbf{q}_2,\mathbf{p}_1+\mathbf{p}_2). \end{aligned} \tag{2.55}$$

The *CCR-algebra* with d degrees of freedom is the closure of the linear span of the Weyl unitaries in $\mathcal{B}(\mathfrak{H})$ in operator norm.

It is not very convenient to work directly in the CCR-algebra for the purpose of quantum mechanics because it turns out that a non-linear dynamics, such as the usual kinetic energy plus a non-quadratic potential, will not map the CCR-algebra into itself. We therefore introduce an $\hbar$-deformation of the classical phase space $\mathbb{R}^{2d}$ that suits our aims much better. We introduce this deformation rather on the level of functions on phase space than on phase space itself and denote by $\mathcal{S}_\hbar(\mathbb{R}^{2d})$ the non-commutative algebra of smooth functions with rapid decrease at infinity on $\hbar$-deformed phase space. This algebra is obtained by modifying the usual pointwise product of the classical algebra $\mathcal{S}(\mathbb{R}^{2d})$. The algebraic structure on $\mathcal{S}_\hbar(\mathbb{R}^{2d})$ is obtained by transporting the algebraic operations in $\mathcal{B}(\mathcal{L}^2(\mathbb{R}^d))$ through the smearing of Weyl operators.

We proceed as follows: as a set, $\mathcal{S}_\hbar(\mathbb{R}^{2d})$ coincides with $\mathcal{S}(\mathbb{R}^{2d})$. Observe in the first place that the Fourier transform

$$g^\wedge(\mathbf{q},\mathbf{p}) := \frac{1}{(2\pi\hbar)^d} \int_{\mathbb{R}^{2d}} d\mathbf{p}' d\mathbf{q}'\, e^{i(\mathbf{q}\cdot\mathbf{p}'-\mathbf{q}'\cdot\mathbf{p})/\hbar}\, g(\mathbf{p}',\mathbf{q}')$$

of a $g \in \mathcal{S}(\mathbb{R}^{2d})$ belongs again to $\mathcal{S}(\mathbb{R}^{2d})$. It can therefore be used to smear the Weyl operators and we obtain a bounded operator

$$W(g^\wedge) := \frac{1}{(2\pi\hbar)^d} \int_{\mathbb{R}^{2d}} d\mathbf{q}\, d\mathbf{p}\, g^\wedge(\mathbf{q},\mathbf{p})\, W(\mathbf{q},\mathbf{p}). \tag{2.56}$$

The mapping $g \mapsto W(g^\wedge)$ is called a *quantization procedure*.

Due to the linearity of the Fourier transform and the integral in (2.56), the addition of elements in $\mathcal{S}_\hbar(\mathbb{R}^{2d})$ and the multiplication by complex scalars remain the same as in the classical algebra. We now consider adjoints and products.

The Hermitian conjugate of $W(g^\wedge)$ is easily seen to be

$$W(g^\wedge)^* = W((g^\wedge)^\dagger)$$

with

$$(g^\wedge)^\dagger(\mathbf{q}, \mathbf{p}) := \overline{g^\wedge(-\mathbf{q}, -\mathbf{p})}.$$

As

$$(g^\wedge)^\dagger(\mathbf{q}, \mathbf{p}) := \frac{1}{(2\pi\hbar)^d} \int_{\mathbb{R}^{2d}} d\mathbf{p}'d\mathbf{q}' \, e^{i(\mathbf{q}\cdot\mathbf{p}'-\mathbf{q}'\cdot\mathbf{p})/\hbar} \, \overline{g(\mathbf{p}', \mathbf{q}')},$$

the adjoint of $g \in \mathcal{S}_\hbar(\mathbb{R}^{2d})$ coincides with its complex conjugate

$$g^* = \overline{g}. \tag{2.57}$$

It remains to construct the product in $\mathcal{S}_\hbar(\mathbb{R}^{2d})$. Using the relations (2.55), we obtain

$$W(g_1^\wedge) \, W(g_2^\wedge) = W(g_1^\wedge * g_2^\wedge),$$

with

$$
\begin{aligned}
&(g_1^\wedge * g_2^\wedge)(\mathbf{q}, \mathbf{p}) \\
&:= \frac{1}{(2\pi\hbar)^d} \int_{\mathbb{R}^{2d}} d\mathbf{q}'d\mathbf{p}' \, e^{i(\mathbf{q}\cdot\mathbf{p}'-\mathbf{q}'\cdot\mathbf{p})/2\hbar} \, g_1^\wedge(\mathbf{q}', \mathbf{p}') \, g_2^\wedge(\mathbf{q} - \mathbf{q}', \mathbf{p} - \mathbf{p}').
\end{aligned}
$$

$g_1^\wedge * g_2^\wedge$ belongs again to $\mathcal{S}(\mathbb{R}^{2d})$ and is therefore the Fourier transform of a unique function in $\mathcal{S}(\mathbb{R}^{2d})$. We define the product in $\mathcal{S}_\hbar(\mathbb{R}^{2d})$ implicitly by

$$(g_1 \cdot_\hbar g_2)^\wedge := g_1^\wedge * g_2^\wedge. \tag{2.58}$$

The operations (2.57) and (2.58), together with the usual addition and multiplication by complex scalars, turn $\mathcal{S}_\hbar(\mathbb{R}^{2d})$ in a non-commutative $*$-algebra, equipped with a norm $\|\cdot\|$ that satisfies the C*-property

$$\|x\|^2 = \|x^*\|^2 = \|x^* x\|.$$

The deformed product $\cdot_\hbar$ in (2.58) can be written out explicitly

$$
\begin{aligned}
&(g_1 \cdot_\hbar g_2)(\mathbf{p}, \mathbf{q}) \\
&= \frac{1}{(\pi\hbar)^{2d}} \int d\mathbf{p}_1 d\mathbf{q}_1 \, d\mathbf{p}_2 d\mathbf{q}_2 \, g_1(\mathbf{p}_1, \mathbf{q}_1) \, g_2(\mathbf{p}_2, \mathbf{q}_2) \\
&\qquad \times e^{2i\left((\mathbf{p}_1-\mathbf{p})\cdot(\mathbf{q}_1-\mathbf{q})-(\mathbf{p}_1-\mathbf{p})\cdot(\mathbf{q}_2-\mathbf{q})\right)/\hbar}.
\end{aligned}
$$

Clearly, the deformed product is no longer local and one can easily see that it tends to the usual pointwise product when $\hbar \to 0$ by using

$$\lim_{n \to \infty} n\, e^{nipq} = \delta(p)\delta(q), \quad p, q \in \mathbb{R}, \tag{2.59}$$

where the convergence in (2.59) is in the sense of distributions.

One may check that the smeared Weyl operators are trace-class with

$$\operatorname{Tr} W(g^\wedge) = g^\wedge(0,0) = \frac{1}{(2\pi\hbar)^d} \int_{\mathbb{R}^{2d}} d\mathbf{p} d\mathbf{q}\, g(\mathbf{p},\mathbf{q}), \quad g \in \mathcal{S}(\mathbb{R}^{2d}).$$

Another useful relation is

$$\operatorname{Tr} W(g^\wedge)^* W(g^\wedge) = \frac{1}{(2\pi\hbar)^d} \int_{\mathbb{R}^{2d}} d\mathbf{q} d\mathbf{p}\, |g^\wedge(\mathbf{q},\mathbf{p})|^2$$
$$= \frac{1}{(2\pi\hbar)^d} \int_{\mathbb{R}^{2d}} d\mathbf{p} d\mathbf{q}\, |g(\mathbf{p},\mathbf{q})|^2.$$

Moreover, $\{W(g^\wedge) \mid g \in \mathcal{S}(\mathbb{R}^{2d})\}$ is a norm-dense subalgebra of the compact operators on $\mathcal{L}^2(\mathbb{R}^d)$. This means that the norm completion of $\mathcal{S}_\hbar(\mathbb{R}^{2d})$ is just the algebra of compact operators on a separable Hilbert space. Therefore, isomorphisms of the norm closure of $\mathcal{S}_\hbar(\mathbb{R}^{2d})$, the compact operators in this case, cannot distinguish between different numbers of degrees of freedom. The useful object to consider is $\mathcal{S}_\hbar(\mathbb{R}^{2d})$ together with its derivations.

Any derivation ∇ of $\mathcal{S}_\hbar(\mathbb{R}^{2d})$, i.e. a linear transformation of $\mathcal{S}_\hbar(\mathbb{R}^{2d})$ that satisfies the conditions

$$\nabla(g^*) = \nabla(g)^*,$$
$$\nabla(g_1 \cdot_\hbar g_2) = \nabla(g_1) \cdot_\hbar g_2 + g_1 \cdot_\hbar \nabla(g_2),$$

can be built in terms of $2d$ fundamental derivations on $\mathcal{S}_\hbar(\mathbb{R}^{2d})$,

$$\nabla_{q_j}(g) := \frac{\partial g}{\partial p_j} \quad \text{and} \quad \nabla_{p_j}(g) := -\frac{\partial g}{\partial q_j}, \quad j = 1, 2, \ldots, d,$$

and suitably chosen infinitely differentiable functions f_j and h_j with slow increase at infinity as

$$\nabla(g) = \sum_j f_j \nabla_{q_j}(g) + h_j \nabla_{p_j}(g).$$

2.5　Composed systems

Forming direct sums and tensor products of Hilbert spaces provides two general ways of building larger spaces from given ones. For example, when describing the decay process of a neutron into a proton, an electron and an antineutrino

$$n \longrightarrow p\, e\, \overline{\nu_e}$$

one has to consider a state space that contains both the states of the unstable particle, the neutron, and of the decay products. This can be done in terms of the direct sum

$$\mathfrak{H} = \mathfrak{H}_n \oplus \mathfrak{H}_{\text{d.p.}}.$$

As the decay products themselves are different particles, a state of the decay products should specify the state of each of them. This is done with the help of the tensor product

$$\mathfrak{H}_{\text{d.p.}} = \mathfrak{H}_p \otimes \mathfrak{H}_e \otimes \mathfrak{H}_{\overline{\nu_e}}.$$

2.5.1 *Direct sums*

Example 2.47 A general column vector $\mathbf{x}$ in $\mathbb{C}^{m+n}$, with entries x_1 up to x_{m+n}, can naturally be identified with two column vectors of lengths m and n obtained by splitting $\mathbf{x}$ between the mth and $(m+1)$th row. These vectors $\mathbf{x}_1$ and $\mathbf{x}_2$ belong to $\mathbb{C}^m$, respectively $\mathbb{C}^n$, and have entries x_1 up to x_m for $\mathbf{x}_1$, and x_{m+1} up to x_{m+n} for $\mathbf{x}_2$. We write $\mathbf{x} = \mathbf{x}_1 \oplus \mathbf{x}_2$ and say that $\mathbf{x}$ is the *direct sum* of $\mathbf{x}_1$ and $\mathbf{x}_2$. It is obvious that

$$(\mathbf{x}_1 + \alpha \mathbf{y}_1) \oplus \mathbf{x}_2 = \mathbf{x}_1 \oplus \mathbf{x}_2 + \alpha(\mathbf{y}_1 \oplus \mathbf{x}_2),$$
$$\mathbf{x}_1 \oplus (\mathbf{x}_2 + \alpha \mathbf{y}_2) = \mathbf{x}_1 \oplus \mathbf{x}_2 + \alpha(\mathbf{x}_1 \oplus \mathbf{y}_2),$$
$$\langle \mathbf{x}_1 \oplus \mathbf{x}_2, \mathbf{y}_1 \oplus \mathbf{y}_2 \rangle = \langle \mathbf{x}_1, \mathbf{y}_1 \rangle + \langle \mathbf{x}_2, \mathbf{y}_2 \rangle.$$

One could say that forming direct sums amounts to extending the range of the index that labels the entries of a vector.

Theorem 2.48 *Given a collection* $\{\mathfrak{H}_1, \mathfrak{H}_2, \dots\}$ *of Hilbert spaces, the set*

$$\{(\varphi_1, \varphi_2, \dots) \mid \sum_j \|\varphi_j\|^2 < \infty\}$$

equipped with the operations

$$(\varphi_j) + \alpha\,(\psi_j) := (\varphi_j + \alpha\,\psi_j),$$
$$\left\langle (\varphi_j), (\psi_j) \right\rangle := \sum_j \langle \varphi_j, \psi_j \rangle$$

is a Hilbert space. (φ_j) *is called the* direct sum *of the vectors* φ_j *and denoted* $\bigoplus_j \varphi_j$. *Similarly, one writes* $\bigoplus_j \mathfrak{H}_j$ *for the Hilbert space of the* (φ_j).

Forming direct sums is commutative and associative in the sense that there are natural identifications of $\mathfrak{H}_1 \oplus \mathfrak{H}_2$ with $\mathfrak{H}_2 \oplus \mathfrak{H}_1$ and of $(\mathfrak{H}_1 \oplus \mathfrak{H}_2) \oplus \mathfrak{H}_3$ with $\mathfrak{H}_1 \oplus (\mathfrak{H}_2 \oplus \mathfrak{H}_3)$ so that we can denote both by $\mathfrak{H}_1 \oplus \mathfrak{H}_2 \oplus \mathfrak{H}_3$. Given orthonormal bases $\{e_{j1}, e_{j2}, \dots\}$ of the $\mathfrak{H}_j$, $\{0 \oplus \dots \oplus 0 \oplus e_{j\ell} \oplus 0 \oplus \dots \mid j, \ell\}$ is an orthonormal basis for $\bigoplus_j \mathfrak{H}_j$. Hence, the dimension of a direct sum is just the sum of the dimensions of the terms. In particular, a countable direct sum of separable spaces is again separable.

Example 2.49 Let S_1 and S_2 be two disjoint measurable subsets of $\mathbb{R}$. The square integrable functions $\mathcal{L}^2(S_1 \cup S_2)$ on $S_1 \cup S_2$ can naturally be identified with $\mathcal{L}^2(S_1) \oplus \mathcal{L}^2(S_2)$ in the following way. Given $f \in \mathcal{L}^2(S_1 \cup S_2)$, let

$$f_1 := f\Big|_{S_1} \quad \text{and} \quad f_2 := f\Big|_{S_2};$$

then

$$f \equiv f_1 \oplus f_2.$$

The *direct integral* is a continuous version of the direct sum construction. Let $(\mathfrak{X}, \Sigma, \mu)$ be a measure space. We say that a Hilbert space $\mathfrak{H}$ is the direct integral of a family $\{\mathfrak{H}_x \mid x \in \mathfrak{X}\}$ of Hilbert spaces if we have a dense subspace $\mathfrak{H}_0$ of $\mathfrak{H}$ such that for $\varphi, \psi \in \mathfrak{H}_0$

(i) φ maps $x \in \mathfrak{X}$ to $\varphi(x) \in \mathfrak{H}_x$,

(ii) $x \mapsto \langle \varphi(x), \psi(x) \rangle$ is measurable,

(iii) $\int_{\mathfrak{X}} \mu(dx) \|\varphi(x)\|^2 < \infty$, and

(iv) $\langle \varphi, \psi \rangle = \int_{\mathfrak{X}} \mu(dx) \langle \varphi(x), \psi(x) \rangle$.

Moreover, if for $x \in \mathfrak{X}$, $\eta(x) \in \mathfrak{H}_x$ is such that all functions $x \mapsto \langle \varphi(x), \eta(x) \rangle$ with $\varphi \in \mathfrak{H}_0$ are measurable and $x \mapsto \|\eta(x)\|^2$ is integrable, then $\eta \in \mathfrak{H}_0$. We use the notations

$$\mathfrak{H} = \int_{\mathfrak{X}}^{\oplus} \mu(dx) \, \mathfrak{H}_x \quad \text{and} \quad \varphi = \int_{\mathfrak{X}}^{\oplus} \mu(dx) \, \varphi(x).$$

2.5.2   *Tensor products*

Example 2.50 The tensor product construction makes $\mathbb{C}^m \otimes \mathbb{C}^n$ isomorphic to $\mathbb{C}^{mn}$. To each pair of vectors $\mathbf{x} \in \mathbb{C}^m$ and $\mathbf{y} \in \mathbb{C}^n$ we associate the vector $\mathbf{x} \otimes \mathbf{y} \in \mathbb{C}^{mn}$ with kth entry $x_{q+1}y_{r+1}$ where $k = qn + r$ with $q, r \in \mathbb{N}$ and $0 \leq r < n$. Vectors of the kind $\mathbf{x} \otimes \mathbf{y}$ are called *elementary tensors* but they do not form a linear subspace of $\mathbb{C}^{mn}$. If $\{\mathbf{e}_1, \mathbf{e}_2, \ldots, \mathbf{e}_m\}$ and $\{\mathbf{f}_1, \mathbf{f}_2, \ldots, \mathbf{f}_n\}$ are the standard bases of $\mathbb{C}^m$ and $\mathbb{C}^n$, then all the entries of $\mathbf{e}_k \otimes \mathbf{f}_\ell$ are zero but for a 1 on row $(k-1)n + \ell + 1$. This means that $\{\mathbf{e}_k \otimes \mathbf{f}_\ell \mid k = 1, 2, \ldots, m, \ \ell = 1, 2, \ldots, n\}$ is just the standard basis of $\mathbb{C}^{mn}$. Therefore, any vector in $\mathbb{C}^{mn}$ is a linear combination of elementary tensors. The scalar product of two elementary tensors is given by

$$\langle \mathbf{x}_1 \otimes \mathbf{y}_1, \mathbf{x}_2 \otimes \mathbf{y}_2 \rangle = \langle \mathbf{x}_1, \mathbf{x}_2 \rangle \langle \mathbf{y}_1, \mathbf{y}_2 \rangle.$$

One may think of the tensor product as creating an additional level of indices for the entries.

Given any two vector spaces $\mathfrak{V}$ and $\mathfrak{W}$, we can form their *algebraic tensor product* $\mathfrak{V} \otimes \mathfrak{W}$. It consists of the finite linear combinations of *elementary tensors* $x \otimes y$ with $x \in \mathfrak{V}$ and $y \in \mathfrak{W}$. These tensors are multilinear in their arguments:

$$(x_1 + \alpha\, x_2) \otimes y = x_1 \otimes y + \alpha\ (x_2 \otimes y),$$
$$x \otimes (y_1 + \alpha\, y_2) = x \otimes y_1 + \alpha\ (x \otimes y_2).$$

If $\mathfrak{V}$ and $\mathfrak{W}$ are normed spaces, then it is generally a delicate problem to turn $\mathfrak{V} \otimes \mathfrak{W}$ into a normed space.

In the Hilbert space case, one shows that

$$\langle \varphi_1 \otimes \psi_1, \varphi_2 \otimes \psi_2 \rangle := \langle \varphi_1, \varphi_2 \rangle \langle \psi_1, \psi_2 \rangle, \tag{2.60}$$

with the φ's in $\mathfrak{H}$ and the ψ's in $\mathfrak{K}$, extends by linearity to an inner product on the algebraic tensor product of $\mathfrak{H}$ and $\mathfrak{K}$. The *Hilbert tensor product* of $\mathfrak{H}$ and $\mathfrak{K}$ is the completion of the algebraic tensor product of the Hilbert spaces $\mathfrak{H}$ and $\mathfrak{K}$ with respect to the norm defined by the inner product (2.60). If $\{e_1, e_2, \dots\}$ and $\{f_1, f_2, \dots\}$ are orthonormal bases of $\mathfrak{H}$ and $\mathfrak{K}$, then $\{e_i \otimes f_j \mid i, j\}$ is an orthonormal basis for $\mathfrak{H} \otimes \mathfrak{K}$.

This leads to the relation $\dim(\mathfrak{H} \otimes \mathfrak{K}) = \dim(\mathfrak{H}) \dim(\mathfrak{K})$. The following property is often useful when computing with tensor products: given an orthonormal basis $\{f_1, f_2, \dots\}$ of $\mathfrak{K}$, there exists for each $\varphi \in \mathfrak{H} \otimes \mathfrak{K}$ a unique set of vectors $\varphi_j \in \mathfrak{H}$ such that

$$\varphi = \sum_j \varphi_j \otimes f_j. \tag{2.61}$$

The mapping $\varphi \mapsto \bigoplus_j \varphi_j$ is an isomorphism between $\mathfrak{H} \otimes \mathfrak{K}$ and $\bigoplus_j \mathfrak{H}$.

There is a natural isomorphism between $\mathfrak{H} \otimes \mathfrak{K}$ and $\mathfrak{K} \otimes \mathfrak{H}$ and between $(\mathfrak{H} \otimes \mathfrak{K}) \otimes \mathfrak{L}$ and $\mathfrak{H} \otimes (\mathfrak{K} \otimes \mathfrak{L})$, which allows us to denote such triple tensor products by $\mathfrak{H} \otimes \mathfrak{K} \otimes \mathfrak{L}$. Moreover, the tensor product is distributive over the direct sum

$$(\mathfrak{H} \oplus \mathfrak{K}) \otimes \mathfrak{L} = (\mathfrak{H} \otimes \mathfrak{L}) \oplus (\mathfrak{K} \otimes \mathfrak{L}).$$

Example 2.51 Let $(\mathfrak{X}_1, \Sigma_1, \mu_1)$ and $(\mathfrak{X}_2, \Sigma_2, \mu_2)$ be two measurable spaces, then there is a natural isomorphism between $\mathcal{L}^2(\mathfrak{X}_1, \mu_1) \otimes \mathcal{L}^2(\mathfrak{X}_2, \mu_2)$ and the space $\mathcal{L}^2(\mathfrak{X}_1 \times \mathfrak{X}_2, \mu_1 \times \mu_2)$ of square integrable functions on the product of the measure spaces. This is explicitly given on elementary tensors by

$$f_1 \otimes f_2 \equiv \Big((x_1, x_2) \mapsto f_1(x_1) f_2(x_2) \Big),$$

$f_1 \in \mathcal{L}^2(\mathfrak{X}_1, \mu_1)$, $f_2 \in \mathcal{L}^2(\mathfrak{X}_2, \mu_2)$, $x_1 \in \mathfrak{X}_1$ and $x_2 \in \mathfrak{X}_2$.

Example 2.52 The Hilbert space of states of a particle with spin s moving in $\mathbb{R}^3$ can be written in a number of different ways using the constructs of above:

$$\mathcal{L}^2(\mathbb{R}^3) \otimes \mathbb{C}^{2s+1} \equiv \bigoplus_{m=-s}^{s} \mathcal{L}^2(\mathbb{R}^3, d\mathbf{x}) \equiv \int_{\mathbb{R}^3}^{\oplus} d\mathbf{x}\, \mathbb{C}^{2s+1}.$$

2.5.3 *Observables and states of composite systems*

Given bounded operators $A \in \mathcal{B}(\mathfrak{H})$ and $B \in \mathcal{B}(\mathfrak{K})$, we can form the *direct sum* $A \oplus B$ by putting

$$(A \oplus B)(\varphi \oplus \psi) := A\varphi \oplus B\psi, \quad \varphi \in \mathfrak{H}, \ \psi \in \mathfrak{K}.$$

As

$$
\begin{aligned}
\|A\varphi \oplus B\psi\|^2 &= \|A\varphi\|^2 + \|B\psi\|^2 \\
&\leq \max\{\|A\|, \|B\|\} \left(\|\varphi\|^2 + \|\psi\|^2\right) \\
&= \max\{\|A\|, \|B\|\} \|\varphi \oplus \psi\|^2,
\end{aligned}
$$

we see that $\|A \oplus B\| \leq \max\{\|A\|, \|B\|\}$. In fact, there is equality.

Similarly, we can form the *tensor product* $A \otimes B$, but this is slightly more delicate. Putting on elementary tensors

$$(A \otimes B)(\varphi \otimes \psi) := A\varphi \otimes B\psi,$$

we have $A \otimes B = (A \otimes \mathbf{1})(\mathbf{1} \otimes B)$. In order to prove the boundedness of $A \otimes B$, it is sufficient to show that $A \otimes \mathbf{1}$ extends to a bounded linear operator. We use (2.61) to get

$$
\begin{aligned}
\|(A \otimes \mathbf{1})\varphi\|^2 &= \left\|(A \otimes \mathbf{1}) \sum_j \varphi_j \otimes f_j\right\|^2 = \left\|\sum_j A\varphi_j \otimes f_j\right\|^2 \\
&= \sum_j \|A\varphi_j\|^2 \leq \|A\|^2 \sum_j \|\varphi_j\|^2 = \|A\|^2 \|\varphi\|^2.
\end{aligned}
$$

Hence, $\|A \otimes B\| \leq \|A\| \|B\|$ and, here too, equality holds. These constructions naturally extend to unbounded operators, taking the proper domains into account.

Example 2.53 For any Hilbert space $\mathfrak{H}$ we have

$$\mathcal{B}(\mathbb{C}^n) \otimes \mathcal{B}(\mathfrak{H}) = \mathcal{B}(\mathbb{C}^n \otimes \mathfrak{H}). \tag{2.62}$$

We can indeed write any vector $\varphi \in \mathbb{C}^n \otimes \mathfrak{H}$ uniquely as

$$\varphi = \sum_{j=1}^{n} e_j \otimes \varphi_j$$

using the standard basis $\{e_1, e_2, \ldots, e_n\}$ of $\mathbb{C}^n$. But then there corresponds to any bounded linear operator A on $\mathbb{C}^n \otimes \mathfrak{H}$ a matrix $[A_{kj}]$ of bounded linear operators on $\mathcal{B}(\mathfrak{H})$ such that

$$A\varphi = \sum_{j,k=1}^{n} e_k \otimes A_{kj}\,\varphi_j.$$

This last equation can be rewritten in terms of the standard matrix units, see Example 2.22, in $\mathcal{M}_n$ as

$$A = \sum_{j,k=1}^{n} |e_k\rangle\langle e_j| \otimes A_{kj},$$

proving (2.62). A completely similar argument shows that the norm closure of $\mathcal{C}(\mathfrak{H}) \otimes \mathcal{C}(\mathfrak{K})$ is dense in $\mathcal{C}(\mathfrak{H} \otimes \mathfrak{K})$.

Example 2.54 The result of Example 2.53 does not extend to the case where both factors in the tensor product are infinite-dimensional, i.e., if $\mathfrak{H}$ is an infinite-dimensional Hilbert space: the finite linear combinations of operators of the form $A \otimes B$ with $A, B \in \mathcal{B}(\mathfrak{H})$ are not dense in norm topology in $\mathcal{B}(\mathfrak{H} \otimes \mathfrak{H})$. Let $\{P_1, P_2, \ldots\}$ be a countably infinite family of pairwise orthogonal one-dimensional projectors and consider the projector

$$P := \sum_{j} P_j \otimes P_j$$

on $\mathfrak{H} \otimes \mathfrak{H}$. The sum is convergent in strong operator topology. Let $X \in \mathcal{B}(\mathfrak{H} \otimes \mathfrak{H})$ and $\varphi \in \mathfrak{H} \otimes \mathfrak{H}$. The estimate

$$\left\| \sum_{k,\ell} (P_k \otimes P_\ell) X (P_k \otimes P_\ell) \varphi \right\|^2 \leq \sum_{k,\ell} \| X (P_k \otimes P_\ell) \varphi \|^2$$

$$\leq \|X\|^2 \sum_{k,\ell} \|(P_k \otimes P_\ell) \varphi\|^2$$

$$\leq \|X\|^2 \|\varphi\|^2$$

shows that the map

$$X \in \mathcal{B}(\mathfrak{H} \otimes \mathfrak{H}) \mapsto \sum_{k,\ell} (P_k \otimes P_\ell) X (P_k \otimes P_\ell)$$

is norm-decreasing. Suppose now that we are able to find for any $\epsilon > 0$ a finite family of operators $A_r, B_r \in \mathcal{B}(\mathfrak{H})$ such that

$$\left\| P - \sum_{r} A_r \otimes B_r \right\| \leq \epsilon,$$

then also

$$\left\| P - \sum_{k,\ell,r} P_k A_r P_k \otimes P_\ell B_r P_\ell \right\| \leq \epsilon. \tag{2.63}$$

As the P_k are one-dimensional, there exist complex numbers $\varphi_k(r)$ and $\psi_k(r)$ such that

$$P_k A_r P_k = \varphi_k(r) P_k \quad \text{and} \quad P_k B_r P_k = \psi_k(r) P_k.$$

Let φ_k and ψ_k be the finite-dimensional vectors with entries $\overline{\varphi_k(r)}$ and $\psi_k(r)$. The smallest norm in (2.63), letting the A_r and B_r vary through the bounded operators, is

$$\inf_{k,\ell} \sup \left| \delta_{k\ell} - \langle \varphi_k, \psi_\ell \rangle \right|, \tag{2.64}$$

where the infimum is taken over sequences $\{\varphi_k \mid k\}$ and $\{\psi_k \mid k\}$ in a finite-dimensional space. Moreover, the elements in these sequences may be chosen in a compact set. Because of the finite dimensionality, we can extract Cauchy subsequences $\{\varphi_{n_k} \mid k\}$ and $\{\psi_{n_k} \mid k\}$. Let φ and ψ be their limits; then

$$\inf_{k,\ell} \sup \left| \delta_{k\ell} - \langle \varphi_k, \psi_\ell \rangle \right| \geq \inf_{\varphi,\psi} \max\{|1 - \langle \varphi, \psi \rangle|, |\langle \varphi, \psi \rangle|\} = \frac{1}{2},$$

contradicting (2.63).

The tensor product construction is essential for describing composed systems. If $\mathfrak{H}_j$ is the Hilbert space of wave functions of the jth component of an n-component system, then the wave functions of the total system belong to $\bigotimes_{j=1}^n \mathfrak{H}_j$. Observables of the global system that pertain only to its k-th component are of the form $A \otimes \bigotimes_{j\neq k} \mathbf{1}$ with $A \in \mathcal{B}(\mathfrak{H}_k)$. More complex observables are built by considering linear combinations of tensor products of single-component observables. If all components are *mutually independent* with component j in the state described by the density matrix ρ_j, then the density matrix of the global system is $\bigotimes_{j=1}^n \rho_j$. Independence manifests itself through

$$\left\langle \bigotimes_{j=1}^n A_j \right\rangle_{\otimes_j \rho_j} = \text{Tr} \bigotimes_{j=1}^n \rho_j \bigotimes_{j=1}^n A_j$$

$$= \prod_{j=1}^n \text{Tr} \, \rho_j A_j = \prod_{j=1}^n \langle A_j \rangle_{\rho_j}.$$

For systems composed of several *indistinguishable particles*, for example, electrons, protons, neutrons, ... , an additional physical requirement is imposed by the *Pauli principle* on the space of admissible wave functions. The consequences will be considered in more detail in Chapter 6.

Sometimes, we need to consider subsystems of a composed system. Let us consider a quantum system composed of two distinguishable parts described by Hilbert spaces $\mathfrak{H}_1$ and $\mathfrak{H}_2$. The Hilbert space of the composed system is the product space $\mathfrak{H}_1 \otimes \mathfrak{H}_2$. Let ρ_{12} be a density matrix on $\mathfrak{H}_1 \otimes \mathfrak{H}_2$ that defines the state

$$\langle A_{12} \rangle_{12} := \text{Tr} \, \rho_{12} A_{12}, \quad A_{12} \in \mathcal{B}(\mathfrak{H}_{12}).$$

If we restrict this state to observables of the type $A_1 \otimes \mathbf{1}_2$ with $A_1 \in \mathcal{B}(\mathfrak{H}_1)$, we obtain a state $\langle \cdot \rangle_1$ on $\mathcal{B}(\mathfrak{H}_1)$ called the *reduction of* $\langle \cdot \rangle_{12}$ *to* $\mathcal{B}(\mathfrak{H}_1)$.

$$\langle A_1 \rangle_1 := \mathrm{Tr}\, \rho_{12}\, A_1 \otimes \mathbf{1}_2. \tag{2.65}$$

It can be shown, using Theorem 2.46 and suitable continuity arguments, that the state in (2.65) is determined by a unique density matrix ρ_1 on $\mathfrak{H}_1$:

$$\mathrm{Tr}\, \rho_1\, A_1 = \mathrm{Tr}\, \rho_{12}\, A_1 \otimes \mathbf{1}_2, \quad A_1 \in \mathcal{B}(\mathfrak{H}_1). \tag{2.66}$$

We explicitly construct ρ_1 in Example 2.55. ρ_1 is the *reduced density matrix of* ρ_{12} and the operation $\rho_{12} \mapsto \rho_1$ is called taking *the partial trace of* ρ_{12} *over* $\mathfrak{H}_2$. It is denoted by

$$\rho_1 = \mathrm{Tr}_{\mathfrak{H}_2}\, \rho_{12}. \tag{2.67}$$

Example 2.55 Let $\{e_1, e_2, \dots\}$ and $\{f_1, f_2, \dots\}$ be orthonormal bases for the Hilbert spaces $\mathfrak{H}_1$ and $\mathfrak{H}_2$; then $\{e_j \otimes f_k \mid j, k\}$ is a basis for $\mathfrak{H}_1 \otimes \mathfrak{H}_2$. Let ρ_{12} be a density matrix on $\mathfrak{H}_1 \otimes \mathfrak{H}_2$. We now define ρ_1 on $\mathfrak{H}_1$ by

$$\langle e_j, \rho_1\, e_k \rangle := \sum_\ell \langle e_j \otimes f_\ell, \rho_{12}\, e_k \otimes f_\ell \rangle. \tag{2.68}$$

The series in (2.68) is convergent because ρ_{12} is a density matrix. Moreover, one shows that ρ_1 is also trace-class. For two arbitrary indices j, k, we compute, using (2.24),

$$\begin{aligned}
\mathrm{Tr}\, \rho_{12}\, (|e_k\rangle\langle e_j| \otimes \mathbf{1}) &= \sum_{\ell r} \langle e_r \otimes f_\ell, \rho_{12}\, (|e_k\rangle\langle e_j| \otimes \mathbf{1})\, e_r \otimes f_\ell \rangle \\
&= \sum_\ell \langle e_j \otimes f_\ell, \rho_{12}\, e_k \otimes f_\ell \rangle \\
&= \langle e_j, \rho_1\, e_k \rangle \\
&= \mathrm{Tr}\, \rho_1\, |e_k\rangle\langle e_j|.
\end{aligned}$$

Example 2.56 In the case of density matrices ρ_{12} on $\mathcal{L}^2(\mathfrak{X}_1, \mu_1) \otimes \mathcal{L}^2(\mathfrak{X}_2, \mu_2)$ given by integral kernels

$$((x_1, x_2), (y_1, y_2)) \mapsto \rho_{12}((x_1, x_2), (y_1, y_2))$$

such as in Example 2.45, we obtain the kernel of the reduced density matrix ρ_1 as

$$(x_1, y_1) \mapsto \rho_1(x_1, y_1) = \int_{\mathfrak{X}_2} \mu_2(dz)\, \rho_{12}((x_1, z), (y_1, z)).$$

In classical probability theory, *marginal probability distributions* correspond to reduced density matrices. On a measurable space of the type $\mathfrak{X}_1 \times \mathfrak{X}_2$ equipped with the product σ-algebra, the marginal probability distribution μ_1 of a probability measure μ_{12} is given by

$$\mu_1(S_1) := \mu_{12}(S_1 \times \mathfrak{X}_2), \quad S_1 \text{ measurable subset of } \mathfrak{X}_1,$$

or in terms of measurable functions

$$\int_{\mathfrak{X}_1} \mu_1(dx_1) \, f_1(x_1) = \int_{\mathfrak{X}_1 \times \mathfrak{X}_2} \mu_{12}(dx_1 \, dx_2) \, f_1(x_1).$$

A deep difference between quantum and classical probability is that in the latter the marginals of pure states are automatically pure. In other words, marginal probability distributions of Dirac measures are again Dirac distributions while reduced density matrices of pure states are seldom pure.

To be more precise: let $\varphi_{12} \in \mathfrak{H}_1 \otimes \mathfrak{H}_2$ be a normalized vector defining the pure state

$$A_{12} \mapsto \langle A_{12} \rangle_{12} := \langle \varphi_{12}, A_{12} \, \varphi_{12} \rangle$$

on $\mathcal{B}(\mathfrak{H}_1 \otimes \mathfrak{H}_2)$. Given an orthonormal basis $\{f_1, f_2, \dots\}$ of $\mathfrak{H}_2$ we write, as in (2.61),

$$\varphi_{12} = \sum_j \varphi_j \otimes f_j, \tag{2.69}$$

with $\varphi_j \in \mathfrak{H}_1$. The reduced density matrix of the state $\langle \cdot \rangle_{12}$ is now given by

$$\langle A_1 \rangle_1 = \sum_j \langle \varphi_j, A_1 \, \varphi_j \rangle = \mathrm{Tr}\left(\sum_j |\varphi_j\rangle\langle\varphi_j| \right) A_1.$$

Clearly, $\langle \cdot \rangle_1$ is pure iff all φ_j are multiples of a same vector in $\mathfrak{H}_1$. In other words, $\langle \cdot \rangle_1$ is pure iff $\langle \cdot \rangle_{12}$ factorizes into the product of two pure states:

$$\langle A_1 \otimes A_2 \rangle_{12} = \langle A_1 \rangle_1 \, \langle A_2 \rangle_2. \tag{2.70}$$

In terms of Hilbert space vectors, $\varphi_{12} = \varphi_1 \otimes f_1$. This property holds in a much more general situation.

The converse construction is also possible. A density matrix ρ_1 on a Hilbert space $\mathfrak{H}_1$ can always be decomposed as

$$\rho_1 = \sum_j |\varphi_j\rangle\langle\varphi_j|, \quad \sum_j \|\varphi_j\|^2 = 1.$$

Therefore, for a suitable choice of the dimension of $\mathfrak{H}_2$, a normalized vector φ in $\mathfrak{H}_1 \otimes \mathfrak{H}_2$ as in (2.69) defines a pure state on $\mathcal{B}(\mathfrak{H}_1 \otimes \mathfrak{H}_2)$ which is an extension of the state on $\mathcal{B}(\mathfrak{H}_1)$ given by ρ_1.

Pure states, defined by elementary tensors $\varphi_{12} = \varphi_1 \otimes \varphi_2$ with $\varphi_j \in \mathfrak{H}_j$ normalized, are called *non-entangled*. *Classical or separable states* are convex combinations of non-entangled mixed states, i.e., of states described by tensor products of density matrices. These states are somehow the trivial states in the quantum computing literature, quantum teleportation, etc.

2.6 Notes

The basic material on mathematical methods for quantum mechanics can be found in many textbooks. In particular, the texts on methods of mathematical physics by Reed and Simon (1972), Richtmyer (1978), Thirring (1981), Ludwig (1985), Blank *et al.* (1994), ... are perfect references for our purposes. They often include background material on measure theory, general topology, and group theory too.

The non-classical probabilistic structure of quantum mechanics was discovered by von Neumann (1932) and settled by Gleason's theorem (Gleason 1957). The book by Peres (1993) pays special attention to probability in quantum mechanics. These results stimulated further research in different directions. One should mention here the so-called axiomatic quantum mechanics presented, for example, in the monographs of Beltrametti and Cassinelli (1981), Ludwig (1983), Mackey (1963), Varadarajan (1968), ... and the vast field of quantum probability, see Quantum Probability and Applications and Quantum Probability & Related Topics. For recent progress, see Alicki *et al.* (1998). Mathematical aspects of the Weyl quantization are treated in Folland (1989). Extending states on subsystems to global states has been considered in (Werner 1990).

Recent experimental progress which allows to observe typical quantum features like superpositions of distinguishable microscopic states and quantum correlations in composite systems, i.e. entanglement, renewed the interest in the foundations of quantum mechanics. These quantum effects may find applications in information processing and quantum computing, see e.g. the recent texts by Gruska (1999) and Nielsen and Chuang (2000).

3

DETERMINISTIC DYNAMICS

This chapter deals with the time evolution of both quantum and classical deterministic systems. By *deterministic system* we understand either a completely isolated system, the dynamics of which is determined by a time-independent Hamiltonian (energy observable), or a very special type of open system. For the later case, it is assumed that the influence of the environment on the system can be fully described in terms of a time-dependent deterministic, i.e. non-random, perturbation of its Hamiltonian. This type of idealization disregards any statistical fluctuations present in the environment and neglects any back reaction of the system on the external world. The absence of irreversibility and randomness in the time evolution of such systems makes them physically and mathematically quite different from the truly irreversible open systems that we shall study in Chapter 8.

The fundamental evolution equation for a quantum system is the *Schrödinger equation* which determines the path $t \mapsto \psi(t)$ of the wave function in the Hilbert space $\mathfrak{H}$ of the system, given an initial condition ψ_0:

$$i\hbar \frac{d\psi}{dt} = H\,\psi, \quad \psi(t_0) = \psi_0. \tag{3.1}$$

H is the *Hamiltonian* of the system, which is a self-adjoint operator on $\mathfrak{H}$. In order to simplify the formulae, we put in the sequel $\hbar = 1$, except for discussing classical limits of course.

The system is *autonomous* if H does not explicitly depend on time, otherwise we speak about time-dependent Hamiltonians. In general, the formal solution of (3.1) is given by a *unitary propagator* $U(t, t_0)$ which has the *groupoid* property

$$U(t, r)\,U(r, s) = U(t, s).$$

For an autonomous system $U(t, r) = U(t - r)$.

The *Schrödinger picture* extends naturally to mixed states, i.e. density matrices ρ, if we identify a wave function ψ with the one-dimensional projector $|\psi\rangle\langle\psi|$

$$\frac{d\rho}{dt} = -i\,[H, \rho], \quad \rho(t_0) = \rho_0.$$

On the level of the observables, we can use the *Heisenberg picture*, which is dual to the Schrödinger picture:

$$\text{Tr}\,\rho(t)\,A_0 = \text{Tr}\,\rho_0\,A(t)$$

and so

$$\frac{dA}{dt} = i\,[H, A], \quad A(t_0) = A_0. \tag{3.2}$$

In terms of the unitary propagator U, the formal solution of Heisenberg's equation is given by

$$A(t_0) = U(t, t_0)^* A_0\, U(t, t_0).$$

The fundamental object for describing a classical dynamical system is the *phase space* whose points completely determine the instantaneous positions and velocities of the particles in the system. A deterministic evolution of such a system is typically obtained as the solution of a, usually non-linear, first order differential equation

$$\frac{dx}{dt} = \mathbf{F}(x, t), \quad x(t_0) = x_0.$$

Here $\mathbf{F}$ is a vector field which specifies the local and instantaneous phase space velocity.

After dealing with deterministic evolutions of quantum systems in Section 3.1, we briefly examine in the second section the connection between quantum and classical dynamics through the *classical limit* and the *mean-field limit*. These limits describe how the temporal evolution of a quantum system approaches a classical evolution when either Planck's constant is considered to be very small, i.e. large quantum numbers, or, for example, the total spin becomes very large. These limits can be treated at different levels such as sharply localized (minimal uncertainty) states evolving for sufficiently short times according to the corresponding classical flow. One can also study the limiting time evolution of quantizations of *classical observables* such as functions of position and momentum when $\hbar$ tends to zero. In Section 3.3, we consider classical evolutions in terms of flows on differentiable manifolds and the fourth section is devoted to free quantum evolutions on compact manifolds.

3.1 Deterministic quantum dynamics

3.1.1 Time-independent Hamiltonians

The fundamental equation which governs the time evolution of an isolated quantum system is Schrödinger's equation

$$i\frac{d\psi}{dt} = H\,\psi. \tag{3.3}$$

H is a self-adjoint Hamiltonian operator which does not depend on t, and ψ is a normalized Hilbert space vector defining a pure state of the system. We write the formal solution of (3.3) as

$$\psi(t) = U(t)\,\psi(0) \quad \text{with } U(t) = \exp(-itH), \ t \in \mathbb{R}. \tag{3.4}$$

The meaning of the exponential in (3.4) can be understood in different but equivalent ways:

Spectral representation

$$U(t) = \int_{\mathrm{Sp}(H)} \exp(-it\epsilon)\ E^H(d\epsilon).$$

Resolvent formula

$$U(t)\,\psi = \lim_{n\to\infty} \left[\mathbf{1} + (it/n)H\right]^{-n}\psi, \quad \psi \in \mathfrak{H}.$$

Power series formula

$$U(t)\,\varphi = \sum_{n=0}^{\infty} \frac{(-it)^n}{n!}\, H^n\varphi. \tag{3.5}$$

In (3.5), φ is an *analytic vector* for H, which means that the series

$$\sum_{n=0}^{\infty} (t^n/n!)\, \|H^n\phi\|$$

converges for all $t \in \mathbb{R}^+$. It follows immediately from the Spectral Theorem, Theorem 2.42, that any self-adjoint operator possesses a dense set of analytic vectors and that $U(t)$ leaves the set of analytic vectors invariant. Therefore, the Schrödinger equation (3.3) is mathematically meaningful for all initial analytic vectors $\psi(0)$ and actually this is true for all $\psi(0) \in \mathrm{Dom}(H)$ as well.

The one-parameter family $\{U(t) \mid t \in \mathbb{R}\}$ of unitary operators enjoys the following properties:

$$\begin{aligned}
U(s)\,U(t) &= U(t+s), \quad s,t \in \mathbb{R}, \\
U(-t) &= U(t)^{-1} = U(t)^*, \quad t \in \mathbb{R}, \\
\lim_{t\to 0} \|U(t)\psi - \psi\| &= 0, \quad \psi \in \mathfrak{H}.
\end{aligned} \tag{3.6}$$

The conditions (3.6) may be viewed as the definition of a *strongly continuous one-parameter group of unitaries*. This provides the most general mathematical setting for quantum reversible, conservative (non-dissipative) evolutions, homogeneous in time. Stone's theorem states that any such evolution is uniquely generated by a Hamiltonian operator.

Theorem 3.1 (Stone) *There exists, for any strongly continuous one-parameter group $\{U(t) \mid t \in \mathbb{R}\}$ of unitary operators on a Hilbert space $\mathfrak{H}$, a unique densely defined self-adjoint operator H such that $U(t) = \exp(-itH)$ for all $t \in \mathbb{R}$.*

3.1.2 *Perturbations of Hamiltonians*

Very often, a Hamiltonian of a quantum system can be formally written as a sum

$$H = H_0 + V,$$

where H_0 is a relatively simple, explicitly tractable, self-adjoint operator and V is another self-adjoint (or only symmetric) operator called *perturbation*. Moreover, typically V can be treated as small compared to H_0. The first problem we meet is to prove the self-adjointness of H for unbounded V. Another physically important property of a Hamiltonian is its *boundedness from below*, i.e. $H + a\mathbf{1} > 0$, for a certain $a \in \mathbb{R}$, which is necessary for thermodynamic stability of the system. In many cases, both these properties can be proved with the help of the *Kato–Rellich criterion*.

Theorem 3.2 (Kato–Rellich)   *Let V be a symmetric operator defined on the domain* $\mathrm{Dom}(H_0)$ *of a densely defined self-adjoint operator H_0 on a Hilbert space $\mathfrak{H}$. Suppose that there exist constants $0 \leq \alpha < 1$ and $\beta \geq 0$ such that for all $\psi \in \mathrm{Dom}(H_0)$*

$$\|V\psi\| \leq \alpha\|H_0\psi\| + \beta\|\psi\|.$$

Then $H = H_0 + V$ is self-adjoint on $\mathrm{Dom}(H) = \mathrm{Dom}(H_0)$. *Moreover, if H_0 is bounded from below, then so is H.*

The next step is to develop approximation schemes for the evolution operator $U(t) = \exp(-itH)$ of a perturbed dynamics. The *Trotter product formula* allows one to express $U(t)$ in terms of a limiting procedure involving the one-parameter unitary groups generated by self-adjoint H_0 and V. Let H_0 and V be self-adjoint such that $H_0 + V$ is essentially self-adjoint on some common dense domain for H_0 and V, then

$$e^{-itH}\psi = \lim_{n \to \infty}\left[e^{-i(t/n)H_0}e^{-i(t/n)V}\right]^n\psi, \quad \psi \in \mathfrak{H},$$

where H is the closure of $H_0 + V$.

A useful notion is the *interaction picture of the dynamics* given in terms of a one-parameter family of unitary operators

$$W(t) := U_0(-t)U(t) \quad \text{with } U_0(t) = \exp(-itH_0).$$

$\{W(t) \mid t \in \mathbb{R}\}$ is not a group but it satisfies the *cocycle relation*

$$W(s + t) = U_0(-t)\, W(s)\, U_0(t)\, W(t), \quad s, t \in \mathbb{R}. \tag{3.7}$$

$W(t)$ describes the deviation of the full evolution from the free evolution governed by U_0 and it is particularly useful in scattering theory and in the theory of open systems.

From (3.7), we obtain the formal differential equation

$$\frac{dW(t)}{dt}\psi = -iV(t)W(t)\psi \tag{3.8}$$

with

$$V(t) = U_0(-t)VU_0(t)$$

which is meaningful, e.g. in the situation of Theorem 3.2 with $\psi \in \mathrm{Dom}(H) = \mathrm{Dom}(H_0)$.

The formal iterative solution of eqn (3.8) yields the *Dyson expansion*

$$W(t)\,\psi = \left(\mathbf{1} + \sum_{n=1}^{\infty}(-i)^n \int_0^t dt_1 \cdots \int_0^{t_{n-1}} dt_n\, V(t_1)\cdots V(t_n)\right)\psi, \tag{3.9}$$

which is mathematically rigorous for bounded V. As usual, unbounded cases need a careful discussion of domain and convergence questions. Introducing the *time-ordering symbol* $\mathbf{T}$, defined by

$$\mathbf{T}(V(t_1)\,V(t_2)\cdots V(t_n)) := V(t_{j_1})\,V(t_{j_2})\cdots V(t_{j_n}),$$

where $t_{j_1} \geq t_{j_2} \geq \cdots \geq t_{j_n}$, we can rewrite formula (3.9) as

$$W(t) = \mathbf{T}\left(\exp\left(-i\int_0^t ds\, V(s)\right)\right).$$

3.1.3   Time-dependent Hamiltonians

As mentioned in the beginning of this chapter, in some cases the influence of the environment can be taken into account by considering a Schrödinger equation with a *time-dependent Hamiltonian*

$$i\frac{d\psi(t)}{dt} = H(t)\psi(t). \tag{3.10}$$

Typical assumptions to make (3.10) meaningful are

- all $H(t)$, $t \in \mathbb{R}$ are self-adjoint operators on a common domain $\mathcal{D}$,
- for any $\psi \in \mathcal{D}$, $t \mapsto H(t)\,\psi$ is continuous on $\mathbb{R}$.

Then, the formal solution $\psi(t)$ of (3.10) with initial condition $\psi(s) = \varphi$ can be expressed in terms of a two-parameter family of unitary operators $\{U(t,s) \mid t, s \in \mathbb{R}\}$, called *unitary propagator*, which has the following properties:

$$U(t,r)\,U(r,s) = U(t,s),$$
$$U(t,t) = \mathbf{1}, \tag{3.11}$$
$$(t,s) \mapsto U(t,s)\psi \text{ is strongly continuous for all } \psi \in \mathfrak{H}.$$

In particular, for $\phi \in \mathcal{D}$ we can give a rigorous meaning to (3.10):

$$i\frac{d}{dt} U(t,s)\,\phi = H(t)\,U(t,s)\,\phi.$$

Similarly to the previous section, the unitary propagator can be formally written as a Dyson series

$$U(t,s)\psi = \left(\mathbf{1} + \sum_{n=1}^{\infty} (-i)^n \int_s^t dt_1 \cdots \int_s^{t_{n-1}} dt_n\, H(t_1) \cdots H(t_n)\right)\psi \qquad (3.12)$$

$$= \mathbf{T}\left(\exp\left(-i\int_s^t dr\, H(r)\right)\right)\psi.$$

For a fixed $r \in \mathbb{R}$, $H(r)$ generates a one-parameter group

$$\{U_r(t) := \exp(-itH(r)) \mid t \in \mathbb{R}\}, \quad t \in \mathbb{R}.$$

Taking a sequence of partitions of a given interval $[s,t]$

$$s \equiv t_0 < t_1 < \cdots < t_{n-1} < t_n \equiv t, \quad \limsup_{n\to\infty} (t_{k-1} - t_k) = 0,$$

we can represent the unitary propagator as a limit

$$U(t,s)\,\psi = \lim_{n\to\infty} U_{t_n}(t_n - t_{n-1})\, U_{t_{n-1}}(t_{n-1} - t_{n-2}) \cdots U_{t_1}(t_1 - t_0)\,\psi.$$

3.1.4 *Periodic perturbations and Floquet operators*

An important class of quantum systems with time-dependent Hamiltonians are those with periodic time dependence, i.e. there is a $T > 0$ such that

$$H(t+T) = H(t), \quad t \in \mathbb{R}. \qquad (3.13)$$

In this context, it is convenient to introduce the time-independent *formalism of Howland* in which the system with time-dependent Hamiltonian $H(t)$ is extended to a larger one described by the Hilbert space $\mathfrak{H}_F$ of $\mathfrak{H}$-valued square integrable functions on $\mathbb{R}$,

$$\mathfrak{H}_F := \left\{ f \,\Big|\, f : \mathbb{R} \mapsto \mathfrak{H},\ \int_{\mathbb{R}} \|f(s)\|^2\, ds < \infty \right\} = \mathcal{L}^2(\mathbb{R}) \otimes \mathfrak{H},$$

and by the extended Hamiltonian

$$H_F := -i\frac{\partial}{\partial t} + H(t). \qquad (3.14)$$

It is easy to check by a direct computation that the solution f_τ of the corresponding extended Schrödinger equation

$$i\frac{df_\tau}{d\tau} = H_F\, f_\tau$$

is given by a strongly continuous one-parameter group $U_F(\tau) := \exp(-i\tau H_F)$ which is related to the propagator $U(t,s)$ of (3.11) by

$$\big[U_F(\tau)f\big](t) = U(t, t - \tau)f(t - \tau).$$

In particular, for periodic $H(t)$ the propagator is also periodic,

$$U(t+T, s+T) = U(t,s),$$

and the group $U_F(\tau)$ acts naturally on the Hilbert space of periodic $\mathfrak{H}$-valued functions

$$\mathfrak{H}_T := \mathcal{L}^2([0,T)) \otimes \mathfrak{H}.$$

The extended Hamiltonian, now denoted by H_T, has the same form as (3.14) but with periodic boundary conditions. It is an important and useful fact that the information about the spectrum of H_T is fully contained in the spectrum of the *Floquet operator* $U(s+T, s)$ which takes the system described by (3.13) through a complete period that starts at s.

Namely, having the solution of the eigenvalue problem for the *Floquet Hamiltonian* H_T

$$(H_T\, f)(t) = \lambda\, f(t), \quad f \in \mathfrak{H}_T. \tag{3.15}$$

and using the explicit form (3.14) of H_T, we can easily solve the corresponding differential equation to obtain

$$U(t,s)\, f(s) = e^{-i\lambda(t-s)}\, f(t).$$

Choosing $t = s + T$, we have

$$U(s+T, s)\, f(s) = e^{-i\lambda T}\, f(s).$$

Conversely, if for a certain $\psi \in \mathfrak{H}$

$$U(s+T, s)\, \psi = e^{-i\lambda T}\, \psi,$$

then $f(t) := e^{i\lambda(t-s)}\, U(t,s)\psi$ is periodic and satisfies eqn (3.15). This argument holds for eigenvalues, but it can be extended to a general spectrum. One should also notice that the spectrum of the Floquet operator $U(s+T, s)$ does not depend on the choice of s.

3.1.5 Kicked dynamics

Kicked systems are relatively simple examples of systems with Hamiltonians periodic in time. They are formally given by

$$H(t) := H_0 + \sum_{n=-\infty}^{\infty} V \, \delta(t - nT) \qquad (3.16)$$

with H_0 and V self-adjoint. These kicked systems can be treated as limits of systems with smooth time dependence. In order to study their spectral properties, we use the fact that the spectrum of the Floquet operator $U(s + T, s)$ is independent on s. Therefore, we analyze the single unitary operator U_T on $\mathfrak{H}$ which can be formally computed by applying (3.12)

$$U_T := \lim_{\epsilon \downarrow 0} U(\epsilon + T, \epsilon) = e^{-iV} \, e^{-iTH_0}. \qquad (3.17)$$

It might be convenient to use another version of the Floquet operator

$$U'_T := \lim_{\epsilon \uparrow 0} U(\epsilon + T, \epsilon) = e^{-iTH_0} \, e^{-iV}.$$

Example 3.3 *(The kicked top)* A periodically *kicked top* has been frequently used as a relatively simple example of a finite quantum system with a chaotic behaviour in the classical limit. This is a quantum spin-j system belonging to a Hilbert space of dimension $2j + 1$. The basic observables are the angular momentum operators J^x, J^y and J^z obeying

$$[J^x, J^y] = iJ^z \text{ and cyclic permutations.} \qquad (3.18)$$

The dynamics is governed by a Hamiltonian periodic in time and of the form (3.16) with

$$H_0 = a \, J^y \quad \text{and} \quad V = b \, (J^z)^2, \quad a, b \in \mathbb{R}.$$

H_0 produces a uniform precession around the y-axis while V describes a nonlinear torsion around the z-axis.

The Floquet operator (3.17), written in the standard parameterization with two real parameters $p, k \in \mathbb{R}$, is

$$U = e^{-ik(J^z)^2/2j} \, e^{-ipJ^y/2}.$$

Example 3.4 *(The baker map of Balacz and Voros)* The *Fourier transform* F_N on the space $\mathbb{C}^N$ of sequences $\mathbf{c} = (c_1, c_2, \dots, c_N)$ of complex numbers of length N is given by

$$(F_N \, \mathbf{c})_k := \frac{1}{\sqrt{N}} \sum_{j=1}^{N} e^{-2\pi ijk/N} \, c_j, \quad k = 1, 2, \dots, N.$$

It is a unitary transformation with inverse

$$(F_N^* \, \mathbf{c})_k := \frac{1}{\sqrt{N}} \sum_{j=1}^{N} e^{2\pi i jk/N} \, c_j, \quad k = 1, 2, \ldots, N.$$

Writing $\mathbb{C}^{2N} = \mathbb{C}^N \oplus \mathbb{C}^N$, the single-step dynamics U of the baker map is

$$U := F_{2N}^* \, (F_N \oplus F_N).$$

3.2   Classical limits

There are two different procedures which relate the quantum description of nature with the classical one. The first one, called *quantization*, is a device which allows to construct the basic elements of a quantum model like states, observables and dynamics from an underlying classical physical system, see Section 2.4.2.

The second procedure, a more pragmatic one, consists in constructing a classical model as a certain kind of approximation to a quantum one valid for high energies or short times or many particles ... This goes under names like *classical approximation (limit), semiclassical or quasi-classical approximation* and *dequantization* and different mathematical techniques are used: WKB approximation, path integrals, stochastic mechanics, coherent states, Moyal brackets, phase space representations ... Several self-consistent approximations, like mean-field or Hartree–Fock, which applied to many-body quantum systems translate into non-linear evolution equations for classical variables or fields, are regarded here as classical limits too. We follow in this book the second point of view, present a general heuristic discussion of the basic ideas and provide some examples.

In the classical theory of deterministic systems, a pure state is identified with a point $\mathbf{x} = (x_1, x_2, \ldots, x_d)$ of a d-dimensional phase space Γ and the most general equation of motion can be written as

$$\frac{d\mathbf{x}}{dt} = \mathbf{F}(\mathbf{x}), \quad \mathbf{F}(\mathbf{x}) = \big(F_1(\mathbf{x}), \ldots, F_d(\mathbf{x})\big). \tag{3.19}$$

To define a classical counterpart of a quantum system, described by a Hilbert space $\mathfrak{H}$ and a Hamiltonian H, we need the following ingredients:

- a *quantization parameter* ϵ which is supposed to be small in a certain sense, formally we consider the limit ϵ tending to 0;
- a set of *relevant quantum observables*, i.e. self-adjoint, not necessarily bounded operators $\mathbf{X} = \{X_1, X_2, \ldots, X_d\}$ whose commutators are of the order ϵ,

$$[X_k, X_l] = O(\epsilon); \tag{3.20}$$

- a family of *classical states* $\{\rho_\mathbf{x} \mid \rho_\mathbf{x} \in \mathcal{T}(\mathfrak{H}), \, \mathbf{x} \in \Gamma\}$ labelled in local coordinates by the elements of the classical phase space. They satisfy the condition

$$\langle X_{j_1}^{k_1} X_{j_2}^{k_2} \cdots X_{j_n}^{k_n} \rangle_{\rho_{\mathbf{x}}} = x_{j_1}^{k_1} x_{j_2}^{k_2} \cdots x_{j_n}^{k_n} + \mathrm{O}(\epsilon),$$

which means that the quantum fluctuations of the observables $\mathbf{X} = \{X_1, X_2, \ldots, X_d\}$ are of the order ϵ. Often, these states are chosen as minimizing an uncertainty relation and are then called *coherent*.

- The choice of observables $\mathbf{X} = \{X_1, X_2, \ldots X_d\}$ must be consistent with the evolution of the system in the sense that

$$i[H, X_k] = F_k[\mathbf{X}] + \mathrm{O}(\epsilon), \quad k = 1, 2, \ldots, d, \tag{3.21}$$

where $F_k[\mathbf{X}]$ is an element of the algebra generated by the observables $\mathbf{X}$ and the identity.

Under the conditions of above, the mean values $x_k := \langle X_k \rangle_{\rho_{\mathbf{x}}}$ evolve according to the approximate equation

$$\frac{d\mathbf{x}}{dt} = \mathbf{F}(\mathbf{x}) + \mathrm{O}(\epsilon) \quad \text{with} \quad \mathbf{F}(\mathbf{x}) = \langle \mathbf{F}[\mathbf{X}] \rangle_{\rho_{\mathbf{x}}},$$

which coincides with (3.19) for $\epsilon \to 0$. We briefly discuss three examples of quantum systems with physically different quantization parameters ϵ.

Example 3.5 For a particle in one dimension, $\mathbf{X} = \{Q, P\}$ with Q and P the position and momentum operators of Examples 2.33 and 2.44. The small parameter ϵ is Planck's constant $\hbar$. If the Hamiltonian is a smooth function of Q and P, then condition (3.21) is fulfilled. As classical states, one can choose coherent states $|q, p\rangle$ which have minimal quantum fluctuations around classical points $(q, p) \in \mathbb{R}^2$. The state $|q, p\rangle$ is the eigenstate of

$$(P - p)^2 + \lambda^2 (Q - q)^2$$

belonging to the lowest eigenvalue. The parameter λ can be chosen according to some natural scale in the Hamiltonian at hand. It turns out that

$$|q, p\rangle = W(q, p) \, \psi_0, \tag{3.22}$$

where ψ_0 is the ground state of $P^2 + \lambda^2 Q^2$; it is the Gaussian

$$\psi_0(x) = \left(\frac{1}{\hbar \pi} \right)^{1/4} e^{-\lambda x^2 / 2\hbar}.$$

The operators W are the Weyl operators of (2.54). The states $|q, p\rangle$ of (3.22) are states of minimal uncertainty as they saturate the Heisenberg uncertainty relation (2.47). The set $\{|q, p\rangle \mid (q, p) \in \mathbb{R}^2\}$ is overcomplete, but it satisfies the following continuous resolution of the unity:

$$\mathbf{1} = \frac{1}{2\pi\hbar} \int_{\mathbb{R}^2} dq\, dp \, |q, p\rangle \langle q, p|.$$

Obviously $\hbar$ is a fundamental physical constant and, therefore, taking the limit $\hbar \to 0$ is a rather dubious procedure. It is more consistent to rescale the canonical variables introducing $Q' := \mu Q$ and $P' := \nu P$; then,

$$[Q', P'] = i\hbar \mu \nu \equiv i\epsilon.$$

The classical limit corresponds now to an approximation that holds for sufficiently small ϵ. For example, for a non-relativistic particle in a periodic potential

$$H = \frac{1}{2m} P^2 + V_0 \cos(kQ).$$

The natural rescaling parameters are here $\mu = k$ and $\nu = 1/\sqrt{m}$.

Example 3.6 A *mean-field ferromagnetic Heisenberg model* is a system of N identical spins with Hilbert space

$$\mathfrak{H} = \bigotimes_{j=1}^{N} \mathbb{C}^{2s+1}$$

and mean-field Hamiltonian

$$H_N := \sum_{\alpha=x,y,z} \sum_{k=1}^{N} h_\alpha S_k^\alpha + \frac{1}{N} \sum_{\alpha,\beta=x,y,z} \sum_{k,j=1}^{N} J(\alpha,\beta)\, S_k^\alpha\, S_j^\beta,$$

where h is an external field and $J \in \mathbb{R}$ describes the coupling between the different spins. It is an essential feature of H_N that all spins interact in the same way with the external field and with one another. As a consequence, the Hamiltonian is invariant under permutations of the spins.

The basic relevant observables (3.20) are mean-field ones,

$$A_N := \frac{1}{N} \sum_{k=1}^{N} a_k,$$

where the a_k are copies of a single-spin observable $a \in \mathcal{M}_{2s+1}$ located at the site k. Their commutators possess the structure

$$[A_N, B_N] = \frac{1}{N} C_N \quad \text{with } c = [a, b]$$

and hence the quantization parameter ϵ is now $1/N$.

The classical phase space of such a mean-field system is the space of $(2s + 1)$-dimensional density matrices. The classical states ρ_N are products of copies of a given single-spin density matrix, i.e. $\rho_N = \bigotimes_N \rho$.

Example 3.7 Similarly to Example 3.5, there is a notion of *spin coherent state* which can be used to describe the classical limit of large spin representation of SU(2). In the following $\mathbf{m}, \mathbf{n}, \mathbf{n}_1, \mathbf{n}_2, \ldots$ denote three-dimensional unit vectors lying on the sphere S^2. The sphere is the *phase space of a classical spin* and by $d\mathbf{n}$ we denote the unique rotation-invariant measure on S^2 normalized to 4π.

The coherent state $|\mathbf{n}\rangle$ is labelled by a unit vector

$$\mathbf{n} = (\sin\theta\cos\phi, \sin\theta\sin\phi, \cos\theta)$$

and given by

$$|\mathbf{n}\rangle := D(\mathbf{n})\,|j, j\rangle,$$

where

$$D(\mathbf{n}) := \exp[-i\theta(\mathbf{m}\cdot\mathbf{J})] \quad \text{and} \quad \mathbf{m} = (\sin\phi, -\cos\phi, 0). \tag{3.23}$$

The operators $\mathbf{J} := (J^x, J^y, J^z)$ in (3.23) are the generators of the $2j + 1$ irreducible representation of SU(2). They satisfy

$$[J^x, J^y] = iJ^z \text{ and cyclic permutations,}$$
$$\mathbf{J}\cdot\mathbf{J} = (J^x)^2 + (J^y)^2 + (J^z)^2 = j(j + 1).$$

The Hilbert space $\mathbb{C}^{2j+1}$ is spanned by the natural orthonormal basis $\{|j, m\rangle \mid m = -j, -j + 1, \ldots, j\}$ with $j \in \frac{1}{2}\mathbb{N}$. The explicit action of the $\mathbf{J}$ reads

$$J^z\,|j, m\rangle = m\,|j, m\rangle,$$
$$(J^x + iJ^y)\,|j, m\rangle = \sqrt{(j - m)(j + m + 1)}\,|j, m + 1\rangle.$$

The coherent states are related to the representation of the groups SU(2) or SO(3). Namely, the operator $\exp(-i\omega\,\mathbf{m}\cdot\mathbf{J})$, $\mathbf{m}^2 = 1$ corresponds to the rotation by the angle ω around the axis directed along $\mathbf{m}$. The rotation $D(\mathbf{n})$ transforms the vector $\mathbf{n}_0 := (0, 0, 1)$, pointing at the north pole, into $\mathbf{n}$. The spin coherent states enjoy the following properties:

- *Resolution of unity or completeness*

$$\int_{S^2} \mu_j(d\mathbf{n})\,|\mathbf{n}\rangle\langle\mathbf{n}| = \mathbf{1} \quad \text{with} \quad \mu_j(d\mathbf{n}) := \frac{2j + 1}{4\pi}\,d\mathbf{n}.$$

- *Scalar product*

$$|\langle\mathbf{n}_1, \mathbf{n}_2\rangle|^2 = \left(\frac{1 + \mathbf{n}_1\cdot\mathbf{n}_2}{2}\right)^{2j}.$$

- The operator $\exp(-i\omega\,\mathbf{m}\cdot\mathbf{J})$ transforms coherent states into coherent states

$$\exp(-i\omega\,\mathbf{m}\cdot\mathbf{J})\,|\mathbf{n}\rangle = e^{i\Phi}\,|\mathbf{n}'\rangle,$$

where $\mathbf{n}'$ is obtained from $\mathbf{n}$ by the corresponding rotation and Φ is a phase.

- Spin coherent states are eigenvectors of $\mathbf{n} \cdot \mathbf{J}$,

$$\mathbf{n} \cdot \mathbf{J} \,|\mathbf{n}\rangle = j \,|\mathbf{n}\rangle, \quad \text{hence} \quad \langle \mathbf{n}, \mathbf{J}\mathbf{n} \rangle = j \,\mathbf{n}.$$

The classical limit is obtained by letting j tend to infinity, and $1/j$ plays the role of the ϵ at the beginning of this section. The basic relevant observables of the system are the rescaled spin operators

$$\mathbf{N} = (N^x, N^y, N^z) := \frac{1}{j}(J^x, J^y, J^z)$$

which describe the position in phase space. From their commutation relations it follows that the quantization parameter is equal to $1/j$ and

$$\langle \mathbf{n}, \mathbf{N}\,\mathbf{n} \rangle = \mathbf{n} \quad \text{and} \quad \langle \mathbf{n}, \mathbf{N}^2\,\mathbf{n} \rangle - \left(\langle \mathbf{n}, \mathbf{N}\,\mathbf{n} \rangle \right)^2 = \frac{1}{j}.$$

$\mathbf{N}$ has a minimal dispersion in coherent states.

To any observable $A \in \mathcal{M}_{2j+1}$ there correspond two classical observables, i.e. functions on S^2, A_{cl} and A^{cl} defined in terms of coherent spin states:

$$A_{cl}(\mathbf{n}) := \langle \mathbf{n}, A\,\mathbf{n} \rangle \quad \text{and} \quad A = \int_{S^2} \mu_j(d\mathbf{n})\, A^{\mathrm{cl}}(\mathbf{n})\, |\mathbf{n}\rangle\langle\mathbf{n}|.$$

The simplest examples are

A	A_{cl}	A^{cl}
$\mathbf{N}$	$\mathbf{n}$	$(1 + \frac{1}{j})\,\mathbf{n}$
$(N^x)^2$	$(1 - \frac{1}{2j})\,(n^x)^2 + \frac{1}{2j}$	$(1 + \frac{1}{j})(1 + \frac{3}{2j})\,(n^x)^2 - \frac{1}{2j}(1 + \frac{1}{j})$

They illustrate the general relation $A_{\mathrm{cl}}(\mathbf{n}) = A^{\mathrm{cl}}(\mathbf{n}) + \mathrm{O}(1/j)$.

3.3 Classical differentiable dynamics

An important category of classical dynamical systems belongs to the phase spaces parametrized by a finite number of real coordinates. In many cases, due to a non-trivial topology of $\mathfrak{X}$, there does not exist a single global coordinate system for the whole of $\mathfrak{X}$. The notion of differentiable manifold provides a natural setting to deal with such systems.

Example 3.8 Consider the circle S^1. Every point of the circle can be identified with an angle $\alpha \in [0, 2\pi[$. Nevertheless, such a parameterization does not describe properly the topological properties of S^1, namely the fact that the points given by the coordinates ϵ and $2\pi - \epsilon$ are close for small ϵ. To amend this default, one can cover the circle with two open arcs

$$\Omega_1 := [0, \tfrac{3}{4}\pi[\bigcup]\tfrac{5}{4}\pi, 2\pi[\quad \text{and} \quad \Omega_2 =]\tfrac{1}{4}\pi, \tfrac{7}{4}\pi[$$

and map these arcs onto the open interval $]-3\pi/4, 3\pi/4[$ using the following local coordinates ω_1 and ω_2 determined by maps ψ_1 and ψ_2:

$$\psi_1 : \Omega_1 \to]-\tfrac{3}{4}\pi, \tfrac{3}{4}\pi[: \omega_1 = \psi_1(\alpha) = \begin{cases} \alpha & \text{if } \alpha \in [0, \tfrac{3}{4}\pi[, \\ \alpha - 2\pi & \text{if } \alpha \in]\tfrac{5}{4}\pi, 2\pi[, \end{cases}$$

$$\psi_2 : \Omega_2 \to]-\tfrac{3}{4}\pi, \tfrac{3}{4}\pi[: \omega_2 = \psi_2(\alpha) = \alpha - \pi.$$

The map which describes the coordinate change $\omega_1 \mapsto \omega_2$ is defined on

$$\psi_1(\Omega_1 \cap \Omega_2) =]\tfrac{1}{4}\pi, \tfrac{3}{4}\pi[\bigcup]-\tfrac{3}{4}\pi, -\tfrac{1}{4}\pi[$$

and given by

$$\psi_2 \circ \psi_1^{-1} : \psi_1(\Omega_1 \cap \Omega_2) \to]-\tfrac{3}{4}\pi, \tfrac{3}{4}\pi[$$

$$: \omega_2 = \psi_2 \circ \psi_1^{-1}(\omega_1) = \begin{cases} \omega_1 - \pi & \text{if } \omega_1 \in]\tfrac{1}{4}\pi, \tfrac{3}{4}\pi[, \\ \omega_1 + \pi & \text{if } \omega_1 \in]-\tfrac{3}{4}\pi, -\tfrac{1}{4}\pi[. \end{cases}$$

Abstract generalization of the construction of above leads to the notion of a d-dimensional *differentiable manifold* $\mathfrak{X}$. We assume that the topological space $\mathfrak{X}$ is covered by a family of open sets $\{\Omega_j\}$. Each set Ω_j is mapped by a continuous map ψ_j into $\mathbb{R}^d$ with continuous inverse on its range. The ψ_j provide *local coordinate systems*. The differentiability of the manifold means that any *coordinate change* $\psi_{j'} \circ \psi_j^{-1}$ is a differentiable map from $\psi_j(\Omega_j \cap \Omega_{j'}) \subset \mathbb{R}^d$ into $\mathbb{R}^d$. Here, differentiability means that such coordinate transformations are r-times continuously differentiable, with $r \in \mathbb{N}$. A pair (Ω_j, ψ_j) is a *local chart* and a collection of charts covering $\mathfrak{X}$ is an *atlas* of $\mathfrak{X}$. Any atlas defines a unique *maximal atlas* by taking all charts compatible with given ones and thereby defines a *differentiable structure*. In practice, we shall apply the following strategy: first we prove a result using a given chart or a given atlas and then we show the invariance of the statement under admissible coordinate changes.

A function $f : \mathfrak{X} \to \mathbb{R}$ is *differentiable* if $f \circ \psi_j^{-1}$ is differentiable on $\psi_j(\Omega_j) \subset \mathbb{R}^d$ for any chart (Ω_j, ψ_j). The set of all functions which are p-times continuously differentiable is denoted by $\mathcal{C}^p(\mathfrak{X})$. Obviously, for a C^r-manifold only the $\mathcal{C}^p$ functions with $p \leq r$ are well-defined.

A fundamental geometrical notion is the *tangent space* $\mathbf{T}_x\mathfrak{X}$ at the point $x \in \mathfrak{X}$ which, in the context of dynamical systems, is a necessary tool for defining continuous-time dynamics. It is known that any differentiable manifold can be embedded into a Euclidean space $\mathbb{R}^m$ with a suitably chosen dimension m, and then the geometrical meaning of the tangent space becomes obvious. There are, however, different equivalent ways to define this notion without referring to the embedding procedure. We follow the definition based on the notion of *derivation*.

For the case $\mathfrak{X} = \mathbb{R}^d$, the tangent space at $x \in \mathfrak{X}$ is obviously again $\mathbb{R}^d$. For a given vector $\mathbf{v} = (v^1, v^2, \dots, v^d)$ of this tangent space, we define the derivation at the point x as a linear functional $\nabla_{(\mathbf{v},x)}$ on the space $\mathcal{C}^1(\mathbb{R}^d)$:

$$\nabla_{(\mathbf{v},x)}(f) := v^j \frac{\partial}{\partial x^j} f(x^1, x^2, \dots, x^d) \tag{3.24}$$

with the Einstein summation convention.

The derivation satisfies the following *Leibnitz rule*:

$$\nabla_{(\mathbf{v},x)}(fg) = g(x)\nabla_{(\mathbf{v},x)}(f) + f(x)\nabla_{(\mathbf{v},x)}(g).$$

Moreover, any linear functional on $\mathcal{C}^1(\mathbb{R}^d)$ satisfying the Leibnitz rule is of the form (3.24). This identification allows us to define the tangent space at a point x for an arbitrary differentiable manifold $\mathfrak{X}$ as the linear space of derivations at the point x, i.e. the space of linear functionals on $\mathcal{C}^1(\mathfrak{X})$ satisfying the Leibnitz rule.

In a given chart (Ω, ψ), the point $x \in \Omega$ is represented by

$$\psi(x) = (x^1, x^2, \dots, x^d) \in \mathbb{R}^d$$

and the function f on the manifold $\mathfrak{X}$ can be treated locally as a function of the coordinates of x. Using such an identification, we can write any derivation in the form (3.24). Hence, the partial derivatives $\partial/\partial x^j$ form a natural basis of the tangent space $\mathbf{T}_x\mathfrak{X}$ associated with a given local coordinate system (Ω, ψ) and $v^1, v^2, \dots, v^d$ are the coefficients of the vector $\mathbf{v} \in \mathbf{T}_x\mathfrak{X}$ with respect to this basis. Changing the coordinate system $(x^1, x^2, \dots, x^d)$ into $(x'^1, x'^2, \dots, x'^d)$, we obtain the associated transformation rule $(v^1, v^2, \dots, v^d) \mapsto (v'^1, v'^2, \dots, v'^d)$ for the coefficients of the tangent vector

$$v^j \frac{\partial}{\partial x^j} = v^j \frac{\partial x'^k}{\partial x^j} \frac{\partial}{\partial x'^k} = v'^k \frac{\partial}{\partial x'^k}$$

and therefore

$$v'^k = \frac{\partial x'^k}{\partial x^j} v^j.$$

The disjoint union $T(\mathfrak{X})$ of tangent spaces $T(\mathfrak{X}) := \bigcup_{x \in \mathfrak{X}} \mathbf{T}_x\mathfrak{X}$ is a *tangent bundle* and it carries the structure of a differentiable manifold.

The physical prototype of a smooth dynamical system is a mechanical system described in the Lagrangian formalism. The dynamics is given by a continuous-time flow on a manifold which has a structure of a tangent bundle over the manifold called configuration space. For a Hamiltonian system, the trajectory of the flow is confined to a submanifold of a fixed energy determined by the initial conditions. These manifolds of constant energy are typically compact differentiable manifolds carrying a natural time-invariant, normalizable, Liouville measure.

More generally, a differentiable dynamical system in continuous time is usually described in terms of a first order differential equation

$$\frac{dx}{dt} = \mathbf{F}(x,t), \quad x = x_0 \text{ at } t = t_0, \tag{3.25}$$

on a compact manifold $\mathfrak{X}$. $\mathbf{F}$ is a *vector field* which associates with each point x of the manifold an element of the tangent space at x and may, moreover, explicitly depend on the time t. We assume that $\mathbf{F}$ depends continuously on x for fixed t and that it is locally Lipshitz in t, which implies that, for any initial condition, there exists a unique global solution

$$t \in \mathbb{R} \mapsto x(t; (x_0, t_0))$$

of (3.25) which is a curve on $\mathfrak{X}$ parametrized by the time t and passing at time t_0 through the point x_0. The *tangent vector* at this curve at time t belongs to the tangent space at $x(t)$ and is the velocity at time t.

The solutions of (3.25) were described in terms of curves passing at a specified time through a given point. We can also look at the dynamical system as a *groupoid* of C^1 transformations $\{T(t,s) \mid s,t \in \mathbb{R}\}$ of $\mathfrak{X}$:

$$T(t,s)(y) := x(t; (y,s)).$$

From the uniqueness of the solution of (3.25) we immediately obtain

$$T(t,r) T(r,s) = T(t,s).$$

The vector field $\mathbf{F}$ is *autonomous* if $\mathbf{F}$ does not explicitly depend on time. In this case, $T(t,s)$ only depends on the difference $t - s$, and we can describe the dynamics in terms of a group $\{T(t) := T(t,0) \mid t \in \mathbb{R}\}$ of C^1 transformations of $\mathfrak{X}$.

Discrete time may appear by taking iterates of the flow map for a given time scale t_0. This is a natural procedure for Lagrangians which depend periodically on time. Another less trivial possibility for obtaining discrete time dynamical systems is the Poincaré first return map.

The abstract generalization of these examples is a *smooth dynamical system* $(\mathfrak{X}, T, \mu)$ where $\mathfrak{X}$ is a compact C^∞ manifold and T is a C^1 homeomorphism of $\mathfrak{X}$ preserving the Borel measure μ. Given an atlas and two charts (Ω_1, ψ_1) and (Ω_2, ψ_2) in that atlas with $x \in \Omega_1$ and $T(x) \in \Omega_2$ and writing

$$(x^1, x^2, \ldots, x^d) = \psi_1(x) \quad \text{and} \quad (t^1, t^2, \ldots, t^d) = \psi_2(T(x)),$$

T is of the form

$$\begin{aligned}(x^1, x^2, &\ldots, x^d) \\ &\mapsto (t^1(x^1, x^2, \ldots, x^d)), t^2(\ldots), \ldots, t^d(x^1, x^2, \ldots, x^d))\end{aligned} \tag{3.26}$$

with all t^j differentiable.

3.4   Self-adjoint Laplacians on compact manifolds

Interesting examples of classical dynamical systems are given by geodesic flows on curved spaces. These describe the free particle motion on a Riemannian configuration space. The corresponding Lagrangian is given by $L(x, \mathbf{v}) := \frac{1}{2} \langle \mathbf{v}, \mathbf{v} \rangle_x$ with resulting motion preserving the length $\|\mathbf{v}\|_x$. The Hamiltonians of their quantum counterparts are Laplacians on compact Riemannian manifolds. Any locally compact differentiable manifold can be equipped with a *Riemannian structure* which allows us to speak about distances between points, angles between intersecting curves, ...

A *Riemannian manifold* is a differentiable manifold with an inner product $\langle \cdot, \cdot \rangle_x$ defined on each tangent space $\mathbf{T}_x \mathfrak{X}$ and which depends smoothly on x. In a local coordinate system, the inner product $\langle \cdot, \cdot \rangle_x$ is given by a positive real matrix $[g_{ij}(x)]$, smoothly depending on x, such that

$$\langle \mathbf{u}, \mathbf{v} \rangle_x := g_{ij}(x) u^i v^j, \quad \mathbf{u} = u^j \frac{\partial}{\partial x^j}, \quad \mathbf{v} = v^j \frac{\partial}{\partial x^j}.$$

Consider a smooth curve $t \in [a, b] \mapsto x(t)$ in $\mathfrak{X}$. The length ℓ of the curve is computed as

$$\ell[a, b] := \int_a^b dt \sqrt{\langle \dot{x}(t), \dot{x}(t) \rangle_{x(t)}},$$

where $\dot{x}(t)$ is the tangent vector $(dx^j/dt)(\partial/\partial x^j)$ in a local coordinate system and $\ell[a, b]$ is invariant with respect to a change $t \mapsto t'$ of the parameterization. The *distance between two points* of the Riemannian space is the infimum of the lengths of curves connecting the points. The topology induced by this metric coincides with the topology of the original manifold .

A Riemannian structure induces also a natural *Riemannian density* given in a local coordinate system by

$$|dx| := \sqrt{|g(x)|} dx^1 dx^2 \cdots dx^d, \quad |g(x)| = \det([g_{ij}(x)]).$$

The volume $\mathrm{Vol}(\mathfrak{X})$ of a compact manifold is the integral of the density over the manifold,

$$\mathrm{Vol}(\mathfrak{X}) := \int_{\mathfrak{X}} |dx|.$$

The Hilbert space $\mathcal{L}^2(\mathfrak{X}, |dx|)$ appears in the description of quantum systems with the Riemannian manifold $\mathfrak{X}$ as configuration space and also in the context of classical dynamical systems with compact finite-dimensional phase spaces. In both cases, the generalized Laplacians on $\mathcal{L}^2(\mathfrak{X}, |dx|)$ play an important role.

A *generalized Laplacian* on $\mathcal{L}^2(\mathfrak{X}, |dx|)$ is a second-order differential operator Δ defined in any local coordinate system by

$$\Delta := -g^{ij}(x) \frac{\partial}{\partial x^i} \frac{\partial}{\partial x^j} + \text{first-order terms}, \tag{3.27}$$

where $[g^{ij}(x)]$ is the inverse of the matrix $[g_{ij}(x)]$.

In particular, we consider generalized Laplacians on a compact manifold which are symmetric operators on the dense domain $\mathcal{C}^\infty(\mathfrak{X})$ in $\mathcal{L}^2(\mathfrak{X}, |dx|)$ like for example the *Beltrami–Laplace operator*

$$[\Delta\varphi](x) := -\frac{1}{\sqrt{|g|}} \frac{\partial}{\partial x^i} \left(\sqrt{|g|}\, g^{ij}(x)\, \frac{\partial}{\partial x^j} \varphi(x) \right).$$

The theory of symmetric generalized Laplacians on compact Riemannian manifolds is well-developed. We recall here some fundamental results which will be used later on.

Theorem 3.9 (Weyl)

(a) *A generalized symmetric Laplacian Δ on a compact d-dimensional Riemannian manifold $\mathfrak{X}$ possesses a unique self-adjoint extension denoted by the same symbol.*

(b) *The operator Δ has a pure point spectrum with finite multiplicity and is bounded from below.*

(c) *The operator $e^{-t\Delta}$, $t > 0$ is trace-class and the following asymptotic formula for $t \to 0$ holds:*

$$\mathrm{Tr}(e^{-t\Delta}) = (4\pi t)^{-d/2}\, \mathrm{Vol}(\mathfrak{X}) + \mathrm{O}(t^{-d/2+1}). \tag{3.28}$$

Theorem 3.9 has a simple physical interpretation. Writing the left-hand side of eqn (3.28) as

$$\mathrm{Tr}(e^{-t\Delta}) = \int_0^\infty dN(\epsilon)\, e^{-t\epsilon},$$

where

$$N(\epsilon) := \#\{\text{eigenvectors of } \Delta \text{ with eigenvalues } \epsilon_k \leq \epsilon\},$$

we obtain from eqn (3.28) for large ϵ

$$N(\epsilon) \approx \left(\frac{1}{4\pi}\right)^{d/2} \frac{1}{\Gamma(1+d/2)}\, \mathrm{Vol}(\mathfrak{X})\, \epsilon^{d/2}.$$

This agrees with the heuristic picture that a single quantum state occupies a volume $(2\pi)^d$ in phase space, putting $\hbar = 1$. Indeed, as the quantum Hamiltonian Δ corresponds to a classical kinetic energy $g^{ij} p_i p_j = p^2$

$$N(\epsilon) \approx (2\pi)^{-d}\, \mathrm{Vol}(\mathfrak{X})\, \mathrm{Vol}(B_{\sqrt{\epsilon}}),$$

where $\mathrm{Vol}(B_p) = \pi^{d/2}\, p^d/\Gamma(1+d/2)$ is the volume of a ball of radius p in momentum space.

Let $\mu(dx) = h^2(x)|dx|$ be a probability measure on $\mathfrak{X}$ with smooth invertible density $h > 0$ with respect to the volume measure $|dx|$. We then consider a generalized Laplacian on the Hilbert space $\mathcal{L}^2(\mathfrak{X}, \mu)$ that is obtained by applying a similarity transformation to the generalized symmetric Laplacian of (3.27):

$$\Delta_h := h^{-1}\Delta h = -g^{ij}(x)\frac{\partial}{\partial x^i}\frac{\partial}{\partial x^j} + \text{first-order terms},$$

This operator is symmetric on the subspace of smooth functions on $\mathfrak{X}$ as

$$\langle \varphi, \Delta_h \psi \rangle_h := \int_{\mathfrak{X}} |dx|\, h^2(x)\, \overline{\varphi(x)}\, (\Delta_h \psi)(x)$$

$$= \int_{\mathfrak{X}} |dx|\, h^2(x)\, \overline{\varphi(x)}\, h^{-1}(x)\, (\Delta\, h\psi)(x)$$

$$= \langle \varphi, h\Delta\, h\psi \rangle$$

$$= \langle h\varphi, \Delta\, h\psi \rangle$$

and Δ is self-adjoint. It then follows that Δ_h has a unique self-adjoint extension and that this extension has the same spectrum as the original Δ characterized by Theorem 3.9.

3.5 Notes

The proofs of Stone's theorem, the Kato–Rellich criterion, the Trotter product formula and the existence of solutions for time-dependent Schrödinger equations can be found in the texts by Kato (1984) and Reed and Simon (1972). The time-independent formalism of (Howland 1974), in particular the case of Hamiltonians periodic in time, is presented in the monograph by Cycon et al. (1987).

Examples of classical and quantum kicked systems such as the kicked top, Example 3.3 and the quantum baker map, Example 3.4, were introduced in (Haake et al. 1987) and (Balazs and Voros 1989). One should also consult the books of Haake (1991) and Stöckmann (1999) and the papers collected in Casati and Chirikov (1995).

A general framework for describing the classical limit, with many references to the literature, can be found in (Werner 1994). Fundamental papers on coherent states are collected by Klauder and Skagerstam (1985). One should also consult the book by Perelomov (1986) and the paper (Perelomov 1972). For a rigorous treatment of quantum mean-field models, see (Duffield and Werner 1992).

For an introduction to differential geometry, see, for example, the texts of Nash and Sen (1983) and Isham (1989). A modern proof of Weyl's theorem, set in a more general framework, can be found in the monograph by Berline et al. (1992).

4

SPIN CHAINS

Spin systems as simplified models for magnetism, in particular for ferromagnetic phase transitions, were already proposed in the twenties, e.g. the Lenz–Ising and Heisenberg models. The detailed rigorous discussion of quantum spin models, in particular the formulation in terms of systems with an infinite number of particles and the proof of the existence of dynamics and equilibrium states in the thermodynamical limit, were developed in the 1960s and 1970s. These results are presented in the fundamental monograph of Bratteli and Robinson (1997), where the references to the original works are given.

We use quantum spin chains for two different purposes. They are, on the one hand, a simple example of an infinite quantum system, used in statistical mechanics to describe linear crystals and, on the other, the prototype of a dynamical system or a quantum stochastic process. In the first case, the points of the chain have the meaning of space while, in the second case, they label discrete time. At each site of the chain, we attach a copy of a spin, which is the simplest possible quantum system described in terms of matrices. The observables belonging to a finite portion of the system are generated by distinguishable copies of such a spin at each point. There is no Pauli principle or effect of quantum statistics involved here. Spins belonging to a finite region again form a discrete system, meaning that they are described by a matrix algebra. It is only in passing to the infinite system that we obtain a fundamentally different object.

We introduce a slightly more general object in that we allow the size of the spin to vary from site to site and that we do not insist on the spins residing on a regular lattice. The description of the observables and states for this type of system is the subject of the first two sections. In the third, we return to spin chains and briefly consider their symmetries and dynamics in the thermodynamic limit.

4.1 Local observables

Interesting dynamical behaviour in statistical mechanics comes about by the interaction of a multitude of relatively simple building blocks. The task of statistical mechanics is to infer the large-scale behaviour of the system in terms of these microscopic building blocks and their interactions. Quantum spin systems on a lattice are probably the simplest examples of collective systems. They mimic crystals. Each molecule in the crystal is modelled by a d-level system attached to a point of a regular lattice in ν dimensions. The simplest case is that of a chain of two-level systems. Spins sitting at different sites of the underlying lattice are, though identical, distinguishable. This means that the full tensor product $\mathbb{C}^{2s+1} \otimes \mathbb{C}^{2s+1}$ is accessible for a system composed of two s-spins and not just its symmetric or antisymmetric part.

Particle systems occurring in real life often consist of huge, though finite, numbers of elementary constituents and one is mainly interested in bulk behaviour, meaning that one wishes to know in first instance the asymptotic behaviour of the system when its number of particles tends to infinity, this is the *thermodynamical limit*. Passing to the thermodynamical limit simplifies the description from a mathematical point of view as it allows to have, for example, lattice translations as a true symmetry of a system or to make sharp distinctions between different coexistent phases of a system in thermal equilibrium.

Describing an infinite quantum spin system in the Hilbert space formalism is not very convenient, as one has to use infinite tensor products of (d-dimensional) Hilbert spaces. This is not easy to deal with. The idea is rather to consider a minimal collection of observables, called the *algebra of quasi-local observables*, that is sufficiently large to specify the nature of the microscopic constituents of the system and their interactions.

The particles on the lattice are described in terms of a few basic observables and their commutation relations such as

$$[S^x, S^y] = iS^z, \text{ and cyclic permutations,}$$
$$(S^x)^2 + (S^y)^2 + (S^z)^2 = s(s+1)$$

for a particle with angular momentum s or

$$c^* c + c c^* = \mathbb{1} \quad \text{and} \quad c^2 = 0$$

for a two-level system.

We now turn to a slightly more general setting, dropping some of the features that are typical for statistical mechanics such as the lattice structure and the identity of the particles at the different sites of the lattice.

Fix for every $j \in \mathbb{N}$ a dimension $d_j \in \{2, 3, \dots\}$. The observables $\mathfrak{A}_j$ of the particle at site j are chosen to be the algebra $\mathcal{M}_{d_j}$ of the complex matrices of dimension d_j. We can think of $\mathfrak{A}_j$ as $\mathcal{B}(\mathfrak{H}_j)$, the linear operators on the d_j-dimensional Hilbert space $\mathfrak{H}_j := \mathbb{C}^{d_j}$. The joint observable A for the spin at site j and B for the spin at k with $j \neq k$ is the elementary tensor

$$A_j \otimes B_k \tag{4.1}$$

acting on $\mathfrak{H}_j \otimes \mathfrak{H}_k$. The subscripts j and k have been added in order to label the sites corresponding to A and B. The observables of the particles at j and k are therefore

$$\mathfrak{A}_{\{i,j\}} = \mathcal{B}(\mathfrak{H}_j \otimes \mathfrak{H}_k) = \mathcal{B}(\mathfrak{H}_j) \otimes \mathcal{B}(\mathfrak{H}_k) = \mathfrak{A}_j \otimes \mathfrak{A}_k.$$

If we are not interested in the particle at k, then we can simply choose $B = \mathbb{1}$ in (4.1). In this way, we have a natural identification of the observables at j with a subalgebra of observables at j and k. Instead of writing $A_j \otimes \mathbf{1}_k$, we simply write A_j. Moreover, observables at j commute with those at k

$$[A_j \otimes \mathbf{1}_k, \mathbf{1}_j \otimes B_k] = [A_j, B_k] = 0, \quad j \neq k.$$

This construction obviously extends to arbitrary finite subset Λ of $\mathbb{N}$,

$$\mathfrak{A}_\Lambda = B(\mathfrak{H}_\Lambda) = \bigotimes_{j \in \Lambda} \mathfrak{A}_j \quad \text{with} \quad \mathfrak{H}_\Lambda = \bigotimes_{j \in \Lambda} \mathfrak{H}_j,$$

and an observable A of particles in $\Lambda \subset M$ is identified with an observable in M through

$$\mathfrak{A}_\Lambda \hookrightarrow \mathfrak{A}_M : A \hookrightarrow A \otimes \Big(\bigotimes_{j \in M \setminus \Lambda} \mathbf{1}_j \Big). \tag{4.2}$$

It is, however, more convenient to forget about these injections and to identify all the identity matrices of the different $\mathfrak{A}_\Lambda$ with a same abstract identity $\mathbf{1}$. We can now consider $\mathfrak{A}_\Lambda$ to be a subalgebra of $\mathfrak{A}_M$ whenever Λ and M are two finite subsets of $\mathbb{N}$ with $\Lambda \subset M$.

The *local observables* $\mathfrak{A}_{\mathrm{loc}}$ are the union of all the algebras $\mathfrak{A}_\Lambda$ with $\Lambda \subset \mathbb{N}$ finite, i.e. local observables are polynomials in the observables of spins in some finite subset of the lattice, this subset depends, of course, on the observable under consideration. Clearly, $\mathfrak{A}_{\mathrm{loc}}$ is a $*$-algebra.

4.2   States of a spin system

Suppose that we have N distinguishable and independent spins at different sites $i_1, i_2, \ldots i_N \in \mathbb{N}$, each of them in the state φ. Computing the overlap of this N-particle state $\varphi_{i_1} \otimes \varphi_{i_2} \otimes \cdots \otimes \varphi_{i_N}$ with a similar state where each spin is in the state ψ, we find the transition probability

$$\mathrm{Prob}\big(\varphi_{i_1} \otimes \varphi_{i_2} \otimes \cdots \otimes \varphi_{i_N}, \psi_{i_1} \otimes \psi_{i_2} \otimes \cdots \otimes \psi_{i_N}\big) = |\langle \varphi, \psi \rangle|^{2N},$$

which tends to zero when N tends to infinity. It is, therefore, not convenient to describe in the thermodynamic limit both type of states simultaneously on Hilbert space level and we shall rather consider states as functionals on the observables.

Returning to the general setting of Section 4.1, suppose that, for each finite $\Lambda \subset \mathbb{N}$, ρ_Λ is a density matrix on $\mathfrak{H}_\Lambda$ which defines an expectation functional or state $\langle \cdot \rangle_\Lambda$ on $\mathfrak{A}_\Lambda$ through

$$\langle A \rangle_\Lambda := \mathrm{Tr}\, \rho_\Lambda A. \tag{4.3}$$

We say that $\{\langle \cdot \rangle_\Lambda \mid \Lambda \subset \mathbb{N}, \text{ finite}\}$ is a *consistent family of local states* if the $\langle \cdot \rangle_\Lambda$ are compatible with the identification (4.2), i.e.

$$\langle A \otimes \mathbf{1}_{M \setminus \Lambda} \rangle_M = \langle A \rangle_\Lambda, \quad A \in \mathfrak{A}_\Lambda \tag{4.4}$$

or, in terms of density matrices (2.66), if ρ_Λ is a partial trace of ρ_M,

$$\text{Tr}_{\mathfrak{H}_{M\setminus\Lambda}} \rho_M = \rho_\Lambda. \tag{4.5}$$

A state on the infinite system is not given by a single density matrix but rather by a consistent family of local states. The ρ_Λ are the *reduced density matrices* of the global state.

Example 4.1 *(Product states)* Let, for any $j \in \mathbb{N}$, ρ_j be a density matrix of dimension d_j and, for any finite $\Lambda \subset \mathbb{N}$, set

$$\rho_\Lambda := \bigotimes_{j \in \Lambda} \rho_j,$$

then $\{\rho_\Lambda \mid \Lambda\}$ is a consistent family of local states. The corresponding state on the infinite system corresponds to all spins being independent with the jth one distributed according to the density matrix ρ_j.

Example 4.2 *(Classical states)* Let $\mathcal{S}_d$ denote the set with d points. It describes a classical model of a spin with d accessible states, not to be confused with a continuous classical spin with phase space S^2, the surface of the unit sphere in $\mathbb{R}^3$. The *configuration space* Ω of a system of classical spins on $\mathbb{N}$ where the spin at j has d_j states is

$$\Omega = \underset{j \subset \mathbb{N}}{\times} \mathcal{S}_{d_j}.$$

It becomes a measurable space if we consider the σ-algebra generated by the *cylinder sets*

$$[\omega_\Lambda] := \{\omega_\Lambda\} \times \left(\underset{k \in \mathbb{N}\setminus\Lambda}{\times} \mathcal{S}_{d_k} \right).$$

Here Λ is a finite subset of $\mathbb{N}$ and ω_Λ a configuration in Λ, which is a Λ-tuple of $s_j \in \mathcal{S}_{d_j}$ with $j \in \Lambda$. We use the notation Ω_Λ for the configurations in Λ.

Let μ be a probability measure on Ω. It is specified by a consistent family of local probability measures $\{\mu_\Lambda \mid \Lambda\}$ similarly to (4.4). To each local configuration $\omega_\Lambda \in \Omega_\Lambda$ we associate the vector

$$\mathbf{e}(\omega_\Lambda) := \otimes_{j \in \Lambda} \mathbf{e}_{s_j}. \tag{4.6}$$

In (4.6), $\{\mathbf{e}_1, \mathbf{e}_2, \dots, \mathbf{e}_d\}$ is the canonical basis in $\mathbb{C}^d$, see (2.2). Then

$$\rho_\Lambda := \sum_{\omega_\Lambda \in \Omega_\Lambda} \mu_\Lambda(\omega_\Lambda) \, |\mathbf{e}(\omega_\Lambda)\rangle\langle\mathbf{e}(\omega_\Lambda)|$$

is a consistent family of local density matrices.

Example 4.3 *(Limiting Gibbs states)*   These states arise naturally in equilibrium statistical mechanics. We sketch the way they are constructed without fur-
nishing any details. For any finite $\Lambda \subset \mathbb{N}$ we are given a *local Hamiltonian* H_Λ
which is a self-adjoint element in $\mathfrak{A}_\Lambda$. The *local Gibbs states* are then determined
by the density matrices

$$\frac{e^{-\beta H_\Lambda}}{\mathcal{Z}_\Lambda}.$$

The parameter $\beta \in \mathbb{R}$ is the inverse temperature and $\mathcal{Z}_\Lambda$, the partition function,
is a normalizing factor

$$\mathcal{Z}_\Lambda := \operatorname{Tr} e^{-\beta H_\Lambda}.$$

Let $\{M_n \mid n\}$ be an absorbing sequence of finite subsets of $\mathbb{N}$, which means that
eventually any finite subset of $\mathbb{N}$ becomes a subset of the M_n. Provided the limit
in (4.7) exists, the reduced density matrices of the *limiting Gibbs state* are given
by

$$\rho_\Lambda^{\text{Gibbs}} := \lim_{n \to \infty} \operatorname{Tr}_{\mathfrak{H}_{M_n \setminus \Lambda}} \frac{e^{-\beta H_{M_n}}}{\mathcal{Z}_{M_n}}. \tag{4.7}$$

This limit is the *thermodynamical limit*. Later on, we shall consider more explicit
situations where the existence of the thermodynamic limit can be proven.

4.3   Symmetries and dynamics

In standard quantum mechanics, both symmetries and dynamics are described
in terms of unitary operators U,

$$\psi \mapsto U\psi.$$

For an infinite system, we need the notion of symmetry on the level of observables, i.e. rather

$$\gamma(X) = U^* X U. \tag{4.8}$$

Maps of the kind (4.8) are called *automorphisms*, because they preserve the full
algebraic structure of the observables:

$$\gamma \text{ is invertible}, \qquad \gamma(\mathbf{1}) = \mathbf{1},$$
$$\gamma(X + \alpha Y) = \gamma(X) + \alpha \gamma(Y),$$
$$\gamma(X Y) = \gamma(X) \gamma(Y), \qquad \gamma(X^*) = (\gamma(X))^*.$$

The notion of automorphism generalizes that of unitarity still in another way, as
many automorphisms are not given in terms of a unitary operator as in (4.8).

Example 4.4 *(Shift-invariant states on a spin chain)* Spin chains are systems of spins with constant dimension d attached at the sites of the chain $\mathbb{Z}$. For such a system there is a natural notion of *translation along the chain*,

$$\gamma(A_\Lambda) = A_{\Lambda+1}.$$

A_Λ is an observable in Λ, and $A_{\Lambda+1}$ is a copy of that observable at the sites of the volume obtained by shifting Λ one unit to the right. The map γ preserves completely the structure of the local observables of the chain and is invertible, the inverse being obviously the left-shift. It is therefore a automorphism. Moreover, γ cannot be written in terms of a local unitary U as in (4.8)

A state is *shift-invariant* if it yields the same expectation for A and $\gamma(A)$ with A an arbitrary local observable. The compatibility conditions (4.4) can be simplified greatly for a shift-invariant state. In order to determine such a state, it is necessary and sufficient to specify a family $\{\sigma(n) \mid n = 1, 2, \ldots\}$ where each $\sigma(n)$ is a density matrix on $\bigotimes_{j=1}^{n} \mathbb{C}^d$ satisfying the compatibility condition

$$\text{Tr}\big(\sigma(n+1)\, A(n) \otimes \mathbf{1}\big) = \text{Tr}\big(\sigma(n)\, A(n)\big) = \text{Tr}\big(\sigma(n+1)\, \mathbf{1} \otimes A(n)\big)$$

for any $A(n) \in \bigotimes_{j=1}^{n} \mathcal{M}_d$. The reduced density matrices ρ_Λ are then given as follows: let $[m, m+n]$ be any interval in $\mathbb{Z}$ containing Λ; then

$$\rho_\Lambda = \text{Tr}_{\{\otimes_j \mathbb{C}^d \mid j+m-1 \notin \Lambda\}}\, \sigma(n).$$

Similarly to the construction of limiting Gibbs states in Example 4.3, the dynamics of an infinite system has to be constructed from local Hamiltonians by applying a thermodynamic limit. The mathematical problems that appear are illustrated by the two following examples.

Example 4.5 *(Nearest-neighbour interactions)* A nearest-neighbour interaction Φ of spins of dimension d is a self-adjoint element of the matrix algebra $\mathcal{M}_d \otimes \mathcal{M}_d$. We assume, for simplicity, that it is symmetric with respect to the permutation of the factors in $\mathcal{M}_d \otimes \mathcal{M}_d$, meaning that

$$\langle \varphi_1 \otimes \psi_1, \Phi\, \varphi_2 \otimes \psi_2 \rangle = \langle \psi_1 \otimes \varphi_1, \Phi\, \psi_2 \otimes \varphi_2 \rangle$$

for any choice of $\varphi_j, \psi_j \in \mathbb{C}^d$. We call the interaction is *non-degenerate* if

$$[\Phi, A \otimes \mathbf{1}] = 0 \text{ implies that } A \text{ is a multiple of } \mathbf{1}.$$

The local Hamiltonian $H_{[-N,N]}$, which yields the interaction energy of the particles in $[-N, N]$, is written in terms of Φ as

$$H_{[-N,N]} = \sum_{j=-N}^{N-1} \Phi_{j\, j+1},$$

where $\Phi_{j\,j+1}$ is a copy of Φ at the neighbouring sites j and $j+1$. The solution of the Heisenberg evolution eqn (3.2) for the local Hamiltonian can be written as a series expansion:

$$A + it\,[H_{[-N,N]}, A] + \frac{(it)^2}{2!}\,[H_{[-N,N]}, [H_{[-N,N]}, A]] + \cdots .$$

Notice that for a fixed local element A the consecutive terms of this series become independent of N for N large enough. This suggests considering the infinite system dynamics:

$$A(t) := A + it\,\delta(A) + \frac{(it)^2}{2!}\,\delta(\delta(A)) + \cdots \tag{4.9}$$

with

$$\delta(A) := \lim_{N \to \infty} [H_{[-N,N]}, A], \quad A \in \mathfrak{A}_{\text{loc}}. \tag{4.10}$$

The map δ in (4.10) is a well-defined linear transformation of $\mathfrak{A}_{\text{loc}}$ that satisfies

$$\delta(A\,B) = \delta(A)\,B + A\,\delta(B) \quad \text{and} \quad \delta(A^*) = -\delta(A)^*.$$

Such maps are called *derivations*.

For A on a single point of the chain,

$$\sum_{n=0}^{\infty} \frac{|t|^n}{n!} \|\delta^n(A)\| \le \left(2\exp\left(2|t|\|\Phi\| + (e^{2|t|\|\Phi\|} - 1)\right) - 1\right)\|A\|. \tag{4.11}$$

This estimate is useful in proving the norm-convergence of the series (4.9) for all $t \in \mathbb{C}$ but for the fact that $\mathfrak{A}_{\text{loc}}$ is not complete. A similar estimate holds for general local elements.

In order to prove (4.11), we observe that $\delta^n(A)$ can be written as a sum of terms t_n from intervals in $\mathbb{Z}$ with lengths ranging from 2 to $n+1$. Suppose, for example, that $A \in \mathfrak{A}_1$; then

$$\delta(A) = \left([\Phi_{01}, A] + [\Phi_{12}, A]\right),$$

$$\delta^2(A) = \left([\Phi_{01}, [\Phi_{01}, A]] + [\Phi_{12}, [\Phi_{12}, A]]\right)$$
$$+ \left([\Phi_{-10}, [\Phi_{01}, A]] + [\Phi_{12}, [\Phi_{01}, A]] + [\Phi_{01}, [\Phi_{12}, A]]\right.$$
$$+ \left.[\Phi_{23}, [\Phi_{12}, A]]\right),$$

where terms from intervals of equal lengths have been grouped together. Each of the terms in $\delta^n(A)$ can be estimated by

$$\|t_n\| \le 2^n \|\Phi\|^n \|A\|. \tag{4.12}$$

We have to count the number of terms t_n appearing in $\delta^n(A)$ from intervals of length $k+1$ with $1 \le k \le n$. Let their number be $2\,c(n,k)$. Using that Φ is a nearest-neighbour interaction, it can be seen that the c's are determined by the recursion

$$c(n,0) = c(n, n + 1) = 0,$$
$$c(n + 1, k) = k\, c(n, k) + 2\, c(n, k - 1),$$

$$(4.13)$$

with initial condition $c(1, 1) = 1$. Solving (4.13) and combining with (4.12) yields (4.11).

We now argue that, for a non-degenerate interaction Φ and A local and not a multiple of $\mathbf{1}$, $A(t)$ as given by (4.9) cannot converge in the local algebra. Suppose to the contrary, by continuity in t, for a non-empty time interval, $A(t)$ must belong to the same algebra $\mathfrak{A}_\Lambda$. Let t_0 be in the interior of such an interval and let Λ be the smallest subset of $\mathbb{Z}$ such that $A(t_0) \in \mathfrak{A}_\Lambda$ and suppose that a is the rightmost point of Λ. Differentiating $A(t)$ with respect to t at t_0, we see that

$$[\Phi_{a\,a+1}, A(t_0)] = 0.$$

As we assumed Φ to be non-degenerate, this implies that $A(t_0)$ belongs to $\mathfrak{A}_{\Lambda \setminus \{a\}}$, contradicting the minimality of Λ. The solution to the non-locality of the dynamics is rather straightforward: we need to consider the completion of the local algebra. This will be considered in the next chapter.

Example 4.6 *(The one-dimensional XY-model)* We now turn to an explicit example of a spin-$\frac{1}{2}$ quantum spin chain. The interaction is of nearest-neighbour type and is explicitly given by

$$\Phi = \sigma^x \otimes \sigma^x + \sigma^y \otimes \sigma^y \in \mathcal{M}_2 \otimes \mathcal{M}_2. \qquad (4.14)$$

The σ's are the usual Pauli matrices:

$$\sigma^x := \begin{pmatrix} 0 & 1 \\ 1 & 0 \end{pmatrix}, \quad \sigma^y := \begin{pmatrix} 0 & -i \\ i & 0 \end{pmatrix}, \quad \sigma^z := \begin{pmatrix} 1 & 0 \\ 0 & -1 \end{pmatrix}. \qquad (4.15)$$

The interaction (4.14) could be generalized to

$$\Phi = \cos\alpha\, \sigma^x \otimes \sigma^x + \sin\alpha\, \sigma^y \otimes \sigma^y + h\,\sigma^z \otimes \mathbf{1} + h\,\mathbf{1} \otimes \sigma^z$$

without affecting much our computations.

The one-dimensional XY-model is simple because it becomes linear when rewritten in terms of suitable operators; this is the point of the *Jordan–Wigner transformation*. Define on an interval $\{1, 2, \ldots, n\}$ operators c_j by

$$c_j := \sigma^z \otimes \cdots \otimes \sigma^z \otimes \sigma^- \otimes \mathbf{1} \otimes \cdots \otimes \mathbf{1}\,,$$

σ^- at the jth place and $\sigma^- := (\sigma^x - i\sigma^y)/2$. The c_j satisfy *canonical anticommutation relations*

$$\{c_j, c_k\} := c_j c_k + c_k c_j = 0, \quad \{c_j, c_k^*\} = \delta_{jk}. \tag{4.16}$$

The relations (4.16) can be expressed in a natural linear fashion by introducing, for a vector $\mathbf{x} \in \mathbb{C}^n$,

$$c^*(\mathbf{x}) := \sum_{j=1}^{n} x_j c_j^*.$$

We obtain

$$c^*(\mathbf{x} + \alpha \, \mathbf{y}) = c^*(\mathbf{x}) + \alpha \, c^*(\mathbf{y}),$$
$$\{c^*(\mathbf{x}), c^*(\mathbf{y})\} = \{c(\mathbf{x}), c(\mathbf{y})\} = 0,$$
$$\{c(\mathbf{x}), c^*(\mathbf{y})\} = \langle \mathbf{x}, \mathbf{y} \rangle \, \mathbb{I}.$$

Putting all these elements together, we can verify that

$$[H_{\{1,2,\dots,n\}}, c^*(\mathbf{x})] = \sum_{j=1}^{n-1} [\Phi_{j\,j+1}, c^*(\mathbf{x})] = c^*(T\,\mathbf{x}), \tag{4.17}$$

where T is the $n \times n$ matrix with entries

$$t_{k\ell} = -2(\delta_{k,\ell+1} + \delta_{k,\ell-1}).$$

Equation (4.17) shows that computation of the dynamics of the XY-model reduces to a linear problem, which simplifies the thermodynamic limit greatly. This is, however, a very particular situation with quantum dynamical systems.

5

ALGEBRAIC TOOLS

In this chapter, we first concentrate on an abstract notion of quantum probability space. A C*-algebra plays the role of continuous functions on the space of elementary outcomes and its expectation functionals are the probability measures. It is always possible, by using the representation theory, to rephrase this probabilistic structure in terms of the usual quantum language: observables acting as linear operators on a Hilbert space and expectations determined by normalized vectors. This is the content of the Gelfand–Naimark–Segal theorem. We present several examples originating in statistical mechanics and quantum dynamical systems.

The main reason to pass from the usual Hilbert space description of quantum mechanics to an algebraic one is that we want to treat infinite systems. A theorem by von Neumann guarantees that the usual momentum and position observables of a system with d degrees of freedom admit a unique Hilbert space representation. A similar result holds for finite systems of fermions. This is no longer true if the system is infinite, i.e. if we want to deal with quantum field theory or with statistical mechanics in the thermodynamical limit. In these cases, a minimal set of observables that make sense in each representation is needed to specify the nature of the system and of its basic interactions. This is precisely the role played by the abstract algebra of observables. A complete specification of a physical situation requires, apart from the algebra, also the knowledge of an expectation functional which determines actually a Hilbert space realization of the system. The couple algebra and expectation functional is a *quantum probability space* and it becomes a *quantum dynamical system* if we specify, moreover, the time evolution of observables or states.

5.1 C*-algebras

The $d \times d$ complex matrices $\mathcal{M}_d$ are basic examples of non-commutative algebras. If we fix any matrix $A \in \mathcal{M}_d$ and consider complex polynomials p in one indeterminate, then we can form the matrix-valued polynomial $p(A)$. These polynomials form an Abelian algebra and satisfy the usual relations

$$p_1(A) + p_2(A) = (p_1 + p_2)(A) \quad \text{and} \quad p_1(A)p_2(A) = (p_1 p_2)(A).$$

We can also find d complex numbers $\gamma_0, \gamma_1, \ldots, \gamma_{d-1}$, which depend on A, such that

$$A^d = \gamma_0 \mathbf{1} + \gamma_1 A + \cdots + \gamma_{d-1} A^{d-1}.$$

Because of this property, one could think of a single $d \times d$ matrix as a function on a discrete set with d points. On the other hand, some matrices have properties very unlike those of complex functions, such as $A^2 = 0$ and, nevertheless, $A \neq 0$.

The structure of a general finite-dimensional Abelian algebra $\mathfrak{A}$ is well-known and given by the *Jordan decomposition*

$$\mathfrak{A} = \bigoplus_{\alpha \in \mathrm{Sp}(\mathfrak{A})} \mathfrak{A}_\alpha, \tag{5.1}$$

where, in our case, $\mathrm{Sp}(\mathfrak{A}) = \mathrm{Sp}(A)$; see (2.31). Each of the $\mathfrak{A}_\alpha$ is a complex algebra generated by an identity element $\mathbb{1}_\alpha$ and a nilpotent N_α, i.e. by an element N_α such that there exists a natural number k_α with $N_\alpha^{k_\alpha} \neq 0$ and $N_\alpha^{k_\alpha + 1} = 0$; k_α is called the order of N_α.

It is useful to think about the full $d \times d$ matrix algebra $\mathcal{M}_d$ as the complex functions on a discrete quantum space with d points and not as much look at it as the union of all Abelian algebras generated by single matrices. When looking for a natural norm on $\mathcal{M}_d$, one could at first try to find a norm that does not depend on the explicit form of the matrix in a given basis, i.e. a norm $\|\cdot\|$ that is invariant under similarity transformations $A \mapsto SAS^{-1}$ with S invertible. The only invariant for general similarity transformations is the Abelian algebra generated by $\mathbb{1}$ and A together with the trace functional. In the Jordan decomposition (5.1) of the algebra generated by $\mathbb{1}$ and A we have

$$A = \bigoplus_{\alpha \in \mathrm{Sp}(A)} (\alpha \mathbb{1}_\alpha + N_\alpha).$$

It is tempting to measure the size of A by putting the norm of A equal to its *spectral radius* $\rho(A)$, i.e. to

$$\rho(A) := \max\{|\alpha| \mid \alpha \in \mathrm{Sp}(A)|\}.$$

Unfortunately, the spectral radius does not provide a proper norm on $\mathcal{M}_d$. It neither satisfies the triangle inequality, nor does $\rho(A) = 0$ imply that $A = 0$. Simple counterexamples can already be found in $\mathcal{M}_2$:

$$1 = \rho\left(\begin{pmatrix} 0 & 1 \\ 1 & 0 \end{pmatrix}\right) \nleq \rho\left(\begin{pmatrix} 0 & 1 \\ 0 & 0 \end{pmatrix}\right) + \rho\left(\begin{pmatrix} 0 & 0 \\ 1 & 0 \end{pmatrix}\right) = 0.$$

The situation can be improved by imposing less invariance on a norm. Hermitian conjugation plays herein a central role. Recall that the Hermitian conjugate A^* of $A \in \mathcal{M}_d$ is given by

$$a_{ij}^* = \overline{a_{ji}}, \quad i, j = 1, 2, \ldots, d.$$

Let S now define a similarity transformation that preserves the Hermitian conjugation: $SA^*S^{-1} = (SAS^{-1})^* = (S^{-1})^* A^* S^*$. Using $(S^{-1})^* = (S^*)^{-1}$, this equality can be rewritten as $A^* S^* S = S^* S A^*$ and, as only multiples of the identity matrix commute with all matrices, we conclude that there exists a complex scalar λ such that $S^*S = \lambda\mathbf{1}$. Because we are working in finite dimensions this means that S is, up to multiplication by a complex scalar, unitary. The invariant, under unitary transformations, is now the algebra $\mathfrak{A}$ generated by A and A^* together with the trace. The elements A^*A and $A\,A^*$ have, up to multiplicities of zero, the same positive eigenvalues, which provides us with the natural choice

$$\|A\| := \rho(A^*A)^{1/2} = \rho(A\,A^*)^{1/2} = \|A^*\|. \tag{5.2}$$

Using the material of Chapter 2, we see that (5.2) indeed defines a norm with the additional property that

$$\|A\|^2 = \|A^*A\|. \tag{5.3}$$

Moreover, (5.2) is the unique norm on $\mathcal{M}_d$ that satisfies (5.3). Suppose $\|\cdot\|_1$ is another. It is clearly sufficient to show that $\|\cdot\|_1$ coincides with $\|\cdot\|$ on matrices of the form A^*A. We use the spectral decomposition

$$A^*A = \sum_\alpha \alpha\,P_\alpha$$

of A^*A, where only non-negative α appear in the sum and

$$P_\alpha = P_\alpha^* = P_\alpha^2 \quad \text{and} \quad P_\alpha P_\beta = \delta_{\alpha\beta}\,P_\alpha.$$

From $\|P_\alpha\|_1^2 = \|P_\alpha\|_1$, we conclude that $\|P_\alpha\|_1 = 1$ whenever $P_\alpha \neq 0$ and, as

$$(A^*A)^k = \sum_\alpha \alpha^k\,P_\alpha,$$

repeated use of (5.3) yields

$$\begin{aligned}
\|A^*A\|_1 &= \lim_{n\to\infty} \|(A^*A)^{2^{2^n}}\|_1^{2^{-2^n}} \\
&= \max_\alpha(\{\alpha\|P_\alpha\|_1\}) \\
&= \max_\alpha(\{\alpha\}) = \|A^*A\| = \|A\|^2.
\end{aligned}$$

We now briefly recall a number of notions and properties from the general theory of C*-algebras and present some examples. We begin with a formal definition.

Definition 5.1 *A C*-algebra $\mathfrak{A}$ is a complex, normed, complete *-algebra such that its norm satisfies*

$$\|x^*x\| = \|x\|^2, \quad x \in \mathfrak{A}.$$

This definition means, in expanded form, that there are three composition laws available:

- multiplication by complex scalars: $(\lambda, x) \in \mathbb{C} \times \mathfrak{A} \to \lambda x \in \mathfrak{A}$,
- addition: $(x, y) \in \mathfrak{A} \times \mathfrak{A} \to x + y \in \mathfrak{A}$,
- multiplication: $(x, y) \in \mathfrak{A} \times \mathfrak{A} \to x y \in \mathfrak{A}$.

These composition laws satisfy the natural requirements for a non-commutative algebra

$$\lambda(x + y) = \lambda x + \lambda y,$$
$$(\lambda x)(\mu y) = \lambda \mu x y,$$
$$x + y = y + x,$$
$$x + (y + z) = (x + y) + z := x + y + z,$$
$$x(y z) = (x y)z := x y z,$$
$$x(y + z) = x y + x z,$$
$$(x + y)z = x z + y z.$$

Moreover, there is an adjoint operation $*$ on $\mathfrak{A}$ that obeys

$$(x^*)^* = x, \qquad (\lambda x)^* = \overline{\lambda} x^*,$$
$$(x + y)^* = x^* + y^*, \qquad (x y)^* = y^* x^*,$$

and the norm satisfies

$$\|x\| \geq 0 \text{ and } \|x\| = 0 \implies x = 0,$$
$$\|\lambda x\| = |\lambda| \|x\|,$$
$$\|x + y\| \leq \|x\| + \|y\|,$$
$$\|x y\| \leq \|x\| \|y\|, \tag{5.4}$$
$$\|x^*\| = \|x\|,$$
$$\|x^* x\| = \|x\|^2.$$

The algebra $\mathfrak{A}$ is *unital* if it contains an element $\mathbf{1}$, called unit, such that $\mathbf{1} x = x \mathbf{1} = x$ for any $x \in \mathfrak{A}$. A unit element is necessarily unique and satisfies $\mathbf{1}^* = \mathbf{1}$. We can, using the adjoint operation, distinguish special classes of elements in a C*-algebra, keeping the terminology of operators on Hilbert spaces:

- self-adjoint: $x = x^*$,
- normal: $x^* x = x x^*$,
- isometry: $x^* x = \mathbf{1}$,
- unitary: $x^* x = x x^* = \mathbf{1}$,
- projector: $x = x^* = x^2$.

Isometries and unitaries make sense only in unital algebras and the existence of non-trivial projectors in not guaranteed in a general C*-algebra. A C*-algebra carries a natural order structure and also the notion of spectrum of an element is available in general. There are a number of deep relations between order, spectrum and norm. We assume that $\mathfrak{A}$ is a unital C*-algebra.

The notions of resolvent, resolvent set and spectrum can be adopted from the Hilbert space theory. So, $\lambda \in \mathbb{C}$ belongs to the resolvent set $\mathrm{Res}(x)$ of x if there exists an element $r_\lambda(x) \in \mathfrak{A}$, called the resolvent of x at λ, such that

$$(x - \lambda \mathbf{1}) \, r_\lambda(x) = r_\lambda(x) \, (x - \lambda \mathbf{1}) = \mathbf{1}.$$

The spectrum, $\mathrm{Sp}(x)$, of x is the complement of $\mathrm{Res}(x)$ in $\mathbb{C}$. $\mathrm{Sp}(x)$ is a closed, non-empty subset of $\mathbb{C}$ contained in $D(0, \|x\|)$, the closed centred disc in $\mathbb{C}$ with radius $\|x\|$.

The location of $\mathrm{Sp}(x)$ can be specified more precisely in terms of the spectral radius $\rho(x)$,

$$\rho(x) := \lim_{n \to \infty} \|x^n\|^{1/n}.$$

$D(0, \rho(x))$ is the smallest centred and closed disc in $\mathbb{C}$ that contains $\mathrm{Sp}(x)$. For a normal element $x \in \mathfrak{A}$, it follows from the C*-property (5.4) of the norm that

$$\|x\|^2 = \|x^* x\| = \lim_{n \to \infty} \|(x^* x)^{2^{2^n}}\|^{2^{-2^n}}$$

$$= \lim_{n \to \infty} \|(x^*)^{2^{2^n}} (x)^{2^{2^n}}\|^{2^{-2^n}}$$

$$= \lim_{n \to \infty} \left(\|(x)^{2^{2^n}}\|^{2^{-2^n}} \right)^2 = \rho(x)^2.$$

The norm of a normal element can therefore directly be read off its spectrum.

For special classes of elements, the location of the spectrum can be sharpened somewhat further:

- x self-adjoint $\implies$ $\mathrm{Sp}(x) \subset [-\|x\|, \|x\|]$,
- x unitary $\implies$ $\mathrm{Sp}(x) \subset \{z \in \mathbb{C} \mid |z| = 1\}$,
- x projector $\implies$ $\mathrm{Sp}(x) \subset \{0, 1\}$.

A property that is sometimes useful in computing the spectrum is

$$\mathrm{Sp}(x\,y) \cup \{0\} = \mathrm{Sp}(y\,x) \cup \{0\}, \tag{5.5}$$

which holds for general $x, y \in \mathfrak{A}$.

An element $x \in \mathfrak{A}$ is positive if there is an element $y \in \mathfrak{A}$ such that $x = y^* y$. A positive element is always self-adjoint and the positive elements form a proper, norm-closed, generating and hereditary cone in $\mathfrak{A}$, i.e.

$0 \leq x \in \mathfrak{A}$ and $0 \leq \lambda \in \mathbb{C}$ $\implies$ $0 \leq \lambda x$,

$0 \leq x \in \mathfrak{A}$ and $0 \leq y \in \mathfrak{A}$ $\implies$ $0 \leq x + y$,

$0 \leq x_n \in \mathfrak{A}$ and $x = \lim_n x_n$ $\implies$ $0 \leq x$,

$0 \leq x \in \mathfrak{A}$ and $-x \geq 0$ $\implies$ $x = 0$,

any $x = x^* \in \mathfrak{A}$ can be written as $x = y - z$ with $0 \leq y$ and $0 \leq z$,

$0 \leq x \in \mathfrak{A}$ and $y \in \mathfrak{A}$ $\implies$ $0 \leq y^* x y$.

We now have

- $x = x^*$ is positive iff $\text{Sp}(x) \subset \mathbb{R}^+$,
- $x = x^*$ is positive iff $\left| \|x\| \mathbf{1} - x \right| \leq \|x\|$,
- $\|x\|$ is the smallest $a \in \mathbb{R}^+$ such that $x^*x \leq a^2 \mathbf{1}$.

We use the term *homomorphism* to denote maps from a C*-algebra $\mathfrak{A}$ to a C*-algebra $\mathfrak{B}$ which transport both the algebraic structure and the Hermitian conjugation. An *isomorphism* is an homomorphism that is one-to-one. An *automorphism* of $\mathfrak{A}$ is an isomorphism from $\mathfrak{A}$ to $\mathfrak{A}$. An automorphism is *inner* if it is of the form

$$x \in \mathfrak{A} \mapsto u\, x\, u^*, \tag{5.6}$$

with u unitary in $\mathfrak{A}$.

5.2 Examples

Example 5.2 Any finite-dimensional C*-algebra $\mathfrak{A}$ can be decomposed into a direct sum of square matrix algebras:

$$\mathfrak{A} = \bigoplus_j \mathcal{M}_{d_j}.$$

Example 5.3 Let $\mathfrak{H}$ be a separable Hilbert space; then the algebra $\mathcal{C}(\mathfrak{H})$ of compact operators on $\mathfrak{H}$, equipped with the operator norm, is a C*-algebra, see Section 2.4.1. If $\mathfrak{H}$ is infinite-dimensional, then $\mathcal{C}(\mathfrak{H})$ is not unital. We can, however, always add a unit to $\mathcal{C}(\mathfrak{H})$, in which case we obtain the algebra $\mathbb{C}\mathbf{1} + \mathcal{C}(\mathfrak{H})$.

$\mathcal{B}(\mathfrak{H})$, the algebra of bounded operators on $\mathfrak{H}$; see Section 2.1.6, is a unital C*-algebra.

In both cases, property (2.25) of the operator norm expresses the C*-property (5.4).

Example 5.4 *(Abelian C*-algebras)* Let $A = A^* \in \mathcal{B}(\mathfrak{H})$; then the set of all continuous and bounded functions of A defined, for example, with the help of the spectral representation (2.40) is an Abelian, or commutative, C*-algebra, isometrically isomorphic to $\mathcal{C}(\text{Sp}(A))$. Commutative algebras provide models of classical physical systems.

Gelfand's theorem shows that Example 5.4 is general.

Theorem 5.5 (Gelfand) *Any commutative unital C*-algebra $\mathfrak{A}$ is isomorphic to the algebra $\mathcal{C}(\mathfrak{X})$ of continuous complex functions on a compact Hausdorff space $\mathfrak{X}$. The space $\mathfrak{X}$, called the spectrum of $\mathfrak{A}$, is uniquely determined up to homeomorphisms.*

The next example furnishes the completion of the algebra of local observables that was considered in Chapter 4. It is known as the *algebra of quasi-local observables*.

Example 5.6 *(UHF-algebras)* Uniformly hyper-finite algebras are norm completions of increasing sequences of matrix algebras.

We first show that a matrix algebra $\mathcal{M}_{n_1}$ is a subalgebra of $\mathcal{M}_{n_2}$ iff n_1 is a divisor of n_2, in which case $\mathcal{M}_{n_2} \cong \mathcal{M}_{n_1} \otimes \mathcal{M}_d$ with $n_2 = n_1 d$ and $\mathcal{M}_{n_1}$ belongs to $\mathcal{M}_{n_2}$ as $\mathcal{M}_{n_1} \otimes \mathbf{1}_d$. Consider a system $\{e_{k\ell}^{n_1} \mid k, \ell = 1, 2, \dots, n_1\}$ of matrix units in $\mathcal{M}_{n_1}$. The projectors $e_k^{n_1} := e_{kk}^{n_1}$ are mutually orthogonal, add up to $\mathbf{1}$ and have equal traces in $\mathcal{M}_{n_2}$ as

$$\operatorname{Tr} e_k^{n_1} = \operatorname{Tr} e_{k\ell}^{n_1} e_{\ell k}^{n_1} = \operatorname{Tr} e_{\ell k}^{n_1} e_{k\ell}^{n_1} = \operatorname{Tr} e_\ell^{n_1}.$$

Hence, n_1 must divide n_2. Let d be the quotient of n_2 by n_1 and choose orthonormal vectors $\{f_i \mid i = 1, 2, \dots, d\}$ in $\mathbb{C}^{n_2}$ such that

$$e_1^{n_1} = \sum_{r=1}^d |f_r\rangle\langle f_r|.$$

We construct matrix units $e_{k\ell}^{n_2}$ in $\mathcal{M}_{n_2}$ by putting for $k_1, \ell_1 = 1, 2, \dots, n_2$ and $k_2, \ell_2 = 1, 2, \dots, d$

$$e_{(k_1 k_2)\,(\ell_1 \ell_2)}^{n_2} := e_{k_1 1}^{n_1} |f_{k_2}\rangle\langle f_{\ell_2}| e_{1\ell_1}^{n_1}.$$

From this construction it is clear that $e_{(k_1 k_2)\,(\ell_1 \ell_2)}^{n_2}$ may be identified with $e_{k_1 \ell_1}^{n_1} \otimes e_{k_2 \ell_2}^d$.

An increasing sequence of matrix algebras is described by specifying a sequence $2 \le n_1 < n_2 < \cdots$ of natural numbers—the dimensions of the corresponding matrix algebras—such that n_k is a divisor of n_{k+1}, i.e. $n_{k+1} = d_{k+1} n_k$. The algebra

$$\bigcup_{k=1}^\infty \mathcal{M}_{n_k} = \bigcup_{k=1}^\infty \bigotimes_{j=1}^k \mathcal{M}_{d_j}$$

is called the algebra of local elements. It is precisely the algebra that was considered in Section 4.1. The local algebra is a $*$-algebra on which there is a unique norm satisfying the C*-property (5.4). The norm of an element x is precisely its norm in all the matrix algebras $\mathcal{M}_{d_j}$ in which it is contained. The UHF-algebra corresponding to the given sequence is the completion of the local algebra with respect to this norm. It will be denoted by $\bigotimes_{j=1}^\infty \mathcal{M}_{d_j}$.

Two UHF-algebras are known to be isomorphic iff they have the same *supernatural number*. Write for each j the prime factorization of d_j as

$$d_j = \prod_{p \text{ prime}} p^{n(p,j)}.$$

The supernatural number of $\mathfrak{A}$ is then $\prod_p p^{n(p)}$, with $n(p) := \sum_j n(p, j)$, infinity being allowed.

Several algebras of common use in mathematical physics are connected with groups and their representations. In physics, groups mostly appear as sets of transformations that preserve the structure of a system. Recall that a *group* is a set $\mathfrak{G}$ equipped with a composition law

$$(g_1, g_2) \in \mathfrak{G} \times \mathfrak{G} \quad \rightarrow \quad g_1 g_2 \in \mathfrak{G}$$

that satisfies the following requirements:
- associativity: $(g_1 g_2) g_3 = g_1 (g_2 g_3) =: g_1 g_2 g_3$,
- existence of an identity element: $\exists e \in \mathfrak{G}$ such that $eg = ge = g$,
- existence of inverse: $\forall g \in \mathfrak{G} \ \exists g^{-1} \in \mathfrak{G}$ such that $gg^{-1} = g^{-1}g = e$.

If we are given a topology on a group $\mathfrak{G}$ such that the composition and the inversion are continuous, then $\mathfrak{G}$ is a *topological group*. A *unitary representation* of a group $\mathfrak{G}$ on a Hilbert space is a map π from $\mathfrak{G}$ to the group $\mathcal{U}(\mathfrak{H})$ of unitary operators on $\mathfrak{H}$ preserving composition and inversion:

$$\pi(g_1 g_2) = \pi(g_1)\pi(g_2) \quad \text{and} \quad \pi(g^{-1}) = \pi(g)^*.$$

If $\mathfrak{G}$ is a topological group and π a unitary representation of $\mathfrak{G}$ on $\mathfrak{H}$, then π is continuous if for any $\varphi \in \mathfrak{H}$ the functions

$$g \in \mathfrak{G} \mapsto \pi(g)\varphi$$

are continuous, i.e.

$$\lim_{g' \to g} \|\pi(g)\varphi - \pi(g')\varphi\| = 0.$$

Example 5.7 *(Group algebras)* Let $\mathfrak{G}$ be a discrete group, i.e. a group with trivial topology: the distance between group elements is either 0 or 1 depending on whether the elements are equal or distinct. There is a natural unitary representation π of $\mathfrak{G}$ on $\ell^2(\mathfrak{G})$ by left-multiplication. If, for $h \in \mathfrak{G}$, e_h denotes the sequence in $\ell^2(\mathfrak{G})$ which is everywhere 0 except at h then $\pi_g e_h := e_{gh}$. The linear span of $\pi_\mathfrak{G}$ is a $*$-algebra and the *reduced group C^*-algebra of $\mathfrak{G}$* is the norm closure of $\mathrm{span}(\pi_\mathfrak{G})$.

Example 5.8 *(Canonical commutation relations)* A well-known example of group algebra is the CCR-algebra, i.e. the algebra of canonical commutation relations, which is built on the *Heisenberg group* or one of its extensions. This can be regarded as an abstract generalization of the construction from Section 2.4.1. Let $\mathfrak{H}$ be a *real symplectic space*, i.e. a vector space over the reals equipped with a symplectic form $\sigma : \mathfrak{H} \times \mathfrak{H} \to \mathbb{R}$:

$$\sigma(\psi, \varphi) = -\sigma(\varphi, \psi),$$
$$\sigma(\psi, \alpha\varphi_1 + \varphi_2) = \alpha\sigma(\psi, \varphi_1) + \sigma(\psi, \varphi_2),$$
$$\text{if } \sigma(\psi, \varphi) = 0 \text{ for all } \varphi \in \mathfrak{H}, \text{ then } \psi = 0.$$

The last condition means that σ is non-degenerate and it can be relaxed. The corresponding Heisenberg group is $\mathbb{T} \times \mathfrak{H}$ with composition law

$$(a_1, \varphi_1) \cdot (a_2, \varphi_2) := (a_1 + a_2 + \tfrac{1}{2}\sigma(\varphi_1, \varphi_2), \varphi_1 + \varphi_2).$$

The addition of the angular part is the addition in $\mathbb{T}$ which is the usual sum modulo 2π. If we identify in the general construction of above $e_{(a,\varphi)}$ with $e^{ia}\varphi$ by dividing $\ell^2(\mathbb{T} \times \mathfrak{H})$ by the closed subspace generated by

$$\{e_{(a,\varphi)} - e^{ia}e_{(0,\varphi)} \mid a \in \mathbb{T},\ \varphi \in \mathfrak{H}\},$$

then $\pi_{(a,\varphi)}$ is of the form $e^{ia}W(\varphi)$ where the $W(\varphi)$ are the *Weyl elements*. These are unitary, i.e.

$$W(\varphi)^* = W(-\varphi),$$
$$W(\varphi)W(\psi) = e^{(i/2)\sigma(\varphi,\psi)}\,W(\varphi + \psi).$$

5.3   States and representations

States on a C*-algebra are general expectation functionals. In the case of Abelian algebras, i.e. continuous functions on compact spaces, states are the functional version of measures such as given in the Banach–Kakutani–Sachs theorem, Theorem 2.36. In the non-commutative situation, they generalize the expectation functionals, either given in terms of measures on the lattice of projections as in Gleason's theorem, Theorem 2.43, or in terms of density matrices; see Theorem 2.46. States on a C*-algebra are an essential ingredient in the probabilistic description of both classical and quantum systems.

Definition 5.9 *A state on a unital C*-algebra $\mathfrak{A}$ is a linear functional ω that is positive, i.e., $\omega(x^*x) \geq 0$ for all $x \in \mathfrak{A}$, and normalized: $\omega(\mathbb{1}) = 1$. The state space $\mathcal{S}(\mathfrak{A})$ of $\mathfrak{A}$ is the set of all states on $\mathfrak{A}$.*

A state ω on $\mathfrak{A}$ defines an inner product $\langle \cdot, \cdot \rangle$ on $\mathfrak{A}$ through the formula

$$\langle x, y \rangle := \omega(x^*y),$$

which implies the *Schwarz inequality for states*:

$$|\omega(x^*y)|^2 \leq \omega(x^*x)\,\omega(y^*y). \tag{5.7}$$

If $x = x^* \in \mathfrak{A}$, then $x \leq \|x\|\mathbb{1}$ and, therefore, $y^*xy \leq \|x\|\,y^*y$ for any $y \in \mathfrak{A}$. Due to positivity,

$$|\omega(y^*x\,y)| \leq \|x\|\,\omega(y^*y). \tag{5.8}$$

These two inequalities allow us to show that a state is continuous and also compute its norm. Indeed, from

$$|\omega(x)| \leq \omega(x^*x)^{1/2} \leq \|x^*x\|^{1/2}\omega(\mathbb{1}) = \|x\|$$

we see that ω is bounded and hence continuous with $\|\omega\| \leq 1$. Moreover, $\|\omega\| = 1$ as $\omega(\mathbb{1}) = 1$. A continuous functional ϕ on $\mathfrak{A}$ is positive iff $\|\phi\| = \phi(\mathbb{1})$. As is the case for signed measures, one shows that each element of the dual of $\mathfrak{A}$ can be decomposed into a linear combination of at most four positive elements.

Example 5.10 Any state ω on a UHF-algebra $\mathfrak{A}$, see Example 5.6, is uniquely determined by a consistent family of local states; see (4.3)–(4.5). Indeed, let ω be a state on $\mathfrak{A}$; then the restrictions $\langle \cdot \rangle_\Lambda$ of ω to a local algebra $\mathfrak{A}_\Lambda$ satisfy the compatibility conditions. Conversely, suppose that we are given a consistent family $\{\langle \cdot \rangle_\Lambda \mid \Lambda \subset \mathbb{N}\}$ of local states. Using compatibility, we define on $\mathfrak{A}_{\text{loc}}$ the functional ω by

$$\omega(x) := \lim_{\Lambda \to \mathbb{N}} \langle x \rangle_\Lambda, \quad x \in \mathfrak{A}_{\text{loc}}.$$

By positivity and normalization it satisfies $|\omega(x)| \leq \|x\|$. Let x be an arbitrary element in $\mathfrak{A}$. For any $n \in \mathbb{N}_0$, we can find a local element x_n such that $\|x - x_n\| < 1/n$. Then $\{\omega(x_n) \mid n\}$ is Cauchy and we put

$$\omega(x) := \lim_{n \to \infty} \omega(x_n).$$

It is now a routine matter to check that ω is well-defined and that it is indeed a state on $\mathfrak{A}$.

We say that a net ϕ_α of continuous functionals on $\mathfrak{A}$ converges in the weak $*$-topology to $\phi \in \mathfrak{A}^*$ if

$$\lim_\alpha \phi_\alpha(x) = \phi(x)$$

for all $x \in \mathfrak{A}$. The state space $\mathcal{S}(\mathfrak{A})$ is a convex subset of $\mathfrak{A}^*$ that is closed both with respect to the norm and the weak $*$-topology and also weak $*$-compact. A state ω is *pure* if it cannot be decomposed into a non-trivial convex combination, i.e. if for $0 < \lambda < 1$ and $\omega_1, \omega_2 \in \mathcal{S}(\mathfrak{A})$

$$\omega = \lambda \omega_1 + (1 - \lambda) \omega_2 \implies \omega_1 = \omega_2 = \omega.$$

The *pure state space* $\mathcal{P}(\mathfrak{A})$ is the weak $*$-closure of the set of pure states on $\mathfrak{A}$. State spaces of general algebras are usually very different from classical ones or from the spaces of density matrices from standard quantum mechanics. They sometimes have surprising properties.

Example 5.11 The pure state space of a UHF-algebra $\mathfrak{A}$ coincides with the entire state space on the algebra, while the pure state space of its commutative counterpart, the continuous functions on a configuration space $\Omega = \times_j S_{d_j}$, is precisely Ω itself. It is also quite different from the standard quantum situation where the pure states on the compact operators are already weak-$*$ closed and equal to the convex boundary of the state space. In order to see this, let ω be an arbitrary state on $\mathfrak{A}$, i.e. a consistent family $\{\langle \cdot \rangle_\Lambda \mid \Lambda \subset \mathbb{N}\}$ of local states and let Λ_0 be any finite subset of $\mathbb{N}$. We modify our consistent family of states as follows. For a suitable choice of a finite $M \subset \mathbb{N}$, we can ensure the existence of a pure state $\langle \cdot \rangle'_M$ on $\mathfrak{A}_M$ that restricts to $\langle \cdot \rangle_{\Lambda_0}$; see the remark after (2.70). We now extend $\langle \cdot \rangle'_M$ by an arbitrary pure product state outside M to obtain a pure state on $\mathfrak{A}$. This finishes the proof because the local algebra is dense in $\mathfrak{A}$.

We are now able to specify one of the fundamental notions that we need.

Definition 5.12 *An* abstract quantum probability space *consists of a unital C*-algebra $\mathfrak{A}$ and a state ω of $\mathfrak{A}$.*

Any abstract quantum probability space $(\mathfrak{A}, \omega)$ has a realization in the form familiar from quantum mechanics. This is the main content of the *Gelfand–Naimark–Segal Representation Theorem* (GNS).

Definition 5.13 *A* representation *of a C*-algebra on a Hilbert space $\mathfrak{H}$ is a linear map π from $\mathfrak{A}$ to $\mathcal{B}(\mathfrak{H})$ satisfying*

$$\pi(x\,y) = \pi(x)\,\pi(y),$$
$$\pi(x^*) = \pi(x)^*,$$
$$\text{if } \mathfrak{A} \text{ is unital, then } \pi(\mathbb{1}) = \mathbb{1}.$$

Theorem 5.14 (Gelfand–Naimark–Segal) *Let ω be a state on a C*-algebra $\mathfrak{A}$, then there exist a Hilbert space $\mathfrak{H}$, a unit vector $\Omega \in \mathfrak{H}$ and a representation π of $\mathfrak{A}$ into $\mathcal{B}(\mathfrak{H})$ such that*

$$\omega(x) = \langle \Omega, \pi(x)\,\Omega \rangle, \quad x \in \mathfrak{A},$$
$$\pi(\mathfrak{A})\,\Omega \text{ is dense in } \mathfrak{H}.$$

The last condition is also cyclicity *of Ω for $\pi(\mathfrak{A})$. Moreover, the triple $(\mathfrak{H}, \Omega, \pi)$ is uniquely determined up to a unitary isomorphism.*

Proof The idea of the proof is to turn $\mathfrak{A}$ into a Hilbert space by introducing the inner product

$$\langle x, y \rangle := \omega(x^*y).$$

This inner product may be degenerate and we therefore divide $\mathfrak{A}$ by its kernel

$$\mathfrak{V}_0 := \{x \in \mathfrak{A} \mid \omega(x^*x) = 0\}.$$

Because ω is norm-continuous, $\mathfrak{V}_0$ is a norm-closed linear subspace of $\mathfrak{A}$ and by inequality (5.8), $\mathfrak{V}_0$ is invariant under left-multiplication by elements of $\mathfrak{A}$: $yx \in \mathfrak{V}_0$ whenever $x \in \mathfrak{V}_0$ and $y \in \mathfrak{A}$. Such a subspace of a C*-algebra is called a *closed left ideal*. The elements of the set $\mathfrak{A}/\mathfrak{V}_0$ consist of the subsets $[x] := x + \mathfrak{V}_0$ of $\mathfrak{A}$ on which we define the vector space operations

$$\lambda\,[x] := [\lambda x] \quad \text{and} \quad [x] + [y] := [x + y].$$

These operations are well-defined because $\mathfrak{V}_0$ is a vector space. Next, we observe that $\omega(x^*y)$ does not change its value when we replace x by $x + z$ with $z \in \mathfrak{V}_0$

and similarly for y. This is a consequence of the Schwarz inequality (5.7) for states,

$$|\omega(z^*y)|^2 \leq \omega(z^*z)\,\omega(y^*y) = 0.$$

Therefore,

$$\langle [x], [y] \rangle := \omega(x^*y) \qquad (5.9)$$

is well-defined on $\mathfrak{A}/\mathfrak{V}_0$. Let $\mathfrak{H}$ be the completion of $\mathfrak{A}/\mathfrak{V}_0$ with respect to the norm induced by the inner product (5.9). The inequality

$$\|[y\,x]\|^2 = \omega(x^*y^*y\,x) \leq \|y\|^2\,\omega(x^*x) = \|y\|^2\,\|[x]\|^2$$

shows that $[x] \mapsto [y\,x]$ extends to a bounded linear operator on $\mathfrak{H}$ which we denote by $\pi(y)$. Clearly, $\pi(y_1 + \alpha y_2) = \pi(y_1) + \alpha\pi(y_2)$ and

$$\begin{aligned}
\langle [x], \pi(y)^*[z] \rangle = \langle \pi(y)\,[x], [z] \rangle &= \langle [y\,x], [z] \rangle \\
&= \omega(x^*y^*z) = \langle [x], [y^*z] \rangle \\
&= \langle [x], \pi(y^*)\,[z] \rangle.
\end{aligned}$$

Hence, π is a representation of $\mathfrak{A}$ into $\mathcal{B}(\mathfrak{H})$.

Finally, we put $\Omega := [\mathbf{1}]$ which yields

$$\langle \Omega, \pi(x)\,\Omega \rangle = \langle [\mathbf{1}], [x] \rangle = \omega(x),$$

and the density of $\pi(\mathfrak{A})\,\Omega = \{[x] \mid x \in \mathfrak{A}\}$ in $\mathfrak{H}$.

Two triplets $(\mathfrak{H}_1, \pi_1, \Omega_1)$ and $(\mathfrak{H}_2, \pi_2, \Omega_2)$ as in the statement of the theorem are unitarily equivalent by mapping $\pi_1(x)\,\Omega_1$ into $\pi_2(x)\,\Omega_2$ for $x \in \mathfrak{A}$. $\qquad \square$

One of the main uses for the GNS-representation is to transport questions from the abstract setting of the C*-algebra to questions phrased in terms of operators on Hilbert spaces. For example, the norm of an element $x \in \mathfrak{A}$ can be obtained as the supremum of the operator norms of all representations of x:

$$\|x\| = \sup_{\pi} \|\pi(x)\|. \qquad (5.10)$$

Explicit representations are also useful in constructing *the universal C*-algebra generated by a number of relations*. One starts by specifying generators of a *-algebra. On these elements and their formal adjoints one imposes relations $\mathcal{R}$. These are a set of equations imposed on the generators and their adjoints. The aim is to construct a C*-algebra $\mathfrak{A}$ that is purely determined by $\mathcal{R}$, in the sense that any other C*-algebra $\mathfrak{B}$, generated by elements which satisfy the same relations as the generators of $\mathfrak{A}$, is a homomorphic image of $\mathfrak{A}$. The idea is that the generators of $\mathfrak{B}$ possibly satisfy additional relations. In order to construct such a $\mathfrak{A}$ one looks first for an explicit realization of $\mathcal{R}$ by bounded operators on some Hilbert space. The norm of a finite linear combination x of products of generators is then chosen as

$$\|x\| = \sup \|\tilde{x}\|,$$

where $\tilde{x}$ is the explicit Hilbert space realization of x and where the supremum is taken over all possible realizations of $\mathcal{R}$.

It is possible now to produce an example that shows an essential difference between finite and infinite systems. It is in fact a discrete version of von Neumann's Uniqueness Theorem for the representation of the canonical commutation relations which we referred to in the introduction to Chapter 5. We consider N identical, distinguishable systems with Hilbert space $\mathbb{C}^d$. For any two pure states ω_1 and ω_2 of such a system, we can always find a unitary operator U_{21} such that

$$\omega_2(X) = \omega_1((U_{21})^* X U_{21}).$$

U_{21} has to rotate the vector ψ_1 that defines ω_1 into ψ_2 which in turn defines ω_2. This no longer holds for infinite systems.

Definition 5.15 *Two states ω_1 and ω_2 on a C*-algebra $\mathfrak{A}$ are equivalent if there exists a unitary operator u_{21} from $\mathfrak{H}_1$ to $\mathfrak{H}_2$ such that*

$$u_{21}\,\pi_1(x) = \pi_2(x)\,u_{21}, \quad x \in \mathfrak{A}. \tag{5.11}$$

$\mathfrak{H}_i$ and π_i in (5.11) are the GNS-spaces and representations of the ω_i.

Example 5.16 Consider two pure product states ω_1 and ω_2 on the UHF-algebra $\mathfrak{A}$ of spins with Hilbert spaces $\mathbb{C}^d$ at the sites of $\mathbb{N}$. Suppose that the single-site reduced states are defined by vectors ψ_1 and ψ_2, i.e.

$$\omega_i(\otimes_{j\in\Lambda} x_j) := \prod_{j\in\Lambda} \langle \psi_i, x_j\,\psi_i \rangle, \quad i = 1, 2, \ x_j \in \mathcal{M}_d.$$

Then ω_1 and ω_2 are equivalent iff they are equal, i.e. $|\langle \psi_1, \psi_2 \rangle| = 1$. Suppose that they are unequal but still equivalent and let u_{21} be as in (5.11). Let $P_1 := |\psi_1\rangle\langle\psi_1|$ be the projector in $\mathcal{M}_d$ on ψ_1 and consider in $\mathfrak{A}$ the local elements

$$P_{1,N} := \bigotimes_{j=1}^{N} P_1.$$

With Ω_i the cyclic vector of ω_i,

$$\|u_{21}\,\pi_1(P_{1,N})\,\Omega_1\|^2 = \|\pi_1(P_{1,N})\,\Omega_1\|^2 = \omega_1(P_{1,N}) = 1. \tag{5.12}$$

For any local element $A \in \mathfrak{A}_{\{1,2,\dots,M\}}$ we also have that

$$\|\pi_2(P_{1,N})\,\pi_2(A)\,\Omega_2\|^2 = \omega_2(A^* P_{1,N}\,A)$$
$$= \omega_2(A^* P_{1,M}\,A)\,\big(\omega_2(P_1)\big)^{N-M}$$
$$= \omega_2(A^* P_{1,M}\,A)\,|\langle \psi_1, \psi_2 \rangle|^{2(N-M)} \xrightarrow{N\to\infty} 0.$$

As $\|P_{1,N}\| = 1$ and $\pi_2(\mathfrak{A}_{\mathrm{loc}})\,\Omega_2$ is dense in $\mathfrak{H}_2$, for any vector $\varphi \in \mathfrak{H}_2$ we have

$$\lim_{N\to\infty} \|\pi_2(P_{1,N})\,\varphi\| = 0.$$

In particular,

$$\lim_{N\to\infty} \|\pi_2(P_{1,N})\,u_{21}\,\Omega_1\| = 0,$$

contradicting (5.12).

Example 5.17 Let $\mathfrak{H}$ be an infinite-dimensional Hilbert space and consider the C*-algebra $\mathfrak{A} := \mathbb{C}\mathbf{1} + \mathcal{C}(\mathfrak{H})$ of the compact operators on $\mathfrak{H}$ extended with a unit. Each state ω on $\mathfrak{A}$ is determined by a positive definite trace-class operator $\rho \in \mathcal{T}(\mathfrak{H})$ satisfying $\mathrm{Tr}\,\rho \le 1$:

$$\omega(\lambda\mathbf{1} + x) := \lambda\left(1 - \mathrm{Tr}\,\rho\right) + \mathrm{Tr}\,\rho\,x, \quad \lambda \in \mathbb{C}, \ \ x \in \mathcal{C}(\mathfrak{H}). \tag{5.13}$$

The pure states on $\mathfrak{A}$ correspond to choosing ρ in (5.13) either 0 or a projector on a one-dimensional subspace of $\mathfrak{H}$. Let, for a general $0 \le \rho \in \mathcal{T}(\mathfrak{H})$, $\{\varphi_i \mid i \in I \subset \mathbb{N}\}$ be an orthonormal family of eigenvectors of ρ corresponding to strictly positive eigenvalues $r_1 \ge r_2 \ge \cdots$ and put, with ω as in (5.13),

$$\mathfrak{H} := \mathbb{C} \oplus \left(\mathfrak{H} \otimes \ell^2(I)\right),$$
$$\Omega := (1 - \mathrm{Tr}\,\rho)^{1/2} \oplus \sum_{i \in I} r_i^{1/2}\,\varphi_i \otimes e_i,$$
$$\pi(\lambda\mathbf{1} + x) := \lambda \oplus (x \otimes \mathbf{1}),$$

then $(\mathfrak{H}, \Omega, \pi)$ is the GNS-triplet of ω.

A general element ϕ in the dual of $\mathfrak{A}$ is of the form

$$\phi(\lambda\mathbf{1} + x) := \mu\lambda + \mathrm{Tr}\,\sigma\,x$$

with $\mu \in \mathbb{C}$ and $\sigma \in \mathcal{T}(\mathfrak{H})$. The norm of ϕ is given by $\max\{\mathrm{Tr}\,|\sigma|, \|\mu| - \mathrm{Tr}\,|\sigma\|\}$.

Example 5.18 *Tracial states* A state τ on a unital C*-algebra $\mathfrak{A}$ is *tracial* if $\tau(a\,b) = \tau(b\,a)$ for all $a, b \in \mathfrak{A}$. As each $a \in \mathfrak{A}$ can be written as a linear combination of at most four unitaries, this is equivalent to $\tau(a) = \tau(u\,a\,u^*)$ for all $a, u \in \mathfrak{A}$ with u unitary.

On $\mathcal{M}_d$, there exists a unique tracial state

$$\tau([a_{ij}]) = \frac{1}{d}\,\mathrm{Tr}[a_{ij}] = \frac{1}{d}\sum_{i=1}^d a_{ii}.$$

UHF-algebras $\bigotimes_{j=1}^\infty \mathcal{M}_{d_j}$ come also with a unique tracial state τ. The expectation of an elementary tensor $\otimes_n x_n$, with each x_n belonging to some $\mathcal{M}_{d_{j_n}}$, is given by:

$$\tau(\otimes_n x_n) = \prod_n \tau_n(x_n),$$

where τ_n is the tracial state on the matrix algebra of x_n.

Similarly to tensor products of Hilbert spaces and operators acting on them (see Section 2.5.3), we have to consider tensor products of abstract C*-algebras to describe composed physical systems. There exists a canonical construction of *spatial tensor product* $\mathfrak{A} \otimes \mathfrak{B}$ of two C*-algebras $\mathfrak{A}$ and $\mathfrak{B}$ with its corresponding *spatial C*-norm*. If we represent faithfully the algebras $\mathfrak{A}$ and $\mathfrak{B}$ on Hilbert spaces $\mathfrak{H}_{\mathfrak{A}}$ and $\mathfrak{H}_{\mathfrak{B}}$, then the spatial tensor product of $\mathfrak{A}$ and $\mathfrak{B}$ is isomorphic to the usual tensor product of the representations of $\mathfrak{A}$ and $\mathfrak{B}$ on $\mathfrak{H}_{\mathfrak{A}}$ and $\mathfrak{H}_{\mathfrak{B}}$; see Section 2.5.3. In the sequel, we shall always use spatial tensor products of C*-algebras.

5.4 Dynamical systems and von Neumann algebras

Symmetries of quantum spin systems and time evolution in Heisenberg picture as in usual quantum mechanics are both instances of automorphisms of an algebra of observables. Moreover, for infinite systems, it often appears that only a limited number of states have physical relevance. The three ingredients, algebra, automorphism and reference state, are the central objects on which we shall focus.

Definition 5.19 *An* abstract dynamical system in discrete time *consists of a unital C*-algebra* $\mathfrak{A}$, *an automorphism* Θ *of* $\mathfrak{A}$ *and a* Θ-*invariant state* ω *of* $\mathfrak{A}$, *called* reference state. *A* dynamical system *in continuous time* is a unital *C*-algebra* $\mathfrak{A}$, *equipped with a strongly continuous one-parameter group* $\{\Theta_t \mid t \in \mathbb{R}\}$ *of automorphisms of* $\mathfrak{A}$ *and a* $\{\Theta_t \mid t \in \mathbb{R}\}$-*invariant reference state* ω.

The introduction of von Neumann algebras allows us to establish the link between dynamical systems in the algebraic version of Definition 5.19 and in the Hilbert space version as in Section 3.1. Given a concrete C*-algebra $\mathfrak{A}$, i.e. a C*-algebra of bounded operators on some given Hilbert space $\mathfrak{H}$, it is natural to consider notions of convergence on $\mathfrak{A}$ weaker than those induced by the operator norm. Specific examples are the weak and strong operator convergences introduced in (2.20) and (2.21). Generally, $\mathfrak{A}$ is not closed with respect to these weaker topologies. Two more topologies are relevant for our purposes since they are related to the description of states in terms of density matrices.

Ultra-strong operator convergence A sequence $\{A_n\}$ of bounded operators on $\mathfrak{H}$ converges to 0 in the *ultra-strong operator sense* if, for any choice of vectors $\varphi_j \in \mathfrak{H}$ such that $\sum_j \|\varphi_j\|^2 \leq \infty$,

$$\lim_{n \to \infty} \sum_j \|A_n \varphi_j\|^2 = 0.$$

Ultra-weak operator convergence A sequence $\{A_n\}$ of bounded operators on $\mathfrak{H}$ converges to 0 in the *ultra-weak operator sense* if, for any choice of vectors $\varphi_j, \psi_j \in \mathfrak{H}$ such that $\sum_j \|\varphi_j\| \|\psi_j\|$ converges,

$$\lim_{n \to \infty} \sum_j \langle \psi_j, A_n \varphi_j \rangle = 0.$$

A fundamental result of von Neumann characterizes the closures of a C*-subalgebra $\mathfrak{A}$ of $\mathcal{B}(\mathfrak{H})$ in terms of the spatial symmetries of $\mathfrak{A}$, i.e. in terms of unitary operators on $\mathfrak{H}$ that commute with each element of $\mathfrak{A}$. Let $\mathfrak{S}$ be a self-adjoint subset of $\mathcal{B}(\mathfrak{H})$ which means that A^* belongs to $\mathfrak{S}$ whenever A does. The *commutant* $\mathfrak{S}'$ of $\mathfrak{S}$ consists of all operators $X \in \mathcal{B}(\mathfrak{H})$ that commute with each element of $\mathfrak{S}$ separately. As

$$[X^*, A] = -[X, A^*]^* \quad \text{and} \quad [XY, A] = X[Y, A] + [X, A]Y,$$

we see that $\mathfrak{S}'$ is a *-algebra containing $\mathbf{1}$. It is not difficult to show that $\mathfrak{S}'$ is closed with respect to all the notions of convergence that were mentioned hitherto: norm, strong, ultra-strong, weak and ultra-weak. Moreover, taking commutants reverses inclusions: if two self-adjoint subsets $\mathfrak{S}_1$ and $\mathfrak{S}_2$ of $\mathcal{B}(\mathfrak{H})$ satisfy $\mathfrak{S}_1 \subset \mathfrak{S}_2$, then $\mathfrak{S}'_1 \supset \mathfrak{S}'_2$. The von Neumann theorem characterizes the bicommutant $\mathfrak{A}'' := (\mathfrak{A}')'$ of a *-subalgebra $\mathfrak{A}$ of $\mathcal{B}(\mathfrak{H})$.

Theorem 5.20 (von Neumann's Bicommutant Theorem) *Let $\mathfrak{A}$ be a unital *-subalgebra of $\mathcal{B}(\mathfrak{H})$, then the strong, ultra-strong, weak and ultra-weak closures of $\mathfrak{A}$ all coincide with the bicommutant $\mathfrak{A}''$ of $\mathfrak{A}$.*

Example 5.21 We return to the algebra $\mathfrak{A} := \mathbb{C}\mathbf{1} + \mathcal{C}(\mathfrak{H})$ of compact operators on $\mathfrak{H}$ extended with the identity operator. We prove that $\mathcal{B}(\mathfrak{H})$ is the ultra-strong closure of $\mathfrak{A}$. Let $A \in \mathcal{B}(\mathfrak{H})$, choose an orthonormal basis $\{e_1, e_2, \dots\}$ in $\mathfrak{H}$ and put

$$A_N := \sum_{k=1}^{N} |e_k\rangle\langle A^* e_k|.$$

Expanding $A\psi$ in this basis, we find

$$A\psi - A_N\psi = \sum_{k=N+1}^{\infty} \langle A^* e_k, \psi \rangle e_k,$$

and therefore

$$\left\| A\psi - A_N\psi \right\|^2 = \sum_{k=N+1}^{\infty} \left| \langle A^* e_k, \psi \rangle \right|^2 = \sum_{k=N+1}^{\infty} \left| \langle e_k, A\psi \rangle \right|^2.$$

Let ψ_j now be a collection of vectors with $\sum_{j=1}^{\infty} \|\psi_j\|^2 \leq \infty$; then

$$\lim_{N \to \infty} \sum_{j=1}^{\infty} \left\| (A - A_N)\psi_j \right\|^2$$

$$= \lim_{N \to \infty} \sum_{j=1}^{\infty} \sum_{k=N+1}^{\infty} \left| \langle e_k, A\,\psi_j \rangle \right|^2$$

$$= \lim_{M \to \infty} \lim_{N \to \infty} \sum_{j=1}^{M} \sum_{k=N+1}^{\infty} \left| \langle e_k, A\,\psi_j \rangle \right|^2 + \lim_{M \to \infty} \lim_{N \to \infty} \sum_{j=M+1}^{\infty} \sum_{k=N+1}^{\infty} \left| \langle e_k, A\,\psi_j \rangle \right|^2$$

$$\leq \lim_{M \to \infty} \sum_{j=M+1}^{\infty} \left\| A\,\psi_j \right\|^2$$

$$\leq \|A\|^2 \lim_{M \to \infty} \sum_{j=M+1}^{\infty} \left\| \psi_j \right\|^2$$

$$= 0.$$

A unital ∗-subalgebra of $\mathcal{B}(\mathfrak{H})$ that is weakly closed is called a *von Neumann algebra*. Because it is closed in strong topology, there is a rich functional calculus in a von Neumann algebra. If X is a self-adjoint element, then the spectral projections of X and even any bounded measurable function of X belong to the algebra. A probability measure μ on a compact space $\mathfrak{X}$, given as a functional on the continuous complex functions on $\mathfrak{X}$, can naturally be extended to a continuous functional on the essentially bounded measurable functions on $\mathfrak{X}$. A similar procedure can be applied to a state ω on a C*-algebra $\mathfrak{A}$ by considering the GNS-triple $(\mathfrak{H}, \pi, \Omega)$ of ω. In this way, we can associate with ω the von Neumann algebra $\pi(\mathfrak{A})''$ together with its state $X \in \pi(\mathfrak{A})'' \mapsto \langle \Omega, X\,\Omega \rangle$. This allows us to characterize the purity of ω: a state ω of $\mathfrak{A}$ is pure iff $\pi(\mathfrak{A})'' = \mathcal{B}(\mathfrak{H})$ or, equivalently, iff $\pi(\mathfrak{A})' = \mathbb{C}\mathbf{1}$, or still iff every non-zero vector $\varphi \in \pi(\mathfrak{A})''\Omega$ is cyclic for $\pi(\mathfrak{A})$.

Theorem 5.22 *Let $(\mathfrak{A}, \Theta, \omega)$ be an abstract discrete time dynamical system and let $(\mathfrak{H}, \pi, \Omega)$ be the GNS-triple of ω, then there exists a unique unitary operator U on $\mathfrak{H}$ such that*

$$\pi(\Theta(x)) = U^* \pi(x)\, U, \quad x \in \mathfrak{A},$$
$$U\,\Omega = \Omega.$$

Moreover, $\pi(x) \mapsto \pi(\Theta(x))$ extends to an automorphism of $\pi(\mathfrak{A})''$.

If $(\mathfrak{A}, \{\Theta_t \mid t \in \mathbb{R}\}, \omega)$ is a dynamical system in continuous time, there exists a unique self-adjoint operator H on $\mathfrak{H}$ such that

$$\pi(\Theta_t(x)) = e^{itH}\, \pi(x)\, e^{-itH}, \quad x \in \mathfrak{A},$$
$$\Omega \in \mathrm{Dom}(H) \quad and \quad H\,\Omega = 0.$$

Moreover, $\pi(x) \mapsto \pi(\Theta_t(x))$ extends to a group of automorphisms of $\pi(\mathfrak{A})''$, strongly operator continuous in t. The operator H is the Hamiltonian *of the system.*

Proof Define on the dense subspace $\pi(\mathfrak{A}) \, \Omega$ of $\mathfrak{H}$ the operator U^* by

$$U^* \, \pi(x) \, \Omega := \pi(\Theta(x)) \, \Omega.$$

By the property of automorphism and the invariance of ω, U^* is an isometry. As Θ is invertible, the range of U^* is dense and therefore U^* is actually unitary. Hence, $U^* \, \Omega = \Omega$ implies $U \, \Omega = \Omega$. Let $X \in \pi(\mathfrak{A})''$; then we can find a (net) $\{x_n\}$ in $\mathfrak{A}$ such that $X = s - \lim_n \pi(x_n)$. But then we also have $U^* X U = s - \lim_n U^* \pi(x_n) \, U$; therefore, $\pi(x) \mapsto U^* \pi(x) \, U$ with $x \in \mathfrak{A}$ extends to an automorphism of $\pi(\mathfrak{A})''$.

The Hamiltonian H in the case of continuous time is obtained through Stone's Theorem, Theorem 3.1. $\qquad\square$

Example 5.23 We apply explicitly Theorem 5.22 to the standard quantum systems with dynamics given by $U(t) = \exp(-itH)$ on a Hilbert space $\mathfrak{H}$ and invariant density matrix ρ: $[\rho, H] = 0$. The Heisenberg evolution provides the automorphism $\Theta_t(A) := U(t)^* A U(t)$ on $\mathcal{B}(\mathfrak{H})$. We suppose for simplicity that $\rho > 0$. The GNS-representation space $\mathfrak{H}_\rho$ of the state $\langle x \rangle := \mathrm{Tr} \, \rho \, x$ is the completion of the vector space

$$\{|A\rangle := A \, \rho^{1/2} \mid A \in \mathcal{B}(\mathfrak{H})\}$$

with respect to the inner product

$$\langle A, B \rangle := \mathrm{Tr} \, \rho \, A^* B = \langle A^* B \rangle. \tag{5.14}$$

Because of the strict positivity of ρ, $\mathfrak{H}_\rho$ is in fact the Hilbert space $\mathcal{T}_2(\mathfrak{H})$ of Hilbert–Schmidt operators on $\mathfrak{H}$, and the inner product (5.14) is the canonical inner product on $\mathcal{T}_2(\mathfrak{H})$; see (2.51). The representation is determined by

$$\pi(X) \, |A\rangle := |X \, A\rangle,$$

and is therefore the left-multiplication on $\mathcal{T}_2(\mathfrak{H})$:

$$\pi(X) \, K := X K.$$

The cyclic vector Ω is $|\mathbf{1}\rangle$, which is also equal to $\rho^{1/2}$.

We now determine the evolution operators $\mathcal{U}(t)$ in GNS-space, using a calligraphic letter type in order to distinguish them from operators on the original Hilbert space $\mathfrak{H}$.

$$\begin{aligned} \mathcal{U}(t) \, (A \, \rho^{1/2}) = \mathcal{U}(t) \, |A\rangle &:= |U(t)^* A U(t)\rangle \\ &= U(t)^* A U(t) \, \rho^{1/2} = U(t)^* A \, \rho^{1/2} U(t). \end{aligned}$$

Hence, on general elements $K \in \mathcal{T}_2(\mathfrak{H})$,

$$\mathcal{U}(t) \, K = e^{it\mathcal{H}} \, K = U(t)^* K \, U(t),$$

where

$$\mathcal{H} K := [H, K].$$

So we see that the evolution obtained in Theorem 5.22 is nothing but the super-operator formalism of quantum optics and that the Hamiltonian $\mathcal{H}$ is in fact the Liouvillian.

Example 5.24 Let $T : \mathfrak{X} \to \mathfrak{X}$ be a measure-preserving transformation of a measure space $(\mathfrak{X}, \Sigma, \mu)$, i.e.

$$\mu(T^{-1}(E)) = \mu(E), \quad E \in \Sigma.$$

The *Koopman operator*

$$U_T : \mathcal{L}^2(\mathfrak{X}, \mu) \to \mathcal{L}^2(\mathfrak{X}, \mu) : (U_T f)(x) := f(T(x)) \qquad (5.15)$$

is an isometry. First remark that the inner product between two indicator functions I_E and I_F of measurable subsets E and F is preserved by U_T. Indeed, $I_E I_F = I_{E \cap F}$ and $T^{-1}(E) \cap T^{-1}(F) = T^{-1}(E \cap F)$. Now, as T preserves μ, we obtain

$$\int_{\mathfrak{X}} \mu(dx) \, (U_T I_E)(x) (U_T I_F)(x) = \mu(T^{-1}(E \cap F))$$

$$= \mu(E \cap F)$$

$$= \int_{\mathfrak{X}} \mu(dx) \, I_E(x) I_F(x).$$

By linearity, U_T extends to a transformation of the simple functions that pre-serves the inner product, but, as the simple functions are dense in $\mathcal{L}^2(\mathfrak{X}, \mu)$, this means that U_T extends, by continuity, to an isometry on $\mathcal{L}^2(\mathfrak{X}, \mu)$. If the map T is invertible and if the inverse is also measure-preserving, i.e.

$$\mu(T^{-1}(E)) = \mu(T(E)) = \mu(E), \quad E \in \Sigma,$$

then U_T as given in (5.15) is in fact unitary because its range is dense in $\mathcal{L}^2(\mathfrak{X}, \mu)$.

Example 5.25 *(Rotation algebra)*   Let u and v be two elements satisfying the relations

$$u u^* = u^* u = \mathbb{1}, \quad v v^* = v^* v = \mathbb{1}, \quad u v = e^{i\theta} \, v \, u, \qquad (5.16)$$

where $0 \leq \theta < 2\pi$ is a *deformation parameter*. The *rotation algebra* $\mathfrak{A}_\theta$ is the universal C*-algebra generated by the relations (5.16). In order to make this a meaningful definition, we have to produce at least one realization of (5.16) by Hilbert space operators.

Consider on $\mathcal{L}^2(\mathbb{T})$ the operators U and V with actions

$$(U\,\varphi)(x) := e^{ix}\varphi(x), \qquad (V\,\varphi)(x) := \varphi(x - \theta),$$

where U and V satisfy (5.16). The constant function 1 is cyclic for the $*$-algebra generated by U and V as powers of U generate the Fourier basis of $\mathcal{L}^2(\mathbb{T})$. Suppose that an operator X on $\mathcal{L}^2(\mathbb{T})$ commutes with U; then it has to be a multiplication operator by a bounded complex function χ on $\mathbb{T}$. Moreover, X commutes with V iff $\chi(x - \theta) = \chi(x)$ for $x \in \mathbb{T}$. If θ is an irrational multiple of π, χ has to be constant, which means that

$$u \mapsto U, \qquad v \mapsto V$$

is the representation of a pure state on $\mathfrak{A}_\theta$.

Consider a second representation on $\mathcal{L}^2(\mathbb{T}^2)$ determined by the operators U_0 and V_0 with actions

$$
\begin{aligned}
(U_0\,\varphi)(x, y) &:= e^{ix}\varphi(x, y), \\
(V_0\,\varphi)(x, y) &:= e^{iy}\varphi(x + \theta, y),
\end{aligned}
\tag{5.17}
$$

where U_0 and V_0 again satisfy (5.16). The constant function 1 is cyclic for the $*$-algebra generated by U_0 and V_0. Applying powers of U_0 and V_0 to 1, we obtain the Fourier basis of $\mathcal{L}^2(\mathbb{T}^2)$ and find

$$\tau(u^m v^n) := \langle 1, U_0^m V_0^n\, 1 \rangle = \delta_{m,0}\,\delta_{n,0}. \tag{5.18}$$

Because of the commutation relations (5.16), we see that 1 generates a tracial state τ on $\mathfrak{A}_\theta$. For irrational multiples θ of π, there is a unique trace on $\mathfrak{A}_\theta$ and the above construction is precisely its GNS-representation. For such θ, the pair of operators U_0' and V_0' given by

$$
\begin{aligned}
(U_0'\,\varphi)(x, y) &:= e^{iy}\varphi(x, y), \\
(V_0'\,\varphi)(x, y) &:= e^{ix}\varphi(x, y + \theta)
\end{aligned}
$$

satisfy the same commutation relations as U_0 and V_0 and generate the commutant of $\{U_0, V_0\}$.

For $t \in \mathbb{R}$, the maps

$$
\begin{aligned}
\gamma_t^{(1)}(u) &:= e^{it}\,u, & \gamma_t^{(1)}(v) &:= v, \\
\gamma_t^{(2)}(u) &:= u, & \gamma_t^{(2)}(v) &:= e^{it}\,v
\end{aligned}
\tag{5.19}
$$

preserve the relations (5.16) and therefore extend to one-parameter groups of automorphisms of $\mathfrak{A}_\theta$ which are, moreover, strongly continuous and preserve the tracial state (5.18). These groups are used to construct *horocyclic actions* and we shall need them in Section 13.2. The triples $(\mathfrak{A}_\theta, \gamma^{(1,2)}, \tau)$ are C*-algebraic dynamical systems in the sense of Definition 5.19. The explicit GNS-representation

(5.17)–(5.18) yields the dynamical systems on the corresponding von Neumann algebras. The dynamics is implemented by the continuous one-parameter groups of unitaries $\{W(t)^{(1,2)} \mid t \in \mathbb{R}\}$:

$$\pi(\gamma_t^{(1,2)}(a)) = (W(t)^{(1,2)})^* \pi(a)\, W(t)^{(1,2)}, \quad a \in \mathfrak{A}_\theta$$

with, for $\varphi \in \mathcal{L}^2(\mathbb{T}^2)$,

$$\big(W(t)^{(1)}\,\varphi\big)(x,y) := \varphi(x - t, y),$$
$$\big(W(t)^{(2)}\,\varphi\big)(x,y) := \varphi(x, y - t).$$

5.5   Notes

A modern comprehensive presentation of the general theory of C*- and von Neumann algebras is presented in the two volumes by Kadison and Ringrose (1986a, b). A presentation of the theory that is directed more towards its application in physics is given in the books by Bratteli and Robinson (1979, 1997). The monograph by Davidson (1991) contains a treasure of information about many of the C*-algebras considered in this book. von Neumann's Uniqueness Theorem appeared in (von Neumann 1931). The CCR-algebra is considered in (Bratteli and Robinson 1997) and in the text by Petz (1990).

6

FERMIONIC DYNAMICAL SYSTEMS

The elementary constituents of matter seem to be fermions, while interactions are mediated by bosons. The models of interacting fermions should suffice to understand the properties of matter in standard conditions. Systems of fermions, even if they are non-interacting, still exhibit some strong correlations due to their statistics. In this chapter, we develop the tools to describe non-interacting systems of finite or infinitely many fermions. They provide dynamical systems which are still manageable but which are also reasonable approximations of real systems. A striking example of this situation is the model of the free-electron gas for a metallic compound. Some quasi-particles in solid state physics are also well-described as fermionic systems.

The first section deals with the Fock space situation where we have a vacuum state out of which essentially a finite number of particles are created. We look into the description of the creation and annihilation processes. In the second section, we develop an algebraic formalism that can also cope with systems with infinitely many particles. This situation is relevant, for example, in statistical mechanics at non-zero temperature. We analyze in particular free states and their associated dynamics.

6.1 Fermions in Fock space

Quite often, one considers systems that are composed of a number of *indistinguishable* particles, called fermions. The *Pauli principle* states that the indistinguishability manifests itself in the symmetry properties of the wave functions of composed systems. If $\mathfrak{H}$ is the Hilbert space of a single particle, then we say that it is a *fermion* if the wave functions of n such particles are totally antisymmetric with respect to permutations of the factors in $\bigotimes^n \mathfrak{H}$.

6.1.1 *Fock space*

For the simplest case of two fermions, each of them described by two levels $|1\rangle$ and $|2\rangle$, the possible wave functions are limited to the one-dimensional subspace of $\mathbb{C}^2 \otimes \mathbb{C}^2$ spanned by $|1, 2\rangle - |2, 1\rangle$, where we used the notation $|1, 2\rangle$ for the vector $|1\rangle \otimes |2\rangle$. In general, the permutation group $\mathfrak{S}_n$ of n objects acts naturally on $\bigotimes^n \mathfrak{H}$:

$$U_\pi \, \varphi_1 \otimes \varphi_2 \otimes \cdots \otimes \varphi_n := \varphi_{\pi(1)} \otimes \varphi_{\pi(2)} \otimes \cdots \otimes \varphi_{\pi(n)}, \quad \pi \in \mathfrak{S}_n.$$

U is a unitary representation of $\mathfrak{S}_n$ on $\bigotimes^n \mathfrak{H}$. A vector $\psi \in \bigotimes^n \mathfrak{H}_i$ is *totally antisymmetric* if

$$U_\pi \psi = \epsilon(\pi) \psi \quad \text{for all } \pi \in \mathfrak{S}_n, \tag{6.1}$$

$\epsilon(\pi) = \pm 1$ according to whether π is an even or odd permutation. The antisymmetric vectors form the closed subspace $\left(\bigotimes^n \mathfrak{H}\right)_{\text{a.s.}}$ of $\bigotimes^n \mathfrak{H}$, also denoted by $\bigwedge^n \mathfrak{H}$.

It is useful to introduce the notation $\varphi_1 \wedge \varphi_2 \wedge \cdots \wedge \varphi_n$ for the antisymmetrized tensor product of n vectors:

$$\varphi_1 \wedge \varphi_2 \wedge \cdots \wedge \varphi_n := \frac{1}{\sqrt{n!}} \sum_{\pi \in \mathfrak{S}_n} \epsilon(\pi)\, \varphi_{\pi(1)} \otimes \cdots \otimes \varphi_{\pi(n)}.$$

As in (6.1), the value of $\epsilon(\pi)$ is ± 1, according to the parity of π. We now easily compute

$$\langle \varphi_1 \wedge \varphi_2 \wedge \cdots \wedge \varphi_n,\ \psi_1 \wedge \psi_2 \wedge \cdots \wedge \psi_n \rangle = \det\left([\langle \varphi_i, \psi_j \rangle] \right). \tag{6.2}$$

The vectors of the form $\varphi_1 \wedge \varphi_2 \wedge \cdots \wedge \varphi_n$ arise by antisymmetrising elementary tensors $\varphi_1 \otimes \varphi_2 \otimes \cdots \otimes \varphi_n$ and, therefore, span the whole of the fermionic n-particle space $\bigwedge^n \mathfrak{H}$.

We can construct an orthonormal base of $\bigwedge^n \mathfrak{H}$ by antisymmetrizing a product basis of $\bigotimes^n \mathfrak{H}$ and removing all linearly dependent vectors. Let $\{e_1, e_2, \dots\}$ be an orthonormal basis of $\mathfrak{H}$. From the computation of the scalar product in (6.2), we find that we should only retain vectors of the type

$$\{e_{i_1} \wedge e_{i_2} \wedge \cdots \wedge e_{i_n} \mid i_1 < i_2 < \cdots < i_n\}.$$

The dimension of $\bigwedge^n \mathfrak{H}$ is therefore given by the binomial coefficient

$$\dim\left(\wedge^n \mathfrak{H}\right) = \binom{\dim(\mathfrak{H})}{n}. \tag{6.3}$$

In particular, the antisymmetric tensor powers of $\mathfrak{H}$ become trivial for $n > \dim(\mathfrak{H})$. If $\mathfrak{H}$ is separable, then $\bigwedge^n \mathfrak{H}$ is also separable.

Often, one has to consider systems composed of many identical particles but without knowing their precise number. The *Fock space* construction has been developed to deal with such a situation.

Definition 6.1 *The fermionic Fock space* $\Gamma_{\text{a.s.}}(\mathfrak{H})$ *over the one-particle Hilbert space* $\mathfrak{H}$ *is*

$$\Gamma_{\text{a.s.}}(\mathfrak{H}) := \mathbb{C} \bigoplus_{n=1}^{\infty} \left(\bigotimes^n \mathfrak{H}\right)_{\text{a.s.}} = \mathbb{C} \bigoplus_{n=1}^{\infty} \bigwedge^n \mathfrak{H}.$$

Fock space has a layered structure, each layer corresponding to a definite number of particles. The zeroth layer, $\mathbb{C}$, describes the state without particles and one calls

$$\Omega := 1 \oplus 0 \oplus \cdots$$

the *vacuum state*.

By (6.3), the dimension of $\Gamma_{\text{a.s.}}(\mathfrak{H})$ with a finite-dimensional one-particle space $\mathfrak{H}$ is

$$\dim\left(\Gamma_{\text{a.s.}}(\mathfrak{H})\right) = \sum_{n=0}^{\dim \mathfrak{H}} \binom{\dim(\mathfrak{H})}{n} = 2^{\dim(\mathfrak{H})}.$$

This formula suggests that

$$\Gamma_{\text{a.s.}}(\mathfrak{H} \oplus \mathfrak{K}) \cong \Gamma_{\text{a.s.}}(\mathfrak{H}) \otimes \Gamma_{\text{a.s.}}(\mathfrak{K}). \tag{6.4}$$

Let $\{e_1, e_2, \dots\}$ be an orthonormal basis for $\mathfrak{H}$ and $\{f_1, f_2, \dots\}$ one for $\mathfrak{K}$; then a basis for the n-particle antisymmetric space $\bigwedge^n(\mathfrak{H} \oplus \mathfrak{K})$ is given by

$$e_{i_1} \wedge \cdots \wedge e_{i_\ell} \wedge f_{j_1} \wedge \cdots \wedge f_{j_{n-\ell}},$$

where $(i_1, \dots, i_\ell)$ and $(j_1, \dots, j_{n-\ell})$ are ordered subsets of $\mathbb{N}_0$. The isomorphism (6.4) is given by

$$e_{i_1} \wedge \cdots \wedge e_{i_\ell} \wedge f_{j_1} \wedge \cdots \wedge f_{j_{n-\ell}} \equiv e_{i_1} \wedge \cdots \wedge e_{i_\ell} \otimes f_{j_1} \wedge \cdots \wedge f_{j_{n-\ell}}.$$

The isomorphism (6.4) is crucial in statistical mechanics. The Hilbert space of a gas of spinless fermions that is enclosed in a finite box $\Lambda \subset \mathbb{R}^3$ is $\Gamma_{\text{a.s.}}(\mathcal{L}^2(\Lambda))$. If we think of Λ as being composed of two disjoint pieces Λ_1 and Λ_2 obtained, for example, by inserting an imaginary wall in Λ, then $\mathcal{L}^2(\Lambda) = \mathcal{L}^2(\Lambda_1) \oplus \mathcal{L}^2(\Lambda_2)$ and consequently $\Gamma_{\text{a.s.}}(\Lambda) = \Gamma_{\text{a.s.}}(\Lambda_1) \otimes \Gamma_{\text{a.s.}}(\Lambda_2)$. It turns out that the equilibrium state of a free gas, at a given temperature and density, enclosed in the box Λ factorizes into a tensor product of equilibrium states in the boxes Λ_1 and Λ_2, up to details involving boundary conditions at the separating wall. The entropy of the gas in the box Λ is as a consequence be equal to the sum of the entropies of the gases in Λ_1 and Λ_2. This implies that the entropy of a free gas in equilibrium is an extensive quantity: it is proportional to the volume occupied by the system. This would fail without the antisymmetrization procedure and therefore lead to the *Gibbs paradox*, namely, that the entropy density is always infinite or zero.

6.1.2 *Creation and annihilation*

The creation and annihilation processes raise or lower the number of particles in a state, therefore, they map the n-particle subspace of Fock space on the $(n+1)$- or $(n-1)$-particle subspace. It will be useful to consider the dense linear subspace Γ_0 spanned by vectors of the type

$$0 \oplus \cdots \oplus 0 \oplus (\varphi_1 \wedge \cdots \wedge \varphi_n) \oplus 0 \oplus \cdots.$$

The creation of a fermion in the mode $\psi \in \mathfrak{H}$ is described by the *creation operator* $a^*(\psi)$ on Γ_0:

$$
\begin{aligned}
a^*(\psi) : \quad & \cdots \oplus 0 \oplus (\varphi_1 \wedge \cdots \wedge \varphi_n) \oplus 0 \oplus \cdots \\
\mapsto \quad & \cdots \oplus 0 \oplus (\psi \wedge \varphi_1 \wedge \cdots \wedge \varphi_n) \oplus 0 \oplus \cdots.
\end{aligned}
\tag{6.5}
$$

It is *a priori* not obvious that $a^*(\psi)$ is bounded and can therefore uniquely be extended to a bounded operator on Fock space. This is precisely the point of the next result.

Theorem 6.2 *For each $\psi \in \mathfrak{H}$, the adjoint of $a^*(\psi)$, as defined in (6.5), is the bounded linear operator $a(\psi)$, uniquely determined by*

$$a(\psi): \ \cdots \oplus 0 \oplus (\varphi_1 \wedge \cdots \wedge \varphi_n) \oplus 0 \oplus \cdots$$

$$\mapsto \sum_{j=1}^{n} (-1)^{j+1} \langle \psi, \varphi_j \rangle \cdots \oplus 0$$

$$\oplus (\varphi_1 \wedge \cdots \wedge \varphi_{j-1} \wedge \varphi_{j+1} \wedge \cdots \wedge \varphi_n) \oplus 0 \oplus \cdots$$

The map $\psi \mapsto a^*(\psi)$ is linear and the operators a and a^* satisfy the relations

$$\begin{aligned}
\{a(\varphi), a(\psi)\} &:= a(\varphi)\,a(\psi) + a(\psi)\,a(\varphi) = 0, \\
\{a(\varphi), a^*(\psi)\} &:= a(\varphi)\,a^*(\psi) + a^*(\psi)\,a(\varphi) = \langle \varphi, \psi \rangle\,\mathbf{1}
\end{aligned} \tag{6.6}$$

on Γ_0. Hence, $a(\psi)$ and $a^*(\psi)$ continuously extend to bounded linear operators on $\Gamma_{\text{a.s.}}(\mathfrak{H})$, which we still denote by $a(\psi)$ and $a^*(\psi)$, and their extensions satisfy (6.6).

The relations (6.6) are known as the *canonical anticommutation relations* (CAR). If the one-particle space is $\mathcal{L}^2(\Lambda)$, with $\Lambda \subset \mathbb{R}^3$ the region where the fermions are, the canonical anticommutation relations are in the physics literature often described in terms of formal pointwise creation and annihilation processes $a^\dagger(\mathbf{x})$ and $a(\mathbf{x})$ related to a^* and a by

$$a^*(\varphi) = \int_\Lambda d\mathbf{x}\, \varphi(\mathbf{x})\, a^*(\mathbf{x}).$$

This leads to the formal relations

$$\{a(\mathbf{x}), a(\mathbf{y})\} = 0 \quad \text{and} \quad \{a(\mathbf{x}), a^*(\mathbf{y})\} = \delta(\mathbf{x} - \mathbf{y}).$$

An important operator on Fock space is the *particle number operator* N which counts the number of particles present in a system:

$$\begin{aligned}
\text{Dom}(N) &= \{\bigoplus_{n=0}^{\infty} \zeta_n \mid \zeta_n \in \wedge^n \mathfrak{H} \text{ and } \textstyle\sum_{n=0}^{\infty} n^2 \,\|\zeta_n\|^2 < \infty\}, \\
N\Big(\bigoplus_{n=0}^{\infty} \zeta_n\Big) &:= \bigoplus_{n=0}^{\infty} n\,\zeta_n \quad \text{on } \text{Dom}(N).
\end{aligned} \tag{6.7}$$

N is a generally unbounded, self-adjoint operator with spectrum $\mathbb{N}$. Each eigenvalue of N is infinitely degenerate unless $\dim(\mathfrak{H}) < \infty$.

Any annihilation operator $a(\varphi)$ kills the vacuum, i.e.

$$a(\varphi)\,\Omega = 0, \quad \varphi \in \mathfrak{H}$$

and if we apply successively n creation operators on Ω, then we end up with an n-particle vector

$$a^*(\varphi_1)a^*(\varphi_2)\cdots a^*(\varphi_n)\,\Omega = \varphi_1 \wedge \varphi_2 \wedge \cdots \wedge \varphi_n. \tag{6.8}$$

Conversely, an annihilation operator $a(\psi)$ removes all components in an n-particle state that are in the mode ψ and returns an $(n-1)$-particle state.

The structure of Fock space can be resumed in the following properties:

(a) There are bounded linear operators $\{a(\varphi) \mid \varphi \in \mathfrak{H}\}$ on $\Gamma_{\mathrm{a.s.}}(\mathfrak{H})$ such that

$$\varphi \mapsto a(\varphi) \quad \text{is } \mathbb{C}\text{-antilinear,}$$
$$\{a(\varphi), a(\psi)\} = 0 \quad \text{and} \quad \{a(\varphi), a^*(\psi)\} = \langle \varphi, \psi \rangle. \tag{6.9}$$

(b) There is a normalized vector Ω in $\Gamma_{\mathrm{a.s.}}(\mathfrak{H})$ such that

$$a(\varphi)\,\Omega = 0 \text{ for } \varphi \in \mathfrak{H},$$
$$\Gamma_{\mathrm{a.s.}}(\mathfrak{H}) = \mathrm{Span}(\{a^*(\varphi_1)\cdots a^*(\varphi_n)\,\Omega \mid \varphi_j \in \mathfrak{H},\ n \in \mathbb{N}\}). \tag{6.10}$$

Properties (6.9) and (6.10) allow us to compute inner products of vectors of the type $a^*(\varphi_1)\cdots a^*(\varphi_n)\,\Omega$. In order to do so, we use the anticommutation relations, moving annihilation operators towards Ω, and use subsequently that annihilation operators kill the vacuum. For example,

$$\langle a^*(\varphi)\,\Omega, a^*(\psi)\,\Omega \rangle = \langle \Omega, a(\varphi)\,a^*(\psi)\,\Omega \rangle$$
$$= \langle \Omega, (\langle \varphi, \psi \rangle - a^*(\psi)a(\varphi))\,\Omega \rangle = \langle \varphi, \psi \rangle.$$

In general, we obtain

$$\langle a^*(\varphi_1)\cdots a^*(\varphi_m)\,\Omega,\ a^*(\psi_1)\cdots a^*(\psi_m)\,\Omega \rangle = \delta_{mn}\,\det\Big([\langle \varphi_i, \psi_j \rangle]\Big).$$

Creation and annihilation operators on the Fock space of a direct sum have a simple relation with those on the Fock spaces of the summands. So, consider the isomorphism (6.4) between $\Gamma_{\mathrm{a.s.}}(\mathfrak{H} \oplus \mathfrak{K})$ and $\Gamma_{\mathrm{a.s.}}(\mathfrak{H}) \otimes \Gamma_{\mathrm{a.s.}}(\mathfrak{K})$ and denote by $a_{\mathfrak{H}}$ the annihilation operators on $\Gamma_{\mathrm{a.s.}}(\mathfrak{H})$; then

$$a_{\mathfrak{H} \oplus \mathfrak{K}}(\varphi \oplus \psi) = a_{\mathfrak{H}}(\varphi) \otimes \mathbf{1}_{\mathfrak{K}} + (-\mathbf{1})^{N_{\mathfrak{H}}} \otimes a_{\mathfrak{K}}(\psi).$$

The operator $(-\mathbf{1})^N$ is the direct sum over the n-particle subspaces of the identity operator on subspaces with an even number of particles and minus the identity operator on subspaces with an odd number of particles.

We shall show that all bounded linear operators on Fock space can be approximated by non-commutative polynomials in the $a(\varphi)$ and $a^*(\varphi)$ in the weak operator topology; see (2.21). By von Neumann's Bicommutant Theorem, Theorem 5.20, this is equivalent to the absence of global symmetries in the system in the sense that the only bounded operators commuting with all the $a(\varphi)$ and $a^*(\varphi)$ are multiples of the identity. Notice that the *von Neumann algebra* generated by $a(\mathfrak{H})$ and $a^*(\mathfrak{H})$ is much larger than the C*-algebra, which is the closure in operator norm of the algebra generated by the creation and annihilation operators. For example, the last one is separable if $\mathfrak{H}$ is, while the former is not, unless $\dim \mathfrak{H} < \infty$.

Theorem 6.3 *An operator $X \in \mathcal{B}(\Gamma_{\mathrm{a.s.}}(\mathfrak{H}))$ which commutes with all the operators $a^*(\psi)$ and $a(\psi)$ is a multiple of the identity. Equivalently, $\mathcal{B}(\Gamma_{\mathrm{a.s.}}(\mathfrak{H}))$ is the von Neumann algebra generated by $\{a(\mathfrak{H}), a^*(\mathfrak{H})\}$.*

Proof We compute the matrix elements of X using (6.8). It suffices to consider vectors in $\Gamma_{\mathrm{a.s.}}(\mathfrak{H})$ of the form $a^*(\varphi_1) \cdots a^*(\varphi_m)\,\Omega$ and $a^*(\psi) \cdots a^*(\psi_n)\,\Omega$:

$$
\begin{aligned}
\langle \varphi_1 \wedge \cdots \wedge \varphi_m \,,\, X\,\psi_1 \wedge \cdots \wedge \psi_n \rangle \\
= \langle a^*(\varphi_1) \cdots a^*(\varphi_m)\,\Omega \,,\, X\,a^*(\psi_1) \cdots a^*(\psi_n)\,\Omega \rangle \\
= \langle \Omega \,,\, a(\varphi_m) \cdots a(\varphi_1)\,X\,a^*(\psi_1) \cdots a^*(\psi_n)\,\Omega \rangle.
\end{aligned}
$$

We now use our assumption that X commutes with creation and annihilation operators:

$$
[X, a(\varphi)] = [X, a^*(\varphi)] = 0, \quad \varphi \in \mathfrak{H}
$$

to get

$$
\langle \varphi_1 \wedge \cdots \wedge \varphi_m \,,\, X\,\psi_1 \wedge \cdots \wedge \psi_n \rangle = \langle \Omega \,,\, X\,a(\varphi_m) \cdots a(\varphi_1)\,a^*(\psi_1) \cdots a^*(\psi_n)\,\Omega \rangle.
$$

Next, we use the anticommutation relations to commute the $a(\varphi)$ over the $a^*(\psi)$. If eventually an $a(\varphi)$ hits Ω, then it returns zero. So, the only possibility to have non-zero contributions is to have $m = n$. Such contributions come from the scalar term in the anticommutation relations. A careful examination reveals that

$$
\begin{aligned}
\langle \Omega, X\,a(\varphi_m) \cdots a(\varphi_1)\,a^*(\psi_1) \cdots a^*(\psi_n)\,\Omega \rangle \\
= \delta_{mn} \langle \Omega, X\,\Omega \rangle \det\left([\langle \varphi_i, \psi_j \rangle] \right) \\
= \delta_{mn} \langle \Omega, X\,\Omega \rangle \langle \varphi_1 \wedge \cdots \wedge \varphi_m, \psi_1 \wedge \cdots \wedge \psi_n \rangle.
\end{aligned}
$$

This precisely means that $X = \langle \Omega, X\,\Omega \rangle \mathbf{1}$.   □

6.1.3 *Second quantization*

Let $X \in \mathcal{B}(\bigotimes^k \mathfrak{H})$ be a k-particle observable that is invariant for permutations of the factors in $\bigotimes^k \mathfrak{H}$. Such observables leave the antisymmetric space $\bigwedge^k \mathfrak{H}$ globally invariant and can be extended to fermionic n-particle systems:

$$X_n := \sum_{\Lambda \subset \{1,2,\dots,n\}} X_\Lambda \otimes \mathbb{1}_{\{1,2,\dots,n\}\backslash\Lambda} \tag{6.11}$$

for $n \geq k$. In this sum, Λ runs through the subsets $\{1, 2, \dots, n\}$ with k elements and X_Λ is a copy of X at the sites belonging to Λ. This is unambiguous because X was assumed to be symmetric.

Second quantization aims at expressing observables like X_n in a concise way for an unspecified number of indistinguishable particles and at using such expressions for global computations. A typical example is the Hamiltonian that often consists of two parts: one-particle contributions such as the kinetic energy and the interaction energy with an external potential and two-particle contributions arising from the interaction between pairs of fermions. The energy H_n of an n-particle system, $n \geq 2$, is of the form

$$H_n = \left(\sum_{j=1}^{n} T_j + \sum_{i<j=1}^{n} V_{ij} \right)\Bigg|_{\mathfrak{H}^{\wedge n}}.$$

T_j is a copy of T at the jth place in $\bigotimes^n \mathfrak{H}$, while V_{ij} is a copy of V at factors i and j. T is a self-adjoint operator on the one-particle Hilbert space $\mathfrak{H}$ and V is self-adjoint and permutation-invariant on $\mathfrak{H} \otimes \mathfrak{H}$. We assume, for simplicity, that both T and V are bounded, otherwise domain problems have to be considered. The global Hamiltonian H on Fock space $\Gamma_{\text{a.s.}}(\mathfrak{H})$ is

$$H = \bigoplus_{n=0}^{\infty} H_n$$

with $H_0 = 0$ and $H_1 = T$. Even if T and V are bounded, H is usually an unbounded self-adjoint operator with domain

$$\{\oplus_n \varphi_n \in \Gamma_{\text{a.s.}}(\mathfrak{H}) \mid \sum_n \|H_n \varphi_n\|^2 < \infty\}.$$

In the notation of second quantization, H is written as $\Gamma(T) + \Gamma(V)$, and $\Gamma(T)$ and $\Gamma(V)$ are expressed in terms of Fock space creation and annihilation operators. We now derive these expressions.

Let $\{e_1, e_2, \dots\}$ be an orthonormal basis for $\mathfrak{H}$ and $\{\varphi_1, \varphi_2, \dots, \varphi_n\}$ an arbitrary collection of vectors in $\mathfrak{H}$:

$$\Gamma(T)\, a^*(\varphi_1)\cdots a^*(\varphi_n)\,\Omega$$

$$= \Gamma(T)\,\cdots 0 \oplus (\varphi_1 \wedge \cdots \wedge \varphi_n) \oplus 0 \oplus \cdots$$

$$= \sum_{j=1}^{n} \cdots 0 \oplus (\varphi_1 \wedge \cdots \wedge \varphi_{j-1} \wedge T\,\varphi_j \wedge \varphi_{j+1} \wedge \cdots \wedge \varphi_n) \oplus 0 \oplus \cdots$$

$$= \sum_{j=1}^{n} a^*(\varphi_1)\cdots a^*(\varphi_{j-1}) a^*(T\,\varphi_j) a^*(\varphi_{j+1})\cdots a^*(\varphi_n)\,\Omega$$

$$= \sum_{j=1}^{n} \sum_{k} \langle e_k, \varphi_j\rangle\, a^*(\varphi_1)\cdots a^*(\varphi_{j-1}) a^*(T\,e_k) a^*(\varphi_{j+1})\cdots a^*(\varphi_n)\,\Omega$$

$$= \sum_{k} a^*(T\,e_k) \sum_{j=1}^{n} (-1)^{j+1}\,\langle e_k, \varphi_j\rangle\, a^*(\varphi_1)\cdots a^*(\varphi_{j-1}) a^*(\varphi_{j+1})\cdots a^*(\varphi_n)\,\Omega$$

$$= \Big(\sum_{k} a^*(T\,e_k) a(e_k) \Big) a^*(\varphi_1)\cdots a^*(\varphi_n)\,\Omega.$$

Therefore

$$\Gamma(T) = \sum_{k} a^*(T\,e_k) a(e_k) = \sum_{i,j} \langle e_i, T\,e_j\rangle\, a^*(e_i) a(e_j).$$

A similar but slightly more tedious computation yields

$$\Gamma(V) = \frac{1}{2} \sum_{i_1,i_2,j_1,j_2} \langle e_{i_1} \otimes e_{i_2}, V\,e_{j_1} \otimes e_{j_2}\rangle\, a^*(e_{i_1}) a^*(e_{i_2}) a(e_{j_2}) a(e_{j_1}).$$

For a general observable as in (6.11), we find

$$\Gamma(X) = \frac{1}{k!} \sum_{i_1,i_2,\ldots,i_k,j_1,j_2,\ldots,j_k} \langle e_{i_1} \otimes e_{i_2} \otimes \cdots \otimes e_{i_k}, V\,e_{j_1} \otimes e_{j_2} \otimes \cdots \otimes e_{j_k}\rangle$$

$$\times\, a^*(e_{i_1}) a^*(e_{i_2})\cdots a^*(e_{i_k}) a(e_{j_k})\cdots a(e_{j_2}) a(e_{j_1}).$$

Example 6.4 A typical example of a many-particle Hamiltonian arises when one considers fermions in a cubic box Λ_L of side L. Often, one chooses $T = -\hbar^2 \Delta_L/2m$ with periodic boundary conditions and for V a translation-invariant two-body potential that satisfies $V(-\mathbf{x}) = V(\mathbf{x})$. It is then useful to consider the orthonormal basis

$$\left\{ e_{\mathbf{k}}(\mathbf{x}) := \frac{1}{L^{3/2}}\, e^{i\mathbf{k}\cdot\mathbf{x}} \;\middle|\; \mathbf{k} \in \frac{2\pi}{L}\, \mathbb{Z}^3 \right\}$$

of eigenfunctions of Δ_L. Using the shorthand notation $a_{\mathbf{k}} := a(e_{\mathbf{k}})$ and the Fourier coefficients v of the potential

$$v(\mathbf{l}) := \frac{1}{L^3} \int_{\Lambda_L} d\mathbf{x}\, V(\mathbf{x})\, e^{-i\mathbf{l}\cdot\mathbf{x}},$$

the Hamiltonian H_L for the box Λ_L becomes

$$H_L = \sum_{\mathbf{k}} \frac{\|\mathbf{k}\|^2}{2m} a_{\mathbf{k}}^* a_{\mathbf{k}} + \frac{1}{2} \sum_{\mathbf{k},\mathbf{k}',\mathbf{l}} v(\mathbf{l}) \, a_{\mathbf{k}}^* a_{\mathbf{k}'}^* a_{\mathbf{k}'-\mathbf{l}} a_{\mathbf{k}+\mathbf{l}}. \tag{6.12}$$

It should be remembered that in (6.12) there is a hidden dependence on the size of the box through the definitions of $a_{\mathbf{k}}$ and v.

The main problem of equilibrium statistical mechanics consists in unravelling the properties of the limiting Gibbs states for Hamiltonians of the type (6.12). The expectation of a local observable X, such as a polynomial in Fock space fields $a^{\#}(\varphi)$ with φ of compact support, is computed as

$$\langle X \rangle_{\beta,\mu} := \lim_{L \to \infty} \frac{\operatorname{Tr} e^{-\beta(H_L - \mu N_L)} X}{\operatorname{Tr} e^{-\beta(H_L - \mu N_L)}}.$$

The parameters $\beta > 0$ and μ are the *inverse temperature* and *chemical potential*. N_L is the Fock space number operator $\sum_{\mathbf{k}} a_{\mathbf{k}}^* a_{\mathbf{k}}$ for the cube Λ_L and the chemical potential has to be adjusted in such a way that the particle density ρ is

$$\rho = \lim_{L \to \infty} \frac{1}{L^3} \langle N_L \rangle_{\beta,\mu}.$$

The *ground state problem* consists in studying the thermodynamic limit, i.e. the limit $L \to \infty$, of the states of minimal energy of H_L on the n-particle subspace with $n = \rho L^3$.

The second-quantization map Γ satisfies $\Gamma(X)^* = \Gamma(X^*)$ and if $X \in \mathcal{B}(\bigotimes^k \mathfrak{H})$ and $Y \in \mathcal{B}(\bigotimes^\ell \mathfrak{H})$ are symmetric k- and ℓ-particle operators, then $\Gamma(X)\Gamma(Y)$ can be expressed as a sum of second-quantized operators arising from symmetrizing contractions of $X \otimes Y$. A simple useful relation is obtained for two one-particle observables $A, B \in \mathcal{B}(\mathfrak{H})$:

$$[\Gamma(A), \Gamma(B)] = \Gamma([A, B]).$$

Example 6.5 On the one-particle space $\mathfrak{H}$, we are given a Hamiltonian

$$H = \sum_k \epsilon_k \, |k\rangle\langle k|,$$

where $\{|1\rangle, |2\rangle, \dots, \}$ is an orthonormal basis for $\mathfrak{H}$ and the energies ϵ_k also contain the chemical potential. We assume that for all $\beta > 0$

$$\sum_k e^{-\beta \epsilon_k} < \infty. \tag{6.13}$$

This situation corresponds to non-interacting fermions essentially confined in a finite volume. The second-quantized Hamiltonian reads

$$\Gamma(H) = \sum_k \epsilon_k \, a_k^* a_k.$$

In this formula, the a_k are a short-hand notation for $a(|k\rangle)$ and $\Gamma(H)$ is a self-adjoint operator on Fock space. The Heisenberg evolution for a single mode is

$$e^{it\Gamma(H)} a_k^* e^{-it\Gamma(H)} = e^{it\epsilon_k} a_k^* = a^*(e^{itH}|k\rangle). \tag{6.14}$$

Formula (6.14) defines in fact the dynamics on the polynomials in a and a^* and, by strong continuity and Theorem 6.3, we can extend it to the full $\mathcal{B}(\Gamma_{\text{a.s.}}(\mathfrak{H}))$. On fermion fields we find

$$e^{it\Gamma(H)} a^*(\varphi) e^{-it\Gamma(H)} = a^*(e^{itH}\varphi), \quad \varphi \in \mathfrak{H}.$$

For $\beta > 0$ we compute

$$\begin{aligned}
\operatorname{Tr} e^{-\beta\Gamma(H)} &= \sum_{n=0}^{\infty} \sum_{k_1 < k_2 < \cdots < k_n} \langle a_{k_1}^* a_{k_2}^* \cdots a_{k_n}^* \Omega, e^{-\beta\Gamma(H)} a_{k_1}^* a_{k_2}^* \cdots a_{k_n}^* \Omega \rangle \\
&= \sum_{n=0}^{\infty} \sum_{k_1 < k_2 < \cdots < k_n} \prod_{j=1}^{n} e^{-\beta\epsilon_{k_j}} \\
&= \prod_{k=1}^{\infty} (1 + e^{-\beta\epsilon_k}) < \infty
\end{aligned}$$

by assumption (6.13). We can therefore define the canonical Gibbs states

$$\langle X \rangle_\beta := \operatorname{Tr} \rho^{\text{Gibbs}} X = \frac{\operatorname{Tr} e^{-\beta\Gamma(H)} X}{\operatorname{Tr} e^{-\beta\Gamma(H)}}.$$

Writing $\mathfrak{H} = \bigoplus_k \mathbb{C}|k\rangle$ (see (6.4)) we have

$$\Gamma_{\text{a.s.}}(\mathfrak{H}) = \bigotimes_k \Gamma_{\text{a.s.}}(\mathbb{C}) \equiv \bigotimes_k \mathbb{C}^2$$

and ρ^{Gibbs} decomposes as

$$\rho^{\text{Gibbs}} = \bigotimes_k \frac{1}{1 + e^{-\beta\epsilon_k}} \begin{pmatrix} e^{-\beta\epsilon_k} & 0 \\ 0 & 1 \end{pmatrix}.$$

Let $\mathbf{k} := (k_1, k_2, \ldots, k_n\}$ and $\mathbf{l} := (l_1, l_2, \ldots, l_m\}$ be two ordered sequences; then a straightforward computation yields

$$\langle a_{k_1}^* a_{k_2}^* \cdots a_{k_n}^* a_{l_m} \cdots a_{l_2} a_{l_1} \rangle_\beta = \delta_{\mathbf{k},\mathbf{l}} \prod_{j=1}^{n} \frac{1}{1 + e^{-\beta\epsilon_{k_j}}}. \tag{6.15}$$

Formula (6.15) extends by antisymmetry and multilinearity to

$$\langle a^*(\varphi_1) a^*(\varphi_2) \cdots a^*(\varphi_n) a(\psi_m) \cdots a(\psi_2) a(\psi_1) \rangle_\beta = \delta_{m,n} \det([\langle a^*(\varphi_i) a(\psi_j) \rangle_\beta]),$$

with

$$\langle a^*(\varphi) a(\psi) \rangle_\beta = \langle \psi, \frac{1}{1 + e^{-\beta H}} \varphi \rangle.$$

6.2   The CAR-algebra

6.2.1   *Canonical anticommutation relations*

We now introduce the CAR-algebra from a more abstract point of view. The algebra that we obtain is isomorphic to the closure in operator norm of the algebra generated by the creation and annihilation operators on the fermionic Fock space as considered in Section 6.1.2. The specific form of the Fock space operators a and a^* is, however, irrelevant in several situations and one often needs representations of the abstract C*-algebra that are different from the Fock space realization of the previous section. In order to avoid confusion, the notations c and c^* are used for abstract creation and annihilation fields which are not tied to any specific representation.

Let $(\mathfrak{H}, \langle \cdot, \cdot \rangle)$ be a separable Hilbert space. The CAR-algebra $\mathfrak{A}(\mathfrak{H})$ is the universal C*-algebra generated by a unit element $\mathbb{1}$ and by $\{c(\varphi) \mid \varphi \in \mathfrak{H}\}$ subject to the following conditions:

$$\varphi \mapsto c(\varphi) \text{ is complex antilinear,}$$
$$\{c(\varphi), c(\psi)\} = 0, \quad \varphi, \psi \in \mathfrak{H}, \quad (6.16)$$
$$\{c(\varphi), c^*(\psi)\} = \langle \varphi, \psi \rangle \, \mathbb{1}, \quad \varphi, \psi \in \mathfrak{H}.$$

Consider the case where $\mathfrak{H}$ is finite-dimensional: $\mathfrak{H} = \mathbb{C}^d$. Using the complex linearity of $\varphi \mapsto c^*(\varphi)$, we can express each $c(\varphi)$ or $c^*(\varphi)$ as a linear combination of c_i or c_i^*, $i = 1, 2, \ldots, d$, where we have fixed an orthonormal basis $\{e_1, e_2, \ldots, e_d\}$ in $\mathbb{C}^d$ and used the notation c_i for $c(e_i)$. The anticommutation relations (6.16) are equivalent to

$$c_i c_j = -c_j c_i \quad \text{and} \quad c_i c_j^* = \delta_{ij} \mathbb{1} - c_j^* c_i. \quad (6.17)$$

Using the relations (6.17), we can express a monomial in c_i's and c_j^*'s as a sum of monomials wherein all factors c_j^* appear in front, if any. Moreover, we can also order the starred and unstarred factors separately and, because $c_i^2 = (c_i^*)^2 = 0$, the indices of the c's and c^*'s can be strictly ordered. Thus, each element in the $*$-algebra $\mathfrak{A}$ generated by $\mathbb{1}$ and the $c(\mathfrak{H})$ is a linear combination of monomials of the type

$$c_{i_1}^* c_{i_2}^* \cdots c_{i_m}^* c_{j_n} \cdots c_{j_2} c_{j_1}$$

with $i_1 < i_2 < \cdots < i_m$, $j_1 < j_2 < \cdots < j_n$ and all i_k and j_ℓ belonging to $\{1, 2, \ldots, d\}$.

The algebra $\mathfrak{A}$ has dimension 2^{2d} as a complex vector space and it is isomorphic to $\mathcal{M}_2^{\otimes d}$. The *Jordan–Wigner transformation* (see Example 4.6) gives an explicit isomorphism. Consider the Pauli matrices as in (4.15). A simple computation shows that

$$(\sigma^z)^2 = \mathbb{1} \quad \text{and} \quad \{\sigma^z, \sigma^+\} = \{\sigma^z, \sigma^-\} = 0,$$

with $\sigma^\pm := (\sigma^x \pm i\sigma^y)/2$. An explicit isomorphism Φ_d from $\mathfrak{A}$ into $\mathcal{M}_2^{\otimes d}$ is defined by

$$\Phi_d(c_j) := \left(\otimes_{k<j}\sigma^z \right) \otimes \sigma^- \otimes \left(\otimes_{\ell>j}\mathbf{1} \right). \tag{6.18}$$

It now remains to consider $\mathfrak{A}(\mathfrak{H})$ with $\mathfrak{H}$ infinite-dimensional and separable. We first compute the norm of $c(\varphi)$. Because of the anticommutation relation $c^*(\varphi)c(\varphi) = -c(\varphi)c^*(\varphi) + \|\varphi\|^2$, we have

$$\mathrm{Sp}\big(c(\varphi)c^*(\varphi)\big) \cup \{0\} = \mathrm{Sp}\big(c^*(\varphi)c(\varphi)\big) \cup \{0\}$$
$$= -\mathrm{Sp}\big(c(\varphi)c^*(\varphi)\big) \cup \{0\} + \|\varphi\|^2.$$

This is only possible if

$$\mathrm{Sp}\big(c(\varphi)c^*(\varphi)\big) \subset \{0, \|\varphi\|^2\};$$

hence,

$$\|c(\varphi)\| \leq \|\varphi\|. \tag{6.19}$$

Let us fix an orthonormal basis $\{e_1, e_2, \dots\}$ in $\mathfrak{H}$ and denote by $\mathfrak{H}_d$ the d-dimensional subspace of $\mathfrak{H}$ spanned by the vectors $\{e_1, e_2, \dots, e_d\}$. Because of the continuity property (6.19), the algebra $\bigcup_d \mathfrak{A}(\mathfrak{H}_d)$ is norm-dense in $\mathfrak{A}(\mathfrak{H})$. We then consider for each d the Jordan–Wigner isomorphism Φ_d as in (6.18). It is clear from the construction of the Φ_d that

$$\Phi_d(x) \otimes \mathbf{1} = \Phi_{d+1}(x), \quad x \in \mathfrak{A}(\mathfrak{H}_d).$$

We can therefore glue all the Φ_d together to obtain a isomorphism Φ between $\bigcup_d \mathfrak{A}(\mathfrak{H}_d)$ and the algebra $\bigcup_d \otimes^d M_2$. As there is a unique C*-norm on the last one, there is also a unique C*-norm on $\bigcup_d \mathfrak{A}(\mathfrak{H}_d)$ and we have shown that

Theorem 6.6 *Let $\mathfrak{H}$ be a separable Hilbert space, then the CAR-algebra $\mathfrak{A}(\mathfrak{H})$ is isomorphic to the UHF-algebra with supernatural number $2^{\dim(\mathfrak{H})}$ (see Example 5.6).*

6.2.2   *Quasi-free automorphisms*

This and the next section are concerned with *quasi-free structures* on $\mathfrak{A}(\mathfrak{H})$. Such structures arise in models of non-interacting fermions, similar to the one presented in Example 6.5. Quasi-free objects exhibit particularly simple combinatorial properties, and are typically characterized by single-particle objects such as operators on one-particle space.

First, we introduce a graded tensor product between CAR-algebras. Let $\mathfrak{H}_1$ and $\mathfrak{H}_2$ be two separable Hilbert spaces with corresponding CAR-algebras $\mathfrak{A}(\mathfrak{H}_1)$ and $\mathfrak{A}(\mathfrak{H}_2)$. There is a natural embedding

$$\jmath_1 : \mathfrak{A}(\mathfrak{H}_1) \hookrightarrow \mathfrak{A}(\mathfrak{H}_1 \oplus \mathfrak{H}_2) \tag{6.20}$$

defined by

$$\jmath_1(c(\varphi_1)) = c(\varphi_1 \oplus 0), \quad \varphi_1 \in \mathfrak{H}_1$$

and, of course, an analogous embedding $\jmath_2$ of $\mathfrak{A}(\mathfrak{H}_2)$. Clearly, $\jmath_1(\mathfrak{A}(\mathfrak{H}_1))$ and $\jmath_2(\mathfrak{A}(\mathfrak{H}_2))$ generate $\mathfrak{A}(\mathfrak{H}_1 \oplus \mathfrak{H}_2)$ but they are not in $\mathfrak{A}(\mathfrak{H}_1 \oplus \mathfrak{H}_2)$ as commuting subalgebras because

$$\{c^{\#}(\varphi_1 \oplus 0), c^{\#}(0 \oplus \varphi_2)\} = 0$$

instead of

$$[c^{\#}(\varphi_1 \oplus 0), c^{\#}(0 \oplus \varphi_2)] = 0.$$

By $c^{\#}$ we mean either c or c^*. Sometimes $\mathfrak{A}(\mathfrak{H}_1 \oplus \mathfrak{H}_2)$ is called the *graded tensor product* of $\mathfrak{A}(\mathfrak{H}_1)$ and $\mathfrak{A}(\mathfrak{H}_2)$ and one uses the notation

$$\mathfrak{A}(\mathfrak{H}_1 \oplus \mathfrak{H}_2) = \mathfrak{A}(\mathfrak{H}_1) \wedge \mathfrak{A}(\mathfrak{H}_2). \tag{6.21}$$

Let W be an *isometry* from the Hilbert space $\mathfrak{H}_1$ into the Hilbert space $\mathfrak{H}_2$; then, because of the construction of CAR-algebras,

$$\alpha_W(c(\varphi_1)) := c(W \varphi_1), \quad \varphi_1 \in \mathfrak{H}_1 \tag{6.22}$$

extends to a homomorphism from $\mathfrak{A}(\mathfrak{H}_1)$ into $\mathfrak{A}(\mathfrak{H}_2)$. If W is *unitary*, then α_W is an isomorphism. Such homomorphisms are called *quasi-free*.

One is often interested in symmetries of a system. For a CAR-algebra $\mathfrak{A}(\mathfrak{H})$ these are typically given in terms of *quasi-free automorphisms*, i.e. by mappings α_U of the type (6.22) with $U \in \mathcal{U}(\mathfrak{H})$ a unitary operator on $\mathfrak{H}$;

$$\alpha : \mathcal{U}(\mathfrak{H}) \to \mathcal{A}ut(\mathfrak{A}(\mathfrak{H})) : U \mapsto \alpha_U$$

is a homomorphism from the group $\mathcal{U}(\mathfrak{H})$ of unitary transformations on $\mathfrak{H}$ into the group $\mathcal{A}ut(\mathfrak{A}(\mathfrak{H}))$ of isomorphisms of $\mathfrak{A}(\mathfrak{H})$. From (6.22) we see that

$$\|\alpha_U(c(\varphi)) - \alpha_V(c(\varphi))\| = \|c(U \varphi) - c(V \varphi)\| = \|(U - V)\varphi\|.$$

Therefore, α is continuous when we consider the strong topologies on $\mathcal{U}(\mathfrak{H})$ and $\mathcal{A}ut(\mathfrak{A}(\mathfrak{H}))$.

Example 6.7 It can be shown that α_U is inner (see (5.6)) iff $U \in \mathbf{1} + \mathcal{T}(\mathfrak{H})$, where $\mathcal{T}(\mathfrak{H})$ are the trace-class operators on $\mathfrak{H}$. Moreover, if $U - V \in \mathcal{T}(\mathfrak{H})$, then

$$\|\alpha_U - \alpha_V\| \leq \|U - V\|_1,$$

$\|\cdot\|_1$ being the trace norm on $\mathcal{T}(\mathfrak{H})$, i.e. $\|X\|_1 := \mathrm{Tr}\,|X|$.

Example 6.8 Let H be a self-adjoint operator on the single-particle space $\mathfrak{H}$ which defines the strongly continuous, unitary, one parameter group $\{U(t) := \exp(-itH) \mid t \in \mathbb{R}\}$ on $\mathfrak{H}$. To this group, there corresponds a one-parameter group $\{\Theta_t \mid t \in \mathbb{R}\}$ of quasi-free automorphisms of $\mathfrak{A}(\mathfrak{H})$:

$$\Theta_t(c^*(\varphi)) := c^*(U(-t)\,\varphi).$$

As the non-commutative polynomials in the $c^\#(\varphi)$ are norm-dense in $\mathfrak{A}(\mathfrak{H})$, we obtain from

$$\lim_{t \to 0} \|c^*(\Theta_t(\varphi)) - c^*(\varphi)\| = \lim_{t \to 0} \|U(-t)\,\varphi - \varphi\| = 0$$

that $\{\Theta_t \mid t \in \mathbb{R}\}$ is strongly continuous in t. If ω is a Θ-invariant state of $\mathfrak{A}(\mathfrak{H})$, then $(\mathfrak{A}(\mathfrak{H}), \{\Theta_t \mid t \in \mathbb{R}\}, \omega)$ is an abstract non-commutative dynamical system in continuous time.

Some quasi-free automorphisms are given special names, such as *parity or grading automorphism* π,

$$\pi : c(\varphi) \mapsto -c(\varphi). \tag{6.23}$$

Elements $x \in \mathfrak{A}(\mathfrak{H})$ that are invariant under the parity automorphism are called *even*, while elements that change sign are *odd*. The subset of even elements in $\mathfrak{A}(\mathfrak{H})$ is a C*-subalgebra $\mathfrak{A}_{\text{even}}(\mathfrak{H})$ of $\mathfrak{A}(\mathfrak{H})$. It is the closure in $\mathfrak{A}(\mathfrak{H})$ of the span of $\mathbf{1}$ and all monomials of the type

$$c^\#(\varphi_1)\,c^\#(\varphi_2) \cdots c^\#(\varphi_{2n})$$

(an even number of factors). Obviously, the odd elements form a closed self-adjoint subspace of $\mathfrak{A}(\mathfrak{H})$ but not an algebra. The product of two even or two odd elements is even, while the product of an even element with an odd element is odd. Writing

$$x = \tfrac{1}{2}(x + \pi(x)) + \tfrac{1}{2}(x - \pi(x)),$$

decomposes each element in $\mathfrak{A}(\mathfrak{H})$ uniquely in a sum of an even and an odd element. One says that $\mathfrak{A}(\mathfrak{H})$ is a $\mathbb{Z}_2$-graded algebra.

The group $\{\gamma_\theta \mid 0 \le \theta < 2\pi\}$ with

$$\gamma_\theta(c(\varphi)) = e^{-i\theta}\,c(\varphi)$$

is the *gauge group* of $\mathfrak{A}(\mathfrak{H})$. An element x in $\mathfrak{A}(\mathfrak{H})$ is *gauge-invariant* if $\gamma_\theta(x) = x$ for all $0 \le \theta < 2\pi$. Again, the gauge-invariant elements of $\mathfrak{A}(\mathfrak{H})$ form a C*-subalgebra of $\mathfrak{A}(\mathfrak{H})$ which is the closure in $\mathfrak{A}(\mathfrak{H})$ of all linear combinations of $\mathbf{1}$ and monomials of the form

$$c^*(\varphi_1) \cdots c^*(\varphi_n)\,c(\psi_n) \cdots c(\psi_1)$$

(as many starred as unstarred factors).

The quasi-free automorphisms in (6.22) are *gauge-invariant*, or rather *gauge-covariant*, in the sense that

$$\gamma_\theta \circ \alpha_U = \alpha_U \circ \gamma_\theta, \quad 0 \le \theta < 2\pi.$$

In many instances, it is useful to consider a wider class of automorphisms, the so-called *Bogoliubov automorphisms*. They mix up creation and annihilation operators and are of the form

$$c(\varphi) \mapsto c(K\varphi) + c^*(L\varphi), \quad \varphi \in \mathfrak{H}. \tag{6.24}$$

The operators K and L are complex linear, respectively antilinear, on $\mathfrak{H}$ and, in order to preserve the anticommutation relations, they must satisfy the conditions

$$\begin{aligned}
0 &= \{c(\varphi), c(\psi)\} \\
&= \{c(K\varphi) + c^*(L\varphi), c(K\psi) + c^*(L\psi)\} \\
&= \langle K\varphi, L\psi \rangle + \langle K\psi, L\varphi \rangle \\
&= \langle \varphi, K^*L\psi \rangle + \langle \varphi, L^*K\psi \rangle
\end{aligned}$$

and

$$\begin{aligned}
\langle \varphi, \psi \rangle &= \{c(\varphi), c^*(\psi)\} \\
&= \{c(K\varphi) + c^*(L\varphi), c^*(K\psi) + c(L\psi)\} \\
&= \langle K\varphi, K\psi \rangle + \langle L\psi, L\varphi \rangle \\
&= \langle \varphi, K^*K\psi \rangle + \langle \varphi, L^*L\psi \rangle.
\end{aligned}$$

Observe that we have taken into account that L is a complex antilinear operator; therefore, the adjoint L^* of L satisfies

$$\langle \varphi, L^*\psi \rangle = \langle \psi, L\varphi \rangle, \quad \varphi, \psi \in \mathfrak{H}.$$

So, in order that (6.24) should define an automorphism of $\mathfrak{A}(\mathfrak{H})$, the operators K and L must satisfy

$$K^*L + L^*K = 0 \quad \text{and} \quad K^*K + L^*L = \mathbf{1}.$$

6.2.3 *Quasi-free states*

These states are the generalization of Example 6.5: canonical Gibbs states for non-interacting systems. Let Q be a bounded linear operator on $\mathfrak{H}$. We define a functional ω_Q on the $*$-algebra generated by $\mathbf{1}$ and $c(\mathfrak{H})$ by the combinatorial rule

$$\begin{aligned}
\omega_Q(\mathbf{1}) &:= 1 \\
\omega_Q\big(c^*(\varphi_1) \cdots c^*(\varphi_m) c(\psi_n) \cdots c(\psi_1)\big) &:= \delta_{mn} \det\big([\langle \psi_i, Q\varphi_j \rangle]_{ij}\big).
\end{aligned} \tag{6.25}$$

Theorem 6.9 *The functional ω_Q defined in (6.25) extends to a state on $\mathfrak{A}(\mathfrak{H})$ iff $0 \leq Q \leq \mathbf{1}$.*

The proof of Theorem 6.9 is based on the explicit construction of the GNS-representation of ω_Q. The germ of the argument can be found in Example 6.13. The states ω_Q are called *quasi-free* and Q the *symbol* of ω_Q. The ω_Q are, by construction, gauge-invariant, but, similarly to the case of quasi-free automorphisms, gauge invariance is not essential.

Theorem 6.10 *Let Q_1, $Q_2 \in \mathcal{B}(\mathfrak{H})$ be the symbols of the gauge-invariant quasi-free states ω_{Q_i} and let $0 < \lambda < 1$. The state $\lambda\omega_{Q_1} + (1 - \lambda)\omega_{Q_2}$ is quasi-free iff $Q_1 - Q_2$ is a rank-one operator on $\mathfrak{H}$, possibly 0.*

Proof   The proof consists in a straightforward computation. If $\lambda\omega_{Q_1}+(1-\lambda)\omega_{Q_2}$ is a quasi-free state ω_Q, then Q must be equal to $\lambda Q_1 + (1 - \lambda)Q_2$ by evaluating two-point functions. We now write $\omega_Q = \lambda\omega_{Q_1} + (1 - \lambda)\omega_{Q_2}$ on a four-point function

$$\omega_Q\big(c^*(\varphi_1)c^*(\varphi_2)c(\varphi_3)c(\varphi_4)\big)$$
$$= \lambda\omega_{Q_1}\big(c^*(\varphi_1)c^*(\varphi_2)c(\varphi_3)c(\varphi_4)\big) + (1 - \lambda)\omega_{Q_2}\big(c^*(\varphi_1)c^*(\varphi_2)c(\varphi_3)c(\varphi_4)\big).$$

Using

$$\omega_Q\big((c^*(\varphi_1)c^*(\varphi_2)c(\varphi_3)c(\varphi_4)\big) = \langle\varphi_4, Q\,\varphi_1\rangle\langle\varphi_3, Q\,\varphi_2\rangle - \langle\varphi_3, Q\,\varphi_1\rangle\langle\varphi_4, Q\,\varphi_2\rangle,$$

similar expressions for Q_1 and Q_2, $Q = \lambda Q_1 + (1-\lambda)Q_2$ and $\lambda(1-\lambda) > 0$, simple algebra leads to

$$\langle\varphi_4, (Q_1 - Q_2)\varphi_1\rangle\langle\varphi_3, (Q_1 - Q_2)\varphi_2\rangle$$
$$= \langle\varphi_3, (Q_1 - Q_2)\varphi_1\rangle\langle\varphi_4, (Q_1 - Q_2)\varphi_2\rangle. \tag{6.26}$$

Now, either $Q_1 = Q_2$ or there exist vectors φ_3 and φ_2 such that $\langle\varphi_3, (Q_1 - Q_2)\varphi_2\rangle > 0$. In this last case, (6.26) states that $Q_1 - Q_2$ is a rank-one operator, and, as both Q_1 and Q_2 lie in the order interval $[0, \mathbf{1}]$, $Q_1 - Q_2$ must in fact be a multiple of a one-dimensional orthogonal projection operator.

It only remains to check that

$$\lambda\omega_{Q_1} + (1 - \lambda)\omega_{Q_2}$$

is quasi-free, whenever $Q_1 = Q_2 + |\zeta\rangle\langle\zeta|$ and Q_2 and ζ are chosen is such a way that $0 \leq Q_2 \leq \mathbf{1}$ and $0 \leq Q_2 + |\zeta\rangle\langle\zeta| \leq \mathbf{1}$. This is most conveniently done by checking the combinatorial relation (6.25) where we take for the φ_j and ψ_j elements in an orthonormal basis of $\mathfrak{H}$ with first element $\zeta/\|\zeta\|$. The verification is straightforward. $\square$

Theorem 6.11 *A gauge-invariant quasi-free state ω_Q is pure iff its symbol is a projector, possibly 0.*

Proof Suppose that $0 \leq Q \leq \mathbb{1}$ is not a projector; then we can always find a non-zero vector $\zeta \in \mathfrak{H}$ such that

$$0 \leq Q \pm |\zeta\rangle\langle\zeta| \leq \mathbb{1}.$$

Then Theorem 6.10 offers an explicit, non-trivial, convex decomposition of ω_Q:

$$\omega_Q = \tfrac{1}{2}\omega_{Q+|\zeta\rangle\langle\zeta|} + \tfrac{1}{2}\omega_{Q-|\zeta\rangle\langle\zeta|}.$$

Conversely, suppose that Q is a projector and let $\omega_Q = \tfrac{1}{2}\omega_1 + \tfrac{1}{2}\omega_2$ be a convex decomposition of ω_Q in, possibly non-quasi-free, states ω_1 and ω_2. For $\varphi \in (\mathbb{1} - Q)\mathfrak{H}$ we have

$$0 \leq \tfrac{1}{2}\omega_1(c^*(\varphi)c(\varphi)) \leq \tfrac{1}{2}\omega_1(c^*(\varphi)c(\varphi)) + \tfrac{1}{2}\omega_2(c^*(\varphi)c(\varphi))$$
$$= \omega_Q(c^*(\varphi)c(\varphi)) = \langle\varphi, Q\varphi\rangle = 0.$$

Similarly, $\omega_1(c(\varphi)c^*(\varphi)) = 0$ whenever $\varphi \in Q\mathfrak{H}$. Both conditions, together with the anticommutation relations, fully determine the state ω_1. Therefore $\omega_1 = \omega_Q$ and ω_Q is pure. $\square$

The case $Q = 0$ corresponds to the Fock state (see (6.9)–(6.10)), whereas the other extreme $Q = \mathbb{1}$ yields the *anti-Fock state*, also called the *Dirac sea*. This state is characterized by the property $\omega(c(\varphi)c^*(\varphi)) = 0$: all available fermion modes are filled and the only possible excitations are annihilations of particles or, equivalently, creations of holes.

Quasi-free states on a CAR-algebra $\mathfrak{A}(\mathfrak{H})$ factorize in suitable (skew) product decompositions of $\mathfrak{A}(\mathfrak{H})$. Suppose that we have two states ω and σ on $\mathfrak{A}(\mathfrak{H})$ and $\mathfrak{A}(\mathfrak{K})$; then we can look for a product $\omega \wedge \sigma$ which yields a state on $\mathfrak{A}(\mathfrak{H}) \wedge \mathfrak{A}(\mathfrak{K}) = \mathfrak{A}(\mathfrak{H} \oplus \mathfrak{K})$. We first identify $x \in \mathfrak{A}(\mathfrak{H})$ and $y \in \mathfrak{A}(\mathfrak{K})$ with their natural embeddings in $\mathfrak{A}(\mathfrak{H} \oplus \mathfrak{K})$ using (6.20). A natural proposal for $\omega \wedge \sigma$ is

$$(\omega \wedge \sigma)(xy) := \omega(x)\sigma(y), \quad x \in \mathfrak{A}(\mathfrak{H}), \ y \in \mathfrak{A}(\mathfrak{K}). \tag{6.27}$$

Such a proposal cannot be correct unconditionally. Indeed, suppose that both x and y are odd; then, if $\omega \wedge \sigma$ is a state, we have that

$$\omega(x)\sigma(y) = (\omega \wedge \sigma)(xy) = \overline{(\omega \wedge \sigma)((xy)^*)}$$
$$= \overline{(\omega \wedge \sigma)(y^*x^*)} = -\overline{(\omega \wedge \sigma)(x^*y^*)}$$
$$= -\overline{\omega(x^*)}\,\overline{\sigma(y^*)} = -\omega(x)\sigma(y).$$

Therefore, at least one of the two states must be even, i.e. it must vanish on odd elements.

Theorem 6.12 *Suppose that ω is an even state on $\mathfrak{A}(\mathfrak{H})$ and that σ is a state on $\mathfrak{A}(\mathfrak{K})$; then there exists a unique state $\omega \wedge \sigma$ on $\mathfrak{A}(\mathfrak{H}) \wedge \mathfrak{A}(\mathfrak{K})$ defined by (6.27).*

Proof As ω is an even state on $\mathfrak{A}(\mathfrak{H})$, it is invariant under the parity automorphism π on $\mathfrak{A}(\mathfrak{H})$. This implies that there is on the GNS-Hilbert space $\mathfrak{H}_\omega$ of ω a unique unitary operator Θ_ω such that

$$\Theta_\omega \Omega_\omega = \Omega_\omega \quad \text{and} \quad \Theta_\omega \pi_\omega(x)\Theta_\omega = \pi_\omega(\pi(x)), \quad x \in \mathfrak{A}(\mathfrak{H}).$$

π_ω is the representation of $\mathfrak{A}(\mathfrak{H})$ and Ω_ω is the cyclic vector in $\mathfrak{H}_\omega$ which generates ω as a vector state

$$\omega(x) = \langle \Omega_\omega, \pi_\omega(x)\Omega_\omega \rangle, \quad x \in \mathfrak{A}(\mathfrak{H}).$$

As $\pi^2 = \mathrm{id}$, with id the identity map, $\Theta_\omega^2 = \mathbf{1}$ and, therefore, Θ_ω is actually a self-adjoint operator. (In the case of the Fock state, Θ_ω is $(-1)^N$ with N the number operator.)

Next, we construct a representation $\tilde{\pi}$ of $\mathfrak{A}(\mathfrak{H}) \wedge \mathfrak{A}(\mathfrak{K})$ on $\mathfrak{H}_\omega \otimes \mathfrak{H}_\sigma$:

$$\tilde{\pi}\big(c(\varphi \oplus \psi)\big) := \pi_\omega(c(\varphi)) \otimes \mathbf{1} + \Theta_\omega \otimes \pi_\sigma(c(\psi)), \quad \varphi \in \mathfrak{H}, \ \psi \in \mathfrak{K}. \tag{6.28}$$

We now claim that

$$(\omega \wedge \sigma)(z) = \langle \Omega_\omega \otimes \Omega_\sigma, \tilde{\pi}(z)\, \Omega_\omega \otimes \Omega_\sigma \rangle, \quad z \in \mathfrak{A}(H) \wedge \mathfrak{A}(\mathfrak{K}).$$

From (6.28) and the properties of Θ_ω, we see that for $x \in \mathfrak{A}(\mathfrak{H})$

$$\tilde{\pi}(xy) = \pi_\omega(x) \otimes \pi_\sigma(y), \quad y \in \mathfrak{A}(\mathfrak{K}) \text{ even,}$$
$$\tilde{\pi}(xy) = \pi_\omega(x)\Theta_\omega \otimes \pi_\sigma(y), \quad y \in \mathfrak{A}(\mathfrak{K}) \text{ odd.}$$

It then suffices to remember that $\Theta_\omega \Omega_\omega = \Omega_\omega$ to prove (6.27). □

Example 6.13 *(Purification of a quasi-free state)* Any given density matrix ρ on a Hilbert space $\mathfrak{H}$ can be recovered as the partial trace of a vector state on $\mathcal{B}(\mathfrak{H} \otimes \mathfrak{H})$; see Section 2.5 and the GNS-construction in Section 5.3. This purification of a state can be worked out explicitly for quasi-free states, preserving the quasi-free structure. We need the GNS-representations of the Fock and anti-Fock states on $\mathfrak{A}(\mathfrak{H})$ and, in order to avoid complicated notations, we denote the Fock representation of $c(\varphi)$ by $c_F(\varphi)$, etc. Let $0 \leq Q \leq \mathbf{1}$ be the symbol of a quasi-free state ω_Q on $\mathfrak{A}(\mathfrak{H})$. We now construct an explicit representation of a quasi-free state on $\mathfrak{A}(\mathfrak{H} \oplus \mathfrak{H})$:

$$\begin{aligned} c(\varphi \oplus 0) &\mapsto c_F((\mathbf{1} - Q)^{1/2}\varphi) \otimes \mathbf{1} + (-1)^{N_F} \otimes c_{AF}(Q^{1/2}\varphi), \\ c(0 \oplus \psi) &\mapsto c_F(Q^{1/2}\psi) \otimes \mathbf{1} - (-1)^{N_F} \otimes c_{AF}((\mathbf{1} - Q)^{1/2}\psi), \end{aligned} \tag{6.29}$$

where N_F is the number operator on Fock space (6.7). The vector state $\Omega_F \otimes \Omega_{AF}$ defines a quasi-free state on $\mathfrak{A}(\mathfrak{H} \oplus \mathfrak{H})$ with symbol

$$\begin{pmatrix} Q & Q^{1/2}(\mathbf{1} - Q)^{1/2} \\ Q^{1/2}(\mathbf{1} - Q)^{1/2} & (\mathbf{1} - Q) \end{pmatrix}.$$

This is an orthogonal projection operator and, therefore, yields a pure quasi-free state on $\mathfrak{A}(\mathfrak{H} \oplus \mathfrak{H})$ and (6.29) is its GNS-representation.

6.3 Notes

There is a vast literature on the CAR-algebra and its quasi-free structures. Some of the main references are the books by Bratteli and Robinson (1997), Plymen and Robinson (1994), and Evans and Kawahigashi (1998).

Early work on representations of the anticommutation relations can be found in (Gärding and Wightman 1954) and (Shale and Stinespring 1964). For an algebraic treatment of the CAR-algebra and its quasi-free states one may consult (Powers 1970) and (Balslev and Verbeure 1968). Quasi-free objects have frequently been used as a laboratory to test various constructions and conjectures and it is often possible to give full characterizations of general properties in terms of operators on the one-particle space. For example, equivalence properties of quasi-free states have been obtained in (Powers and Størmer 1970). Our presentation of pure quasi-free states is based on Theorem 6.10, due to Wolfe (1975).

Ergodic and chaotic properties of quasi-free dynamical systems will be considered later on. A number of non-trivial physical models have also been treated in terms of quasi-free states on CAR-algebras. We refer, for example, to the solution of the ground state problem for the XY-model (Araki and Matsui 1985) and of the two-dimensional Ising model (Evans and Lewis 1986).

7

ERGODIC THEORY

During its time evolution, a dynamical system traces out a path in the space of pure states, starting at some given initial condition. Time averages of instantaneous expectations of observables define an invariant state. Ergodic theory is concerned with the (in)dependence of such temporal averages on initial conditions and with their properties. For example, in statistical mechanics the ergodic hypothesis claims that for almost all relevant initial conditions, time averages coincide with the average defined by a natural measure, e.g. the micro-canonical ensemble. It finds herein the justification for replacing the time average of the expectation of an observable over the phase space trajectory by a static ensemble average. Proving with some degree of generality that the ergodic hypothesis holds is, however, an intractable problem.

In some simplified models, one can explicitly study various degrees of ergodicity. We recall first some notions from the classical theory and then turn in Section 7.2 to quantum systems. After considering asymptotically Abelian systems, we consider multitime correlation functions and the statistics of fluctuations of observables around their temporal averages. This leads to different notions of independence with their associated statistics. The third section is concerned with a geometrical description of random behaviour in terms of Lyapunov exponents.

7.1 Ergodicity in classical systems

The geometrical picture of ergodicity in classical dynamical systems expresses that no sizeable part of the phase space can be globally invariant under the dynamics. Stronger notions of randomness ask that the probability distribution in the whole phase space can be reconstructed by observing the system locally. In view of comparing these notions with their quantum counterparts, we translate them into properties of phase space functions. It is even instructive to pass to the Hilbert space formalism of classical dynamical systems as given by the Koopman formalism of Example 5.24.

Let $T : \mathfrak{X} \to \mathfrak{X}$ be a measure-preserving transformation of a probability space $(\mathfrak{X}, \Sigma, \mu)$; then T is *ergodic* if the only sets in Σ that are globally T-invariant are either of full or of zero measure. A classical result by Birkhoff shows that for a stationary dynamical system

$$\lim_{N \to \infty} \frac{1}{N} \sum_{n=0}^{N-1} f(T^n(x)) \tag{7.1}$$

exists for almost all $x \in \mathfrak{X}$. In (7.1), f is an integrable function. Moreover, if the system is ergodic, then the sets on which the limit (7.1) is constant are T-invariant and measurable and, therefore, they are either of full or zero measure. Expressed in a slightly different manner: if T is ergodic, then

$$\mu(f) = \lim_{N \to \infty} \frac{1}{N} \sum_{n=0}^{N-1} f(T^n(x)) \tag{7.2}$$

μ-a.e. or

$$\lim_{N \to \infty} \frac{1}{N} \sum_{n=0}^{N-1} \mu(E \cap T^{-n}(F)) = \mu(E) \, \mu(F)$$

for measurable subsets E and F of $\mathfrak{X}$. An equivalent characterization of ergodicity is extremal invariance among the invariant measures: a T-invariant probability measure is ergodic iff it cannot be decomposed into a non-trivial convex combination of T-invariant measures.

A stronger form of randomness, called *mixing*, allows to reconstruct the measure of arbitrary sets, not by averaging, but by taking limits:

$$\lim_{n \to \infty} \mu(E \cap T^{-n}(F)) = \mu(E) \, \mu(F) \tag{7.3}$$

for any two measurable subsets E and F of $\mathfrak{X}$. In terms of phase space functions, mixing becomes

$$\lim_{n \to \infty} \mu(f \, (g \circ T^n)) = \mu(f) \, \mu(g).$$

The notation $g \circ T$ denotes the function $x \mapsto g(T(x))$.

Example 7.1 Ergodicity means that any measurable subset E of $\mathfrak{X}$, of non-zero measure, sweeps through the entire space in the course of time. In particular, if we choose for f in (7.2) the indicator function I_E of E, Birkhoff's theorem implies that for almost all initial conditions $x_0 \in \mathfrak{X}$

$$\mu(E) = \lim_{N \to \infty} \frac{1}{N} \sum_{n=0}^{N-1} I_E(T^n x_0)$$

$$= \lim_{N \to \infty} \frac{1}{N} \#\{n \mid T^n x_0 \in E \text{ for } n \in \{0, 1, \ldots, N-1\}\},$$

i.e. the measure of a set is given by the asymptotic relative frequency of visiting the set, irrespective of the initial condition.

Mixing is much stronger: any measurable set spreads out through the whole system with a local density determined by the measure of that set. Writing the condition (7.3) in terms of relative frequencies, we find for two non-trivial measurable subsets E and F

$$\mu(F)$$

$$= \lim_{m \to \infty} \lim_{N \to \infty} \frac{\#\{n \mid T^{n+m} x_0 \in F \text{ and } T^n x_0 \in E, \; n \in \{0, 1, \ldots, N-1\}\}}{\#\{n \mid T^n x_0 \in E, \; n \in \{0, 1, \ldots, N-1\}\}}.$$

On the unit circle, rotation over an irrational multiple of π is an ergodic transformation with respect to the normalized Lebesgue measure. This transformation is, however, not mixing. Indeed, for a small arc E, $\mu(E T^{-n}(E))$ always vanishes unless the set E has been rotated over approximately an integer multiple of 2π. This excludes the existence of the limit in (7.3). The phase-doubling map $z \mapsto z^2$ on $\{z \in \mathbb{C} \mid |z| = 1\}$ is an example of a dynamical map that is mixing with respect to the Lebesgue measure.

Ergodicity and mixing can be characterized in terms of the spectrum of the Koopman propagator U_T, which acts on $\mathcal{L}^2(\mathfrak{X}, \mu)$ as $(U_T f)(x) := f(T(x))$; see Example 5.24. This relies on *von Neumann's Ergodic Theorem*.

Theorem 7.2 (von Neumann's Ergodic Theorem) *Let U be a unitary operator on a Hilbert space $\mathfrak{H}$; then for any $\varphi \in \mathfrak{H}$*

$$\lim_{N \to \infty} \frac{1}{N} \sum_{n=0}^{N-1} U^n \varphi = P \varphi,$$

where P is the projector on the subspace of eigenvectors of U belonging to the eigenvalue 1.

Proof Let E^U be the spectral measure of U (see (2.41)) and φ an arbitrary vector in $\mathfrak{H}$. We can write

$$\left\| P\varphi - \frac{1}{N} \sum_{n=0}^{N-1} U^n \varphi \right\|^2 = \int_{\mathbb{T}} \left| \delta_{\lambda,0} - \frac{1}{N} \sum_{n=0}^{N-1} e^{in\lambda} \right|^2 \| E^U(d\lambda)\varphi \|^2.$$

Moreover, for any $\lambda \in \mathbb{T}$

$$\lim_{N \to \infty} \frac{1}{N} \sum_{n=0}^{N-1} e^{in\lambda} = \delta_{\lambda,0} \quad \text{and} \quad \left| \delta_{\lambda,0} - \frac{1}{N} \sum_{n=0}^{N-1} e^{in\lambda} \right| \le 2.$$

We now use the dominated convergence theorem to finish the proof. $\qquad \square$

Theorem 7.3 *A dynamical map T on a probability space $(\mathfrak{X}, \Sigma, \mu)$ is ergodic iff the eigenvalue 1 of the corresponding Koopman evolution operator U_T is non-degenerate. It is mixing iff U_T has an absolutely continuous spectrum, except for the single non-degenerate eigenvalue 1.*

Proof We write the definition of ergodicity in the Koopman formalism. Denoting by P the projector on the eigenspace of U_T corresponding to the eigenvalue 1 and using Theorem 7.2,

$$\lim_{N\to\infty} \frac{1}{N} \sum_{n=0}^{N-1} \mu(f\,(g\circ T^n)) = \frac{1}{N} \sum_{n=0}^{N-1} \langle \overline{f}, (U_T)^n g\rangle$$
$$= \langle \overline{f}, Pg\rangle$$
$$= \langle \overline{f}, 1\rangle\,\langle 1, g\rangle.$$

As this holds for any choice of f and g bounded, we have $P = |1\rangle\langle 1|$.

From its definition, we conclude that T is mixing iff

$$w - \lim_{n\to\infty} (U_T)^n = |1\rangle\langle 1|.$$

Let f and g in $\mathcal{L}^2(\mathfrak{X}, \mu)$ be orthogonal to the constant function 1; then by mixing

$$\lim_{n\to\infty} \langle \overline{f}, (U_T)^n g\rangle = \int_{\mathbb{T}} e^{in\lambda}\, \langle \overline{f}, E^U(d\lambda)g\rangle = 0.$$

But this means that the Fourier coefficients of the bounded signed measure $\langle \overline{f}, E^U(d\lambda)g\rangle$ tend to 0 at infinity or, equivalently, that $\langle \overline{f}, E^U(d\lambda)g\rangle$ is absolutely continuous with respect to the Lebesgue measure $d\lambda$. □

7.2 Ergodicity in quantum systems

7.2.1 *Asymptotic Abelianness*

In order to link for a quantum dynamical system, the probabilistic expectation ω of observables with the time evolution, it is convenient to return to its GNS-representation. Assuming ω to be time-invariant, we know that there exists a unique unitary U on the GNS-Hilbert space $\mathfrak{H}$ of ω such that

$$\pi(\Theta(x)) = U^*\pi(x)\,U \quad \text{and} \quad U\,\Omega = \Omega, \tag{7.4}$$

where Ω and π are the cyclic vector and the representation of ω. Knowledge of the correlation functions

$$a, b \in \mathfrak{A} \mapsto \langle \pi(a)\,\Omega, U^*\pi(b)\,\Omega\rangle = \omega(a^*\Theta(b))$$

is sufficient to reconstruct the evolution. Ergodic theory, however, aims at finding a simplified behaviour of dynamical correlation functions when observables are largely separated in time.

The trajectory of an initial wave function $\psi \in \mathfrak{H}$ is given by a sequence $\{\psi, U\psi, U^2\psi, \dots\}$ and as every $\psi \in \mathfrak{H}$ can be arbitrarily well approximated by vectors of the type $\pi(x)\,\Omega$ with $x \in \mathfrak{A}$, we may restrict our attention to correlation functions of the type

$$n \in \mathbb{N} \mapsto \omega(a^*\Theta^n(x)b) = \langle U^n\pi(a)\,\Omega, \pi(x)\,U^n\pi(b)\,\Omega\rangle, \qquad (7.5)$$

with $a, b, x \in \mathfrak{A}$. The simplest notion is that of ergodicity, which deals with the large n behaviour of correlation functions as in (7.5). Unlike the classical case, it is not sufficient to restrict attention to

$$n \mapsto \omega(a\,\Theta^n(x)).$$

This is due to the non-commutativity of the system. For quantum dynamical systems, several degrees of ergodicity can be discerned, according to the behaviour of the time-dependent correlation functions (7.5).

The state ω is *ergodic* for the evolution Θ if for any choice of $a, b, x \in \mathfrak{A}$

$$\lim_{N\to\infty} \frac{1}{N} \sum_{n=0}^{N-1} \omega(a\,\Theta^n(x)\,b) = \omega(ab)\,\omega(x).$$

In terms of the GNS-representation, ergodicity means that the time average of an observable $x \in \mathfrak{A}$ converges weakly to its expectation value

$$w - \lim_{N\to\infty} \frac{1}{N} \sum_{n=0}^{N-1} (U^*)^n\pi(x)\,U^n = \omega(x)\,\mathbf{I}.$$

The state ω is *mixing or weakly clustering* if for any choice of $a, b, x \in \mathfrak{A}$

$$\lim_{n\to\infty} \omega(a\,\Theta^n(x)\,b) = \omega(ab)\,\omega(x).$$

This corresponds in GNS-space to observables weakly converging to their expectation values when they evolve for a long time. The degree of clustering can be specified more precisely in terms of the speed of convergence to 0 of the function $\omega(a\,\Theta^n(x)b) - \omega(ab)\,\omega(x)$. Gentle perturbations of a state ω by an element $p \in \mathfrak{A}$ with $\omega(p^*p) > 0$ are often chosen as

$$\omega_p(x) := \frac{\omega(p^*x\,p)}{\omega(p^*p)}. \qquad (7.6)$$

The state ω is mixing iff

$$\lim_{n\to\infty} \omega_p(\Theta^n(x)) = \omega(x),$$

which means that there is a *return to equilibrium* for any perturbed state of the form (7.6).

For many-particle quantum dynamical systems in statistical mechanics such as quantum spins on a lattice, one expects that local observables will be smeared out in the course of time. As spatially separated observables commute, one can therefore expect the system to be asymptotic Abelian in time. More precisely, $(\mathfrak{A}, \Theta)$ is *norm asymptotic Abelian* whenever

$$\lim_{n \to \infty} \|[\Theta^n(x), y]\| = 0, \quad x, y \in \mathfrak{A}. \tag{7.7}$$

The continuous-time version of this property is expected to hold for quantum spin lattice systems with truly non-commutative interactions, e.g. systems with nearest-neighbour interactions $h \in \mathfrak{A} \otimes \mathfrak{A}$ (see Example 4.5) such that

$$x \in \mathfrak{A} \quad \text{and} \quad [x \otimes \mathbf{1}, h] = 0 \quad \text{implies that } x \text{ is a multiple of } \mathbf{1}.$$

We call $(\mathfrak{A}, \Theta, \omega)$ *weakly asymptotic Abelian* if

$$\lim_{n \to \infty} \omega(a[\Theta^n(x), y]b) = 0$$

for all choices of $a, b, x, y \in \mathfrak{A}$. Using the identity

$$\omega(a[\Theta^n(x), y]b) = \omega(a[\Theta^n(x), yb]) - \omega(ay[\Theta^n(x), b])$$

weak asymptotic Abelianness is equivalent to

$$\lim_{n \to \infty} \omega(a[\Theta^n(x), y]) = 0.$$

For time-asymptotic Abelian systems, the ergodic properties of ω can be expressed in terms of the spectrum of the unitary U that generates the dynamics in precisely the same way as in the classical case.

Theorem 7.4 *A weakly asymptotic Abelian dynamical system* $(\mathfrak{A}, \Theta, \omega)$ *is ergodic iff the eigenvalue 1 of the evolution operator U which implements the dynamics (7.4) is non-degenerate. It is mixing iff U has an absolutely continuous spectrum, except for the single non-degenerate eigenvalue 1.*

Proof Because of the weak asymptotic Abelianness, ergodicity and mixing translate into

$$\lim_{N \to \infty} \frac{1}{N} \sum_{n=0}^{N-1} \omega(a^* \Theta^n(x)) = \lim_{N \to \infty} \frac{1}{N} \sum_{n=0}^{N-1} \langle \pi(a) \, \Omega, U^n \, \pi(x) \, \Omega \rangle$$

$$= \omega(a^*) \, \omega(x)$$

$$= \langle \pi(a) \, \Omega, \Omega \rangle \, \langle \Omega, \pi(x) \, \Omega \rangle$$

and

$$\lim_{n \to \infty} \omega(a^* \Theta^n(x)) = \lim_{n \to \infty} \langle \pi(a) \, \Omega, U^n \pi(x) \, \Omega \rangle$$

$$= \omega(a^*) \, \omega(x)$$

$$= \langle \pi(a) \, \Omega, \Omega \rangle \, \langle \Omega, \pi(x) \, \Omega \rangle.$$

We have now reached precisely the same situation as in Theorem 7.3. $\qquad\square$

Example 7.5 Let $\tilde{U}$ be a unitary with absolutely continuous spectrum acting on a Hilbert space $\tilde{\mathfrak{H}}$ and extend it to a unitary U on $\mathfrak{H} := \mathbb{C} \oplus \tilde{\mathfrak{H}}$ by $U(\alpha \oplus \varphi) := \alpha \oplus \tilde{U}\varphi$. We choose as observables the compact operators $\mathcal{C}(\tilde{\mathfrak{H}})$ extended with the constants $\mathbb{C}\mathbf{1}$, as dynamics $\Theta(x) := U^* x\, U$ and as reference state $\omega(x) := \langle 1 \oplus 0, x\, 1 \oplus 0 \rangle$. This dynamical system is weakly asymptotic Abelian. In order to check this, we may limit ourselves, in the computation of

$$\lim_{n \to \infty} \omega(a^*[\Theta^n(x), c]),$$

to rank-one operators for x and c. We have to compute for arbitrary $\alpha, \alpha_1, \alpha_2, \beta_1, \beta_2 \in \mathbb{C}$ and $\varphi, \varphi_1, \varphi_2, \psi_1, \psi_2 \in \tilde{\mathfrak{H}}$

$$\lim_{n \to \infty} \Big\langle \alpha \oplus \varphi, \Big[\Theta^n(|\alpha_1 \oplus \varphi_1\rangle\langle\beta_1 \oplus \psi_1|), |\alpha_2 \oplus \varphi_2\rangle\langle\beta_2 \oplus \psi_2|\Big] 1 \oplus 0 \Big\rangle. \qquad (7.8)$$

Because of the assumption on $\tilde{U}$,

$$\lim_{n \to \infty} \langle \alpha \oplus \varphi, (U^*)^n \beta \oplus \psi \rangle = \overline{\alpha}\beta. \qquad (7.9)$$

Combining (7.9) with (7.8), we obtain weak asymptotic Abelianness. A simple computation on rank-one operators shows that there is no asymptotic Abelianness in norm in this case. We can now immediately apply Theorem 7.4 and conclude that the system is mixing. The mixing property, however, does not extend to the level of the von Neumann algebra. There exist indeed many constants of the motion X in $\mathcal{B}(\mathfrak{H})$ different from $\mathbf{1}$ and obviously one cannot, for such an observable X, have that $\Theta^n(X)$ weakly tends to a multiple of the identity.

Example 7.6 *(Quasi-free fermion automorphisms)* Let U be a unitary operator on a one-particle space $\mathfrak{H}$ which defines a discrete time evolution through $\Theta(c(\varphi)) = c(U\,\varphi)$. A quasi-free state ω_Q is time-invariant iff its symbol Q satisfies $[U, Q] = 0$. The state ω_Q is ergodic iff

$$\lim_{N \to \infty} \frac{1}{N} \sum_{n=0}^{N-1} \omega_Q(c^*(\varphi)\Theta^n(c(\psi))) = \lim_{N \to \infty} \frac{1}{N} \sum_{n=0}^{N-1} \langle U^n \psi, Q\varphi \rangle = 0.$$

By von Neumann's Ergodic Theorem, Theorem 7.2, this is equivalent to asking that $PQ = 0$ where P is the orthogonal projector on the eigenvectors of U belonging to the eigenvalue 1. The state is mixing iff

$$\lim_{n \to \infty} \omega_Q(c^*(\varphi)\Theta^n(c(\psi))) = \lim_{n \to \infty} \langle U^n \psi, Q\varphi \rangle = 0,$$

which amounts to requiring that the spectrum of U restricted to the range of Q is absolutely continuous.

In view of the anticommutation relations, it is natural to consider a graded form of asymptotic Abelianness. We have to distinguish between even and odd elements. The notion is the same as the usual one, except for the case of pairs of odd elements x^{odd} and y^{odd} where one imposes the restriction

$$\lim_{n \to \infty} \|\{x^{\text{odd}}, \Theta^n(y^{\text{odd}})\}\| = 0.$$

As

$$\{c^*(\varphi), \Theta^n(c(\psi))\} = \{c^*(\varphi), c(U^n\psi)\} = \langle U^n\psi, \varphi \rangle \, \mathbf{I},$$

there is no difference between weak and norm graded asymptotic Abelianness, which amounts to the condition that U must have a purely absolutely continuous spectrum. For such systems any time-invariant state ω_Q is mixing.

Example 7.7 (*Highly anticommutative systems*) A system $(\mathfrak{A}, \Theta)$ is *highly anticommutative* if there exists a subset S of $\mathfrak{A}$ such that $\text{Span}\{\mathbf{I}, S\}$ is dense in $\mathfrak{A}$ and for any $x \in S$, $\epsilon > 0$ and $N \in \mathbb{N}$ there exist times $0 < n_1 < n_2 < \cdots < n_N$ with

$$\|\{\Theta^{n_i}(x^*), \Theta^{n_j}(x)\}\| \leq \epsilon, \quad i \neq j. \tag{7.10}$$

For such a system, there exists but one invariant state ω_0 determined by the property

$$\omega_0(\mathbf{I}) = 1, \quad \omega_0(x) = 0, \ x \in S.$$

Let x and y be two positive elements in a C*-algebra, then

$$\|x + y\| \geq \|x\|.$$

This follows from (2.27) and (5.10). In particular, $\|x\|^2 \leq \|\{x^*, x\}\|$. We apply this inequality to the situation of (7.10) to obtain that

$$\left\| \frac{1}{N} \sum_{i=0}^{N-1} \Theta^{n_i}(x) \right\|^2 \leq \frac{1}{N^2} \left((N^2 - N)\epsilon + 2N\|x\|^2 \right), \tag{7.11}$$

which becomes arbitrarily small for suitable N and ϵ. The Θ-invariance of ω leads to

$$|\omega(x)| = \left| \frac{1}{N} \sum_{i=0}^{N-1} \omega(\Theta^{n_i}(x)) \right| \leq \left\| \frac{1}{N} \sum_{i=0}^{N-1} \Theta^{n_i}(x) \right\|.$$

But (7.11) implies that $\omega(x) = 0$.

Example 7.8 *(Powers–Price shifts)*   Consider a simple model of random commutation or anticommutation relations described in terms of a universal C*-algebra $\mathfrak{A}$ generated by $\mathbf{1}$ and elements $\{e_j \mid j \in \mathbb{N}_0\}$ subject to the conditions

$$(e_j)^2 = \mathbf{1}, \quad e_j^* = e_j \text{ and } e_j e_k = c(j,k)\, e_k e_j, \tag{7.12}$$

where c is a sequence of ± 1's which satisfies

$$c(j,k) = c(k,j) \quad \text{and} \quad c(j,j) = 1.$$

The existence of such an algebra can be shown in a similar way as for the UHF- or CAR-algebras by constructing an explicit realization of the relations (7.12) in terms of the Pauli matrices

$$\sigma^x = \begin{pmatrix} 0 & 1 \\ 1 & 0 \end{pmatrix} \quad \text{and} \quad \sigma^z = \begin{pmatrix} 1 & 0 \\ 0 & -1 \end{pmatrix}.$$

We write $c(j,k) = (-1)^{\epsilon(j,k)}$ with ϵ equal to 0 or 1. The matrices

$$\tilde{e}_k := \left(\bigotimes_{j=1}^{k-1} (\sigma^z)^{\epsilon(j,k)} \right) \otimes \sigma^x \otimes \left(\bigotimes_{j=k+1}^{N} \mathbf{1} \right)$$

with $k = 1, 2, \ldots, N$ realize the relations (7.12) between the first N generators. The algebra generated by the first N $\tilde{e}_k$ can then be embedded in that generated by the first $N+1$ generators by tensoring with an extra $\mathbf{1}$ on the right. The algebra $\mathfrak{A}$ is the completion of the linear span of words

$$w_{\mathbf{j}} := e_{j_1} e_{j_2} \ldots e_{j_\ell}, \quad \mathbf{j} = (j_1, j_2, \ldots, j_\ell), \tag{7.13}$$

and the empty word $\mathbf{1}$. Words are multiplied by concatenation and subsequent simplification, using the relations (7.12). Hence, each word w can be brought, up to a sign, in a canonical form

$$w_{\mathbf{j}} = e_{j_1} e_{j_2} \ldots e_{j_\ell}, \quad j_1 < j_2 < \cdots < j_\ell. \tag{7.14}$$

We call $\mathbf{j}$ the *support of the word* $w_{\mathbf{j}}$ if $w_{\mathbf{j}}$ is written in the simplified form (7.14). The adjoint w^* of a word w is obtained by rewriting its letters in reverse order and equals by (7.12) either w or $-w$. There exists a canonical tracial state τ on $\mathfrak{A}$ determined by

$$\tau(w) := \delta_{w,\mathbf{1}}.$$

Moreover, this τ is the unique tracial state iff the following condition holds: for all choices of $j_1 < j_2 < \cdots < j_N$ there exists a k such that

$$c(j_1, k)\, c(j_2, k) \cdots c(j_N, k) = -1. \tag{7.15}$$

This is easily seen from the tracial property

$$\tau(e_{j_1}e_{j_2}\cdots e_{j_N}) = \tau(e_k e_k e_{j_1}e_{j_2}\cdots e_{j_N})$$
$$= \tau(e_k e_{j_1}e_{j_2}\cdots e_{j_N}e_k)$$
$$= c(j_1,k)\,c(j_2,k)\cdots c(j_N,k)\,\tau(e_k e_k e_{j_1}e_{j_2}\cdots e_{j_N})$$
$$= c(j_1,k)\,c(j_2,k)\cdots c(j_N,k)\,\tau(e_{j_1}e_{j_2}\cdots e_{j_N}).$$

Moreover, if condition (7.15) would fail for a certain choice of indices $j_1 < j_2 < \cdots < j_N$, then $w := e_{j_1}e_{j_2}\cdots e_{j_N}$ would commute with all elements in $\mathfrak{A}$ and

$$x \in \mathfrak{A} \mapsto \tau((\mathbf{1}+w)x)$$

would define another tracial state on $\mathfrak{A}$.

The *Powers–Price shift* is a particular instance of this construction where the indices of the generators are labelled by $\mathbb{Z}$ and

$$c(j,k) = a(|j-k|).$$

a is called the *bit-stream*, it is a sequence of ± 1's indexed by the natural numbers with $a(0) = 1$. Condition (7.15) amounts to the requirement that the set $\{n \in \mathbb{Z} \mid a(|n|) = -1\}$ is not periodic.

The dynamics Θ is determined by shifting the indices of the letters to the right: $\Theta(e_i) = e_{i+1}$. This Θ preserves the commutation relations and the properties $e_i^* = e_i$ and $e_i^2 = \mathbf{1}$ and extends, therefore, to a homomorphism of $\mathfrak{A}$. Obviously, Θ is invertible and, therefore, an automorphism. A word w such as in (7.13) evolves, after t time-steps, into

$$w_{\mathbf{j}}(t) = w_{\mathbf{j}+t} \quad \text{with } \mathbf{j}+t := (j_1+t, j_2+t, \dots, j_\ell+t).$$

Finally, the reference state ϕ is chosen to be the unique tracial state ϕ on $\mathfrak{A}$. Notice that $\phi(w_{\mathbf{i}}w_{\mathbf{j}}) = 0$ unless the two supports $\mathbf{i}$ and $\mathbf{j}$ coincide. The Powers–Price shifts satisfy the generalized weak cluster property (7.20) and are, therefore, weakly asymptotic Abelian. They are asymptotic Abelian in norm iff $a(n) = 1$ for n sufficiently large. It can also be shown that they are highly anticommutative under suitable assumptions on the bit stream a, implying that the trace is their unique invariant state.

Example 7.9 *(Quantum cat maps)* We return to the rotation algebra $\mathfrak{A}_\theta$, introduced earlier in Example 5.25, and consider the Weyl unitaries

$$W(\mathbf{n}) := e^{-i\theta n_1 n_2}\, u^{n_1}\, v^{n_2},$$

indexed by two-dimensional vectors $\mathbf{n}$ with integral components

$$\mathbf{n} = (n_1, n_2) \in \mathbb{Z}^2.$$

The operators $W(\mathbf{n})$ obey the relations

$$W(\mathbf{n})^* = W(-\mathbf{n}),$$
$$W(\mathbf{m})\,W(\mathbf{n}) = e^{(i/2)\theta\sigma(\mathbf{m},\mathbf{n})}\,W(\mathbf{m}+\mathbf{n}) = e^{i\theta\sigma(\mathbf{m},\mathbf{n})}\,W(\mathbf{n})\,W(\mathbf{m}), \qquad (7.16)$$

with σ the *symplectic form*

$$\sigma(\mathbf{m},\mathbf{n}) := m_1 n_2 - m_2 n_1.$$

Note that products $\prod_j W(\mathbf{m}_j)$ reduce to just a single $W(\sum_j \mathbf{m}_j)$ multiplied by a phase. Any 2×2 matrix T with integer entries and determinant 1 defines a map

$$\Theta_T : W(\mathbf{n}) \mapsto W(T^{\mathsf{T}}\mathbf{n}).$$

It is readily checked that Θ_T preserves the commutation relations (7.16). Because $\mathfrak{A}_\theta$ is a universal C*-algebra, this implies that Θ_T extends to a homomorphism of $\mathfrak{A}_\theta$. Because T is invertible and T^{-1} has properties similar to T, Θ_T is in fact an automorphism. Such automorphisms are called *cat maps*. We now consider the dynamical system $(\mathfrak{A}_\theta, \Theta_T, \tau)$, where τ is the tracial state on $\mathfrak{A}_\theta$ (see (5.18)), and we assume that $\mathrm{Tr}\,T > 2$. This last condition ensures that T has two irrational eigenvalues $\lambda^\pm$ such that $0 < \lambda^- < 1 < \lambda^+ = 1/\lambda^-$. This dynamical system is weakly clustering. It suffices to verify that

$$\lim_{n\to\infty} \tau(W(\mathbf{k})\,\Theta_T^n(W(\mathbf{l}))\,W(\mathbf{m})) = \tau(W(\mathbf{k})\,W(\mathbf{m}))\,\tau(W(\mathbf{l})).$$

The case $\mathbf{l} = 0$ is trivial. If $\mathbf{l} \neq 0$, then for n sufficiently large we always find $\mathbf{k} + \mathbf{m} + T^n \mathbf{l} \neq 0$.

For a subset of measure one of deformation parameters θ, the cat maps are highly anticommutative dynamical systems. Nevertheless, for particular values of the deformation parameter, they turn out to be norm asymptotic Abelian.

7.2.2 *Multitime correlations*

Additional information on algebraic relations between observables widely separated in time is encoded in the asymptotic behaviour of multitime correlation functions

$$n_1, n_2, \ldots \mapsto \omega\big(\Theta^{n_{i_1}}(x_1)\Theta^{n_{i_2}}(x_2)\cdots\Theta^{n_{i_j}}(x_j)\big), \qquad (7.17)$$

when all differences between the different times n_k appearing in the correlation function become large, a given time can appear more than once. In model systems, the limit of (7.17) does not always exist and one should rather consider consecutive ergodic averages

$$\lim_{N_k\to\infty}\frac{1}{N_k}\sum_{n_k=0}^{N_k-1}\cdots\lim_{N_1\to\infty}\frac{1}{N_1}\sum_{n_1=0}^{N_1-1}.$$

A suitable framework to deal with this situation is provided by the infinite free product $\bigstar_{i=1}^{\infty} \mathfrak{A}_i$ of copies of the algebra $\mathfrak{A}$ equipped with the state

$$\omega_{\infty}(x_{i_1}^{(1)} x_{i_2}^{(2)} \cdots x_{i_j}^{(j)})$$

$$:= \lim_{N_k \to \infty} \cdots \lim_{N_1 \to \infty} \frac{1}{N} \sum_{n_k=0}^{N_k-1} \cdots \frac{1}{N} \sum_{n_1=0}^{N_1-1} \omega\big(\Theta^{n_{i_1}}(x^{(1)}) \cdots \Theta^{n_{i_j}}(x^{(j)})\big) \quad (7.18)$$

whenever the limits exist. The superscript (k) in $x_{i_k}^{(k)}$ labels an element $x^{(k)}$ in $\mathfrak{A}$ and the subscript i_k refers to which copy of $\mathfrak{A}$ in $\bigstar_{i=1}^{\infty} \mathfrak{A}_i$ $x^{(k)}$ belongs.

The algebra $\bigstar_{i=1}^{\infty} \mathfrak{A}_i$ encodes completely unspecified relations between observables widely separated in time. It is generated by an identity $\mathbf{1}$ and words, which are monomials $w = x_{i_1}^{(1)} x_{i_2}^{(2)} \cdots x_{i_j}^{(j)}$ with all $x^{(k)} \in \mathfrak{A}$ and $i_1 \neq i_2 \neq i_3 \neq \cdots$, repetitions among the i_k are allowed whenever they are separated by another index. Then the algebraic relations between the words are for $x, y \in \mathfrak{A}$, $\lambda \in \mathbb{C}$, $j \in \mathbb{N}_0$ and w, w' two generic words:

$$\begin{aligned} w\mathbf{1}_j w' &= ww', \\ w(x_j + \lambda y_j)w' &= wx_j w' + \lambda wy_j w', \\ wx_j y_j w' &= w(xy)_j w'. \end{aligned} \quad (7.19)$$

Concatenation, together with the above simplification rules, defines the product of words. The product xy in (7.19) is not concatenation, but the usual operator product in the algebra $\mathfrak{A}$. The adjoint w^* of a word $w = x_{i_1}^{(1)} x_{i_2}^{(2)} \cdots x_{i_j}^{(j)}$ equals $(x^{(j)\,*})_{i_j} (x^{(j-1)\,*})_{i_{j-1}} \cdots (x^{(1)\,*})_{i_1}$.

Two extreme instances of $(\bigstar_{i=1}^{\infty} \mathfrak{A}_i, \omega_{\infty})$ are furnished by *commutative and free independence*.

Example 7.10 *(Commutative independence)* Suppose that $(\mathfrak{A}, \Theta, \omega)$ is a norm asymptotic Abelian system and that ω satisfies the following, slightly strengthened, version of mixing:

$$\lim \omega\big(\Theta^{n_{i_1}}(x^{(1)}) \cdots \Theta^{n_{i_j}}(x^{(j)})\, y\, \Theta^{n_{i_{j+1}}}(x^{(j+1)}) \cdots \Theta^{n_{i_k}}(x^{(k)})\big) = 0 \quad (7.20)$$

for all elements y with $\omega(y) = 0$, the limit in (7.20) being taken for all times and separations between different times going to infinity. In this case, it is easy to compute the asymptotic state ω_{∞} of (7.18) and one obtains

$$\omega_{\infty}(x_{i_1}^{(1)} x_{i_2}^{(2)} \cdots x_{i_j}^{(j)}) = \prod_m \omega\Big(\overset{\rightarrow}{\prod_{i_k=m}} x^{(k)}\Big).$$

We can form the quotient of $\bigstar_{i=1}^{\infty} \mathfrak{A}_i$ by the closed two-sided ideal generated by $[x_i, y_j]$ with $i \neq j$. In the proof of the GNS-Theorem 5.14, we used left ideals. If we divide an algebra by a two-sided ideal, then the quotient is again an algebra.

As the state ω_∞ vanishes on the two-sided ideal, it defines a state on the quotient. Doing so, we end up the product of a countable number of copies of ω on the tensor product $\bigotimes_{i=1}^{\infty} \mathfrak{A}_i$.

Consider unital C*-algebras $\mathfrak{A}$ and $\mathfrak{B}$ and their free product $\mathfrak{A} \star \mathfrak{B}$. Given a state ω on $\mathfrak{A}$ and a state σ on $\mathfrak{B}$, the equations

$$(\omega \star \sigma)(\tilde{a}_1 \, \tilde{b}_1 \, \tilde{a}_2, \dots) = 0$$

whenever $\tilde{a}_j \in \mathfrak{A}$ and $\tilde{b}_j \in \mathfrak{B}$ are such that $\omega(\tilde{a}_j) = \sigma(\tilde{b}_j) = 0$ define a linear functional on the dense subalgebra $(\mathfrak{A} \star \mathfrak{B})_0$ of $\mathfrak{A} \star \mathfrak{B}$ spanned by words. The tilde on the elements a and b refers to the fact that they are centred. For example, for general $a_1, a_2 \in \mathfrak{A}$ and $b_1, b_2 \in \mathfrak{B}$

$$\begin{aligned}(\omega \star \sigma)(a_1 \, b_1 \, a_2 \, b_2) = &\,\omega(a_1)\,\omega(a_2)\,\sigma(b_1 \, b_2) + \omega(a_1 \, a_2)\,\sigma(b_1)\,\sigma(b_2)\\ &- \omega(a_1)\,\omega(a_2)\,\sigma(b_1)\,\sigma(b_2).\end{aligned}$$

It can be shown that $\omega \star \sigma$ is positive on $(\mathfrak{A} \star \mathfrak{B})_0$ and extends by continuity to a state on $\mathfrak{A} \star \mathfrak{B}$, called *the free product* ω and σ and again denoted by $\omega \star \sigma$. This construction can be extended to free products of an arbitrary number of states on C*-algebras.

Example 7.11 *(Free independence)*   Consider the universal C*-algebra $\mathfrak{A}$ generated by $\mathbf{1}$ and elements $\{e_i \mid i \in \mathbb{Z}\}$ obeying

$$e_i^* = e_i \quad \text{and} \quad e_i^2 = \mathbf{1}. \tag{7.21}$$

The algebra $\mathfrak{A}_0$ spanned by simple words is norm-dense in $\mathfrak{A}$. A simple word w of length n is a monomial $e_{i_1} e_{i_2} \cdots e_{i_n}$ that cannot be simplified by means of the relations (7.21), i.e. $i_1 \neq i_2$, $i_2 \neq i_3, \dots i_{n-1} \neq i_n$. The identity $\mathbf{1}$ is considered as the empty word.

The *free shift* Θ is the automorphism of $\mathfrak{A}$ determined by

$$\Theta(e_i) = e_{i+1}, \quad i \in \mathbb{Z}.$$

Clearly, $\mathfrak{A}_0$ is globally invariant under Θ. There is a tracial state τ on $\mathfrak{A}$ determined by

$$\tau(w) := \delta_{w,\emptyset}.$$

It is obvious that the state τ is invariant under Θ and $(\mathfrak{A}, \Theta, \tau)$ is a dynamical system. The multiple averages in (7.18) can be replaced by limits and the state ω_∞ turns out to be the free product of copies of τ.

It is possible to show the existence of the asymptotic state ω_∞ for several model systems such as the Powers-Price shifts and the cat map. In general, depending on the bit-stream or the deformation parameter, a wide variety of structures may arise somehow between the extremes of two examples of above.

7.2.3 *Fluctuations around ergodic means*

In the von Neumann algebra picture of a dynamical system $(\mathfrak{A}, \Theta, \omega)$ ergodicity and mixing mean that time averages or long-time limits of elements of the C*-algebra weakly converge to multiples of the identity, e.g.

$$\lim_{N \to \infty} \frac{1}{N} \sum_{n=0}^{N-1} \langle \varphi, (U^*)^n \pi(x) U^n \, \psi \rangle = \omega(x) \, \langle \varphi, \psi \rangle,$$

for any $x \in \mathfrak{A}$.

Although for mixing systems time averages of observables tend to scalars, such averages still fluctuate when considered at the right time-scale. In mixing systems, fluctuation theory studies the asymptotic behaviour of quantities of the type

$$F_N(x) := \frac{1}{\sqrt{N}} \sum_{n=0}^{N-1} \Big(\Theta^n(x) - \omega(x) \mathbf{1} \Big).$$

While mixing is on the level of the law of large numbers, fluctuations are on the level of central limit theorems. For studying the asymptotic behaviour of F_N, one may consider limits of correlations such as

$$\Phi\Big(F(x_1) F(x_2) \cdots F(x_r) \Big) := \lim_{N \to \infty} \omega \Big(F_N(x_1) F_N(x_2) \cdots F_N(x_r) \Big). \qquad (7.22)$$

One may then try to reconstruct the algebraic relations of global fluctuations F by studying the functional Φ. We restrict our attention to the two extreme cases of commutative and free independence.

For sufficiently well time-clustering dynamical systems $(\mathfrak{A}, \Theta, \omega)$, it is meaningful to consider the *truncated correlation functions* or *cumulants* ω^{T}. These are recursively defined through

$$\omega(x_1 x_2 \cdots x_j) := \sum_{\mathcal{P}} \prod_{\ell} \omega^{\mathrm{T}}(x_{i_1}, x_{i_2}, \dots, x_{i_{m_\ell}}), \qquad x_k \in \mathfrak{A}. \qquad (7.23)$$

The sum in (7.23) runs over the set $\mathcal{P}$ of all ordered partitions $(\Lambda_1, \Lambda_2, \dots, \Lambda_r)$ of $\{1, 2, \dots, j\}$ in non-empty sets $\Lambda_\ell = \{i_1, i_2, \dots, i_{m_\ell}\}$. The first few truncated correlation functions are explicitly given by

$$\begin{aligned}
\omega^{\mathrm{T}}(x) &= \omega(x), \\
\omega^{\mathrm{T}}(x, y) &= \omega(xy) - \omega(x)\,\omega(y), \\
\omega^{\mathrm{T}}(x, y, z) &= \omega(xyz) - \omega(x)\,\omega(yz) - \omega(y)\,\omega(xz) \\
&\quad - \omega(z)\,\omega(xy) + 2\omega(x)\,\omega(y)\,\omega(z).
\end{aligned} \qquad (7.24)$$

The expectations of products of fluctuations are given by

$$\omega\Big(F_N(x_1)F_N(x_2)\cdots F_N(x_r)\Big)$$
$$= \frac{1}{N^{r/2}} \sum_{n_1=0}^{N-1} \cdots \sum_{n_r=0}^{N-1} \sum_{\mathcal{P}} \prod_{\ell} \omega^{\mathrm{T}}\big(\Theta^{n_1}(x_{i_1}), \Theta^{n_2}(x_{i_2}), \ldots, \Theta^{n_\ell}(x_{i_{m_\ell}})\big). \quad (7.25)$$

The summation in (7.25) is over all ordered partitions of $\{1, 2, \ldots, r\}$ in subsets $\Lambda_\ell = \{i_1, i_2, \ldots, i_{m_\ell}\}$ containing at least two elements. Imposing

$$\sum_{n\in\mathbb{Z}} \big|\omega(x\Theta^n(y)) - \omega(x)\,\omega(y)\big| < \infty \quad (7.26)$$

and appropriate conditions on the vanishing of averages of truncated correlated functions of order larger than two for elements belonging to a dense subalgebra $\mathfrak{A}_0$ of $\mathfrak{A}$ that is globally invariant under the dynamics, the limit in (7.22) exists and yields a functional Φ that satisfies for $x_i \in \mathfrak{A}_0$

$$\Phi(F(x_1)F(x_2)\cdots F(x_{2n+1})) = 0,$$
$$\Phi(F(x_1)F(x_2)\cdots F(x_{2n})) = \sum_{\mathcal{P}} \prod_{\ell=1}^{n} \Phi(F(x_{i_\ell})F(x_{j_\ell})). \quad (7.27)$$

In (7.27), we have to consider in the summation *all ordered pair partitions* of $\{1, 2, \ldots, 2n\}$, i.e. all partitions of the type $(\{i_1, j_1\}, \{i_2, j_2\}, \ldots, \{i_n, j_n\})$ with $i_1 < i_2 < \cdots < i_n$ and $i_k < j_k$. Moreover,

$$\Phi(F(x)F(y)) = \sum_{n\in\mathbb{Z}} \big(\omega(x\Theta^n(y)) - \omega(x)\,\omega(y)\big).$$

The structure of the functional Φ is completely similar to that of the correlation functions of *Bose fields in a quasi-free state*. Such Bose fields are described by unbounded operators $\{b(\varphi) \mid \varphi \in \mathfrak{H}\}$ on the bosonic Fock space $\Gamma_s(\mathfrak{H})$. The Hilbert space $\mathfrak{H}$ is the one-particle space and, quite similarly to the fermionic Fock space,

$$\Gamma_s(\mathfrak{H}) := \mathbb{C} \bigoplus_{n=1}^{\infty} \Big(\bigotimes{}^{\!n} \mathfrak{H}\Big)_s,$$

where $\Big(\bigotimes{}^{\!n} \mathfrak{H}\Big)_s$ is the fully symmetric part of $\bigotimes{}^{\!n} \mathfrak{H}$. It is useful to introduce the notation $\varphi_1 \times \varphi_2 \times \cdots \times \varphi_n$ for the symmetrized tensor product of n vectors:

$$\varphi_1 \times \varphi_2 \times \cdots \times \varphi_n := \frac{1}{\sqrt{n!}} \sum_{\pi\in\mathfrak{S}_n} \varphi_{\pi(1)} \otimes \cdots \otimes \varphi_{\pi(n)}, \quad (7.28)$$

$\mathfrak{S}_n$ being the permutation group of n elements.

On the dense subspace Γ_0 of $\Gamma_s(\mathfrak{H})$ spanned by vectors of the type (7.28), we consider the Bose creation fields $b^*(\psi)$, $\psi \in \mathfrak{H}$:

$$b^*(\psi)\,\varphi_1 \times \varphi_2 \times \cdots \times \varphi_n := \sqrt{n+1}\,\psi \times \varphi_1 \times \varphi_2 \times \cdots \times \varphi_n.$$

It can be shown that the creation operators are closable and have densely defined adjoints, called annihilation operators. These operators satisfy on Γ_0 the following canonical commutation relations:

$$b(\lambda\varphi_1 + \varphi_2) = \overline{\lambda}b(\varphi_1) + b(\varphi_2),$$
$$[b(\varphi), b(\psi)] = 0,$$
$$[b(\varphi), b^*(\psi)] = \langle \varphi, \psi \rangle.$$

Γ_0 is a space of analytic vectors for $\overline{b(\varphi) + b^*(\varphi)}$ and

$$W(\varphi) := \exp \frac{i}{\sqrt{2}}\left(\overline{b(\varphi) + b^*(\varphi)}\right)$$

satisfy the canonical relations in Weyl form (cf. Example 5.8)

$$\begin{aligned} W(\varphi)^* &= W(-\varphi), \\ W(\varphi)\,W(\psi) &= e^{-i\Im m\langle \varphi, \psi \rangle/2}\,W(\varphi + \psi). \end{aligned} \tag{7.29}$$

The CCR-algebra $\Delta(\mathfrak{H}_0)$ is the uniform closure of the algebra generated by $\{W(\varphi) \mid \varphi \in \mathfrak{H}_0\}$. The space $\mathfrak{H}_0$ is a dense subspace of $\mathfrak{H}$ which can be chosen according to the problem at hand.

A *quasi-free state* ω_Q on the C*-algebra generated by the Weyl operators (7.29) is a state of the form

$$\omega_Q(W(\varphi)) = \exp -\tfrac{1}{4}\|\varphi\|^2 \ \exp -\tfrac{1}{2}\langle \varphi, Q\,\varphi \rangle.$$

Such functionals yield a state iff the symbol Q satisfies $Q \geq 0$ and

$$\omega_Q\big(b^*(\varphi_1)\cdots b^*(\varphi_n)b(\psi_n)\cdots b(\psi_1)\big) = \sum_{\pi \in \mathfrak{S}_n} \prod_{\ell=1}^{n} \langle \psi_\ell, Q\,\varphi_{\pi(\ell)} \rangle. \tag{7.30}$$

In (7.30), the sum is over all permutations π of $\{1, 2, \ldots, n\}$.

The functional (7.27) yields precisely the correlation functions of a *quasi-free state on Bose fields* with the following identifications. Let $(\mathfrak{H}, \pi, \Omega)$ be the GNS-triple of the reference state ω of a dynamical system $(\mathfrak{A}, \Theta, \omega)$ (see Theorem 5.14); then

$$\begin{aligned} \mathfrak{H}_0 &= \pi(\mathfrak{A}_0)\,\Omega \quad \text{and} \quad \langle \pi(x)\Omega, \pi(y)\Omega \rangle = \omega(x^*y), \\ F(x) &= b(\pi(x^*)\Omega) + b(\pi(x)\Omega), \end{aligned}$$

and the symbol Q is given by

$$\langle \pi(x)\Omega, Q\,\pi(y)\Omega \rangle = \sum_{n \in \mathbb{Z}} \big(\omega(x^*\Theta^n(y)) - \omega(x^*)\,\omega(y)\big).$$

Example 7.12 If we impose the strong summability condition

$$\sum_{n_1,\dots,n_j\in\mathbb{Z}} \left|\omega^{\mathrm{T}}(x_0,\Theta^{n_1}(x_1),\dots,\Theta^{n_j}(x_j))\right| < \infty \qquad (7.31)$$

on all truncated correlation functions, then the limits in (7.22) exist. The limiting correlation functions can be reconstructed as those of Bose fields in a quasi-free state. The structure of canonical commutation relations for fluctuations, in Weyl form, can be recovered under much milder conditions than those in (7.31).

A very different situation arises when, for sufficiently large times n, $\Theta^n(x)$ has more or less random commutation relations with y. A possible way to express this is

$$\lim_{N\to\infty} \frac{1}{N} \sum_{n=0}^{N-1} \omega(a\,\Theta^n(x)\,b\,\Theta^n(y)\,c) = 0$$

for centred observables b, x and y. In such cases, it is appropriate to consider the *non-crossing truncated correlation functions* or *non-crossing cumulants* ω^{NC} defined through

$$\omega(x_1 x_2 \cdots x_j) := \sum_{\mathcal{P}} \prod_{\ell} \omega^{\mathrm{NC}}\left(x_{i_1}, x_{i_2}, \dots, x_{i_{m_\ell}}\right). \qquad (7.32)$$

The sum in (7.32) runs over the set $\mathcal{P}$ of all ordered non-crossing partitions $(\Lambda_1,\Lambda_2,\dots,\Lambda_r)$ of $\{1,2,\dots,j\}$ in non-empty sets $\Lambda_\ell = \{i_1,i_2,\dots,i_{m_\ell}\}$. Non-crossing means that for four indices $i_1 < i_2 < i_3 < i_4$ it never happens that i_1 and i_3 belong to an element of the partition and i_2 and i_4 to another. The first few non-crossing correlation functions coincide with those of (7.24). The first difference appears in the function $\omega^{\mathrm{NC}}(w,x,y,z)$ of order four, where the term $\omega(wy)\,\omega(xz)$ is missing in ω^{NC}.

The expectations of products of fluctuations are again given by

$$\omega\left(F_N(x_1)F_N(x_2)\cdots F_N(x_r)\right)$$
$$= \frac{1}{N^{r/2}} \sum_{n_1=0}^{N-1} \cdots \sum_{n_r=0}^{N-1} \sum_{\mathcal{P}} \prod_{\ell} \omega^{\mathrm{NC}}\left(\Theta^{n_1}(x_{i_1}),\Theta^{n_2}(x_{i_2}),\dots,\Theta^{n_\ell}(x_{i_{m_\ell}})\right),$$

where the summation is now over all non-crossing ordered partitions of the set $\{1,2,\dots,r\}$ in subsets $\Lambda_\ell = \{i_1,i_2,\dots,i_{m_\ell}\}$ containing at least two elements. If we impose, besides condition (7.26), that the multitime averages of non-crossing truncated correlation functions vanish sufficiently rapidly, we obtain for the correlations of the limiting fluctuations

$$\Phi(F(x_1)F(x_2)\cdots F(x_{2n+1})) = 0,$$
$$\Phi(F(x_1)F(x_2)\cdots F(x_{2n})) = \sum_{\mathcal{P}} \prod_{\ell=1}^{n} \Phi(F(x_{i_\ell})F(x_{j_\ell})), \qquad (7.33)$$

with

$$\Phi(F(x)F(y)) = \sum_{n \in \mathbb{Z}} \left(\omega(x\Theta^n(y)) - \omega(x)\,\omega(y) \right).$$

In (7.33), we must consider in the summation *all ordered non-crossing pair partitions* of $\{1, 2, \ldots, 2n\}$. The correlations in (7.33) are those of semicircularly distributed random variables.

Considering a single fluctuation with $x = x^*$, we find

$$\Phi(F(x)^{2k+1}) = 0,$$

$$\Phi(F(x)^{2k}) = \frac{1}{k+1} \binom{2k}{k} \left(\Phi(F(x)^2) \right)^k.$$

These are exactly the moments of the semicircular distribution

$$\rho(x) = \begin{cases} \frac{2}{\pi r^2}\sqrt{r^2 - x^2} & \text{for } |x| \leq r, \\ 0 & \text{elsewhere,} \end{cases}$$

with $r^2 = 4\Phi(F(x)^2)$.

Example 7.13 Let a be a generic bit-stream for the unbiased Bernoulli measure on $\{0,1\}^{\mathbb{N}}$. This means that a is a sequence of ± 1's such that for any given sequence $(\epsilon_1, \epsilon_2, \ldots, \epsilon_{k_0})$ of ± 1's of arbitrary length k_0

$$\lim_{N \to \infty} \frac{1}{N} \#\left(\{ n \in \mathbb{N} \mid 0 \leq n \leq N - 1 \quad \text{and} \quad a(n+k) = \epsilon_k \right.$$

$$\forall k = 1, 2, \ldots, k_0 \}\big) = 2^{-k_0}.$$

Consider now the Powers–Price shift determined by such a bit map. A careful counting argument shows that the limiting fluctuations of any generator e_j satisfy (7.33) and that they are, therefore, semicircularly distributed with variance

$$\Phi(F(e)^2) = 1.$$

7.3 Lyapunov exponents

The ergodic properties in the previous section dealt with probabilistic aspects of the dynamics. Classical dynamical systems as they appear in physics have, however, often a relevant geometrical structure. Chaotic behaviour, which is rather typical for smooth dynamical systems, is usually understood as a sensitive dependence of the motion on initial conditions. It means that phase space points that are initially very close become separated during the evolution by a distance that increases exponentially in time. A similar approach to quantum dynamical systems still offers many difficulties and is in a very preliminary state. The lack of a notion of phase space poses a fundamental problem for a geometrical description and one has to look for suitable concepts on an algebraic level. We present a rather special example at the end of the section. We shall also consider a relation between statistical and geometrical notions of chaos in Section 13.2.

7.3.1 *Classical dynamics*

Heuristically, one can study sensitive dependence on initial conditions by linearizing the solution

$$(T(x))^j = t^j(x)$$

of the evolution equation. Using, as in (3.26), coordinates given by some chart, we write for the coordinates of $T^n(x)$

$$(T^n(x))^j = t^j(n; x).$$

Identifying x with its coordinates $(x^1, x^2, \ldots, x^d)$, a small variation δx on the initial condition leads, after n time steps, to a variation

$$\delta(T^n(x))^j = \frac{\partial t^j(n; x)}{\partial x^i} \delta x^i. \tag{7.34}$$

In order to formulate this idea rigorously, we have to replace the infinitesimal vectors δx and $\delta T^n(x)$ by elements of the tangent spaces in x and $T^n(x)$. Therefore, we consider the transport of a vector $\mathbf{u} \in \mathbf{T}_x\mathfrak{X}$ into a vector $\mathbf{v} \in \mathbf{T}_{T(x)}\mathfrak{X}$. Consider a curve $s \mapsto \gamma(s)$ passing at $s = 0$ through x with s-velocity $d\gamma/ds$ equal to $\mathbf{u}$ at $s = 0$. The dynamical map T transforms γ into the curve $s \mapsto T(\gamma(s))$ which has s-velocity $\mathbf{v} := dT(\gamma)/ds$ at $s = 0$. The vector $\mathbf{v}$ does not depend on the explicit choice of the curve γ and it is given by the action of a linear transformation $D(T, x)$ of $\mathbf{u}$:

$$D(T, x) : \mathbf{T}_x\mathfrak{X} \to \mathbf{T}_{T(x)}\mathfrak{X} : \mathbf{u} \mapsto \mathbf{v} = D(T, x)\mathbf{u}.$$

$D(T, x)$ is explicitly given by

$$\left(D(T, x)\mathbf{u}\right)^j = \frac{\partial t^j}{\partial x^k}(x) u^k$$

and one has, upon iteration,

$$D(T^{n+1}, x) = D(T, T^n(x)) D(T^n, x).$$

The rigorous version of (7.34) becomes

$$\mathbf{u} \mapsto \mathbf{u}(n) := D(T^n, x)\mathbf{u} = D(T, T^{n-1}(x)) D(T, T^{n-2}(x)) \cdots D(T, x) \mathbf{u}.$$

For chaotic systems, we expect that for some directions of the vector $\mathbf{u}$, the length of $\mathbf{u}(n)$ increases exponentially with n. In other directions, it may decrease. Before we define rigorously the associated Lyapunov exponents, we consider two examples.

Example 7.14 *(Expanding maps)* These are non-invertible maps T_k (endo-morphisms) of the circle S^1 labelled by $k = 2, 3, \ldots$ In multiplicative notation they are given by

$$T_k(z) := z^k, \quad z \in \mathbb{C}, \ |z| = 1,$$

or, in additive notation, by

$$T_k(x) := k\, x \bmod 1, \quad x \in [0, 1[.$$

The distance between two infinitesimally close points increases by a factor k with each step of the evolution and, hence, the Lyapunov exponent is $\log k$.

Example 7.15 *(Arnold cat maps)* These are invertible diffeomorphisms (au-tomorphisms) of the torus $\mathbb{T}^2$ preserving the Lebesgue measure $dx^1\, dx^2$:

$$T \begin{pmatrix} x^1 \\ x^2 \end{pmatrix} := \begin{pmatrix} t_{11} & t_{12} \\ t_{21} & t_{22} \end{pmatrix} \begin{pmatrix} x^1 \\ x^2 \end{pmatrix} \bmod 1$$

where $x^1, x^2 \in [0, 1[$ and $[t_{ij}]$ is an integer-valued matrix with determinant 1 and trace larger than 2. As T is essentially a linear transformation, stretching and contracting directions are given by the eigenvectors of the matrix $[t_{ij}]$ and the corresponding eigenvalues $\lambda^+ \geq 1, \lambda^- = (\lambda^+)^{-1}$ determine the stretching ratio. Therefore, we obtain two Lyapunov exponents $\pm \log \lambda^+$.

In order to define the *Lyapunov exponents* in the general case of a smooth dynamical system $(\mathfrak{X}, T, \mu)$, we equip the manifold with a Riemannian structure which allows us to define the distance between infinitesimally close points of $\mathfrak{X}$ or, more rigorously, to define the length $\|\mathbf{u}\|$ of a vector $\mathbf{u}$ of the tangent space $\mathbf{T}_x\mathfrak{X}$, see Section 3.3. For a pair $(x, \mathbf{u})$, $x \in \mathfrak{X}$, $\mathbf{u} \in \mathbf{T}_x\mathfrak{X}$, we define the *upper Lyapunov exponent of* $(x, \mathbf{u})$ as the, possibly infinite, number

$$\chi(x, \mathbf{u}) := \limsup_{n \to \infty} \frac{1}{n} \log \|D(T^n, x)\, \mathbf{u}\|.$$

Some properties of $\chi(x, \mathbf{u})$ can be easily proven:

Theorem 7.16 (a) *The value of* $\chi(x, \mathbf{u})$ *does not depend on the Riemannian metric,*

(b) *T-invariance:* $\chi(x, \mathbf{u}) = \chi(T(x), D(T, x)\, \mathbf{u})$,

(c) $\chi(x, \alpha \mathbf{u}) = \chi(x, \mathbf{u})$ *for all* $0 \neq \alpha \in \mathbb{R}$,

(d) $\chi(x, \mathbf{u} + \mathbf{v}) \leq \max\{\chi(x, \mathbf{u}), \chi(x, \mathbf{v})\}$,

(e) *if* $\chi(x, \mathbf{u}) \neq \chi(x, \mathbf{v})$, *then* $\chi(x, \mathbf{u} + \mathbf{v}) = \max\{\chi(x, \mathbf{u}), \chi(x, \mathbf{v})\}$.

Proof Property (a) follows from the fact that on a compact differentiable manifold all Riemannian metrics are equivalent, i.e. for any two metrics there exists
an $\epsilon > 0$ such that for $x \in \mathfrak{X}$ and $\mathbf{u} \in \mathbf{T}_x \mathfrak{X}$

$$\epsilon \|\mathbf{u}\|_1 \leq \|\mathbf{u}\|_2 \leq \frac{1}{\epsilon} \|\mathbf{u}\|_1.$$

Parts (b) and (c) are obvious. Part (d) follows from

$$\|D(T^n, x)(\mathbf{u} + \mathbf{v})\| \leq 2 \max\{\|D(T^n, x)\mathbf{u}\|, \|D(T^n, x)\mathbf{v}\|\}.$$

In order to prove (e), suppose that $\chi(x, \mathbf{u}) < \chi(x, \mathbf{v})$; then

$$\chi(x, \mathbf{u} + \mathbf{v}) \leq \chi(x, \mathbf{v}) = \chi(x, \mathbf{v} - \mathbf{u} + \mathbf{u})$$
$$\leq \max\{\chi(x, \mathbf{u} + \mathbf{v}), \chi(x, -\mathbf{u})\} = \chi(x, \mathbf{u} + \mathbf{v}).$$

The last equality holds because we know that $\chi(x, -\mathbf{u}) = \chi(x, \mathbf{u}) < \chi(x, \mathbf{v})$.
□

The following picture emerges from Theorem 7.16. For each $x \in \mathfrak{X}$ and each
real number χ, let us define a linear subspace of the tangent space $\mathbf{T}_x \mathfrak{X}$:

$$E_\chi(x) := \{\mathbf{u} \in \mathbf{T}_x \mathfrak{X} \mid \chi(x, \mathbf{u}) \leq \chi\},$$

which is T-invariant in the sense of property (b) of Theorem 7.16. Obviously, if
$\chi_1 \geq \chi_2$, then $E_{\chi_1}(x) \supset E_{\chi_2}(x)$. Because all tangent spaces are d-dimensional,
we obtain a collection of numbers

$$\chi_1(x) < \chi_2(x) < \cdots < \chi_{k(x)}(x), \quad k(x) \leq d,$$

and associated linear subspaces

$$\{0\} \subset E_{\chi_1}(x) \subset E_{\chi_2}(x) \subset \cdots \subset E_{\chi_{k(x)}}(x) = \mathbf{T}_x \mathfrak{X}$$

such that if $\mathbf{u} \in E_{\chi_j}(x) \setminus E_{\chi_{j-1}}(x)$, then $\chi(x, \mathbf{u}) = \chi_j(x)$. The numbers $\{\chi_j(x)\}$
are the *Lyapunov exponents* at x and the number

$$l_j(x) := \dim E_{\chi_j}(x) - \dim E_{\chi_{j-1}}(x)$$

is the *multiplicity* of the Lyapunov exponent $\chi_j(x)$.

Stronger results are provided by the *Oseledec Multiplicative Ergodic Theorem*.
Here we present the most important statements which hold for reversible smooth
dynamical systems.

• There exists a decomposition

$$\mathbf{T}_x \mathfrak{X} = \bigoplus_{j=1}^{k(x)} H_j(x),$$

with $\dim(H_j(x)) = l_j(x)$, which is invariant under the map

$$(x, \mathbf{u}) \mapsto (T(x), D(T, x)\mathbf{u}).$$

- The T-invariant Lyapunov exponents $\{\chi_j(x) \mid j = 1, 2, \ldots, k(x)\}$ exist as limits and

$$\lim_{n \to \pm \infty} \frac{1}{|n|} \log \frac{\|D(T^n, x)\,\mathbf{u}\|}{\|\mathbf{u}\|} = \pm \chi_j(x)$$

  uniformly in $\mathbf{u} \in H_j(x) \setminus \{0\}$.
- For ergodic systems, the Lyapunov exponents are constant almost everywhere, i.e.

$$\chi_j(x) = \chi_j, \quad l_j(x) = l_j \quad a.e.$$

There is a whole hierarchy of classical dynamical systems according to their degree of mixing. One of the strongest requirements is to ask for a decomposition of a differentiable manifold into invariant submanifolds with strictly positive and strictly negative Lyapunov exponents: Anosov systems.

7.3.2 Quantum dynamics

There are only few examples of quantum dynamical systems for which geometrical structures have been defined which allow to introduce quantum Lyapunov exponents. One of the possibilities are *horocyclic actions* of automorphisms $\{\gamma_t^{(j)} \mid t \in \mathbb{R}\}$ on the algebra of observables $\mathfrak{A}$ leaving the reference state invariant. They satisfy the relations

$$\Theta_t \, \gamma_s^{(j)} \, \Theta_{-t} = \gamma_{e^{-\chi_j t} s}^{(j)} \tag{7.35}$$

with respect to the dynamical group $\{\Theta_t \mid t \in \mathbb{R}\}$, where

$$\chi_1 \leq \cdots \leq \chi_k < 0 \leq \chi_{k+1} \leq \cdots \leq \chi_m$$

are the *quantum Lyapunov exponents*. The classical intuition behind this structure is that the horocyclic actions describe the shifts in phase space along the directions of pure stretching or squeezing. There exists, of course, a version of this situation for discrete horocyclic actions and/or discrete time evolutions. We shall consider another type of Lyapunov exponents in Section 13.2.

Example 7.17 (*Quantum cat maps*) For the dynamical system $(\mathfrak{A}_\theta, \Theta_T, \tau)$ defined in Example 7.9, there exist two horocyclic actions

$$\gamma_t^{(\pm)} := \gamma_{u_1^\pm t}^{(1)} \, \gamma_{u_2^\pm t}^{(2)},$$

where $\gamma_t^{(1,2)}$ are defined in (5.19) and $\mathbf{u}^\pm = (u_1^\pm, u_2^\pm)$ are the eigenvectors of T^T. These horocyclic actions satisfy the relation (7.35) with Lyapunov exponents $\chi^\pm := \log \lambda^\pm$, where $\lambda^\pm$ are, as in Example 7.9, the eigenvalues of T, $\chi^+ > 0$ and $\chi^- = -\chi^+$.

7.4   Notes

There are many texts and monographs devoted to classical ergodic theory such
as the books by Arnold and Avez (1967), Cornfeld *et al.* (1982), Walters (1982),
Petersen (1983), and the comprehensive exposition by Katok and Hasselblatt
(1996).

C*-dynamical systems are considered in the books of Emch (1972), Bratteli
and Robinson (1979, 1997), Thirring (1983), and Sewell (1986) where one can
also find basic results on ergodic properties of asymptotically Abelian systems.
For further developments of the general quantum ergodic theory we refer to
the book by Benatti (1993) and to (Narnhofer and Thirring 1994). See also the
book by Schmidt (1990) for C*- and von Neumann algebra techniques applied
to classical dynamical systems.

Free probability has been developed by Voiculescu and one can consult the
books by Voiculescu *et al.* (1992) and Hiai and Petz (2000) on this topic.

The application of free product algebras to the characterization of quantum
dynamical systems has been proposed in (Benatti and Fannes 1998). The free
shift of Example 7.11 was studied in (Størmer 1992) and the Powers–Price shifts
were introduced in (Price 1987) and (Powers 1988). The non-commutative torus
and its automorphisms has been widely investigated. Particular attention to
quantum cat maps as models of chaotic dynamical systems was paid in the papers
(Benatti *et al.* 1991; Narnhofer 1992; Emch *et al.* 1994). The notion of highly
anticommutative systems was introduced in (Narnhofer and Thirring 1994) and
illustrated by the examples of the Powers–Price shifts and quantum cat maps
(Narnhofer and Thirring 1995).

The first examples of non-commutative central limit theorems can be found
in the papers (Cushen and Hudson 1971; Hudson 1973). The applications of
quantum limit theorems to Markovian descriptions of quantum open systems
were reviewed in (Accardi *et al.* 1992).

Ergodic properties of quasi-free fermionic dynamical systems have been considered in (Lanford and Robinson 1972; Haag *et al.* 1973) and in a generalized
sense in (Kishimoto 1979). Recent developments for interacting systems were
considered, e.g. in (Bach *et al.* 1998; Maassen *et al.* 1999; Fidaleo and Liverani
1999).

The systematic study of quantum fluctuations as Bose fields in infinite quantum systems has been undertaken in the paper (Goderis *et al.* 1989); see also
(Verbeure and Zagrebnov 1993).

The bosonic model of fluctuations appears when independence of temporally
(or spatially) separated observables implies commutativity, i.e.

$$y\Theta^n(x) - \Theta^n(x)y \to 0 \quad \text{for} \quad n \to \infty.$$

In quantum probability, different notions of independence like free independence
(Voiculescu *et al.* 1992) or that related to the q-deformed commutation relations
as discussed, e.g. in (Biedenharn 1990), were studied in the past decade. In
particular, the corresponding central limit type theorems have been proved in

(Bożejko and Speicher 1996). Fluctuations in Power–Price dynamical system were discussed in (Andries *et al.* 2000) for a more general case than that presented in Example 7.13.

In the presentation of classical Lyapunov exponents we follow the book by Katok and Hasselblatt (1996). The proofs of the different formulations and generalizations of Oseledec's theorem can be found in (Oseledec 1968; Ledrappier 1984). The idea of quantum exponents defined in terms of horocyclic actions is developed in (Emch *et al.* 1994). Another proposal was made in (Majewski and Kuna 1993).

8

QUANTUM IRREVERSIBILITY

8.1 Measurement theory

The status of measurement theory in quantum mechanics is still a topic of debate. Even the fundamental measurement problem: 'How is it possible to obtain objective pointer readings?', is unsolved. We do not know whether a new theory of macrosystems is needed or whether standard quantum mechanics can solve this problem or whether the problem is unsolvable within the standard frame of quantum mechanics and the interpretation of a quantum measurement should be added as an additional postulate.

A deep discussion of the foundations of quantum mechanics is not within the scope of this book and we restrict ourselves to the presentation of the mathematical formalism used in the operational approach to quantum physics in a modern form.

The standard description of the measurement of an observable A with a discrete set of possible values $\{\lambda_j\}$ is due to von Neumann. Associate with such an observable is a self-adjoint operator

$$A := \sum_j \lambda_j P_j. \tag{8.1}$$

$\{P_j \mid j\}$ is a spectral family, i.e. a collection of projectors such that

$$P_i P_j = \delta_{ij} P_i \quad \text{and} \quad \sum_j P_j = \mathbb{1}. \tag{8.2}$$

The *state reduction postulate* states, assuming that the initial state of the system is given by a density matrix ρ, that, after performing the measurement of A and selecting the outcome λ_k, the final state ρ' of the system is

$$\rho \mapsto \rho' = \frac{P_k \rho P_k}{\operatorname{Tr} \rho P_k}. \tag{8.3}$$

According to the standard interpretation, the probability of obtaining the outcome λ_k is equal to $\operatorname{Tr} \rho P_k$ and the mean value of the observable is $\operatorname{Tr} \rho A$. The map

$$\rho \mapsto \sum_j P_j \rho P_j \tag{8.4}$$

describes the state of the system after measurement without recording the result. The transformations (8.3) and (8.4) satisfy the *repeatability condition*, which means that a second measurement, performed immediately after the first one, produces the same outcome and does not change the state of the system anymore.

This theory is not completely satisfactory in several aspects. First of all, the maps (8.3) and (8.4) perturb very strongly the initial state of the observed system and this does not seem to be the case in many real experiments. There are also long standing problems of the existence of a phase observable in quantum optics or a time delay observable in scattering theory, which cannot be solved within the standard scheme of self-adjoint operators. Last but not least, joint measurements of non-commuting observables cannot be described within this scheme, although there exist experimental settings for approximate measurements of such observables.

The modern approach to measurement theory does not rely so strongly on the von Neumann state reduction postulate for the microscopic system under observation. Instead, in operational terms one discusses a measurement as an act of interaction of an object system with another quantum system: the *apparatus*. We discuss a simple model of a measurement leading to the notion of an *unsharp observable* E with a discrete set of outcomes $\{\mu_j\}$. The scheme is very flexible and covers much more general situations, including continuous observables with values in multidimensional spaces, e.g. joint measurements of non-commuting observables, coarse-graining and convolutions with confidence measures, etc.

In our model, the apparatus is described by a Hilbert space $\mathfrak{H}_A$ spanned by the orthonormal set of vectors $\{\varphi_k \mid k \in I\}$. We assume that the set of indices can be decomposed into disjoint subsets, each of them corresponding to a possible outcome μ_j of the measurement, i.e.

$$I = \bigcup_j I_j, \quad I_j \iff \mu_j.$$

The apparatus must be macroscopic, in the sense that the *macrostates* $\{\varphi_k \mid k \in I_j\}$ are distinguishable. Why this should happen is an instance of the measurement problem. The initial state of the composite system, object + apparatus, is assumed to be a product

$$\rho_{in} = \rho \otimes |\varphi_0\rangle\langle\varphi_0|$$

where ρ is an initial state of the object and the initial state of the apparatus is, for simplicity, chosen to be pure with φ_0 taken from the basis $\{\varphi_k \mid k = 0, 1, 2, \ldots\}$. The interaction between both systems is described by a unitary operator U on the tensor product $\mathfrak{H} \otimes \mathfrak{H}_A$ of the Hilbert space of the object with that of the apparatus. We can always write

$$U = \sum_{k,\ell} U_{k\ell} \otimes |\varphi_k\rangle\langle\varphi_\ell|, \quad U_{k\ell} \in \mathcal{B}(\mathfrak{H}) \tag{8.5}$$

and the unitarity reads

$$\sum_{k \in I} U_{mk} (U_{nk})^* = \delta_{mn} \mathbf{I}.$$

The final state of object + apparatus is

$$\rho_{\text{fin}} = U \rho_{\text{in}} U^* = \sum_{k,\ell \in I} U_{k0}\, \rho\, (U_{\ell 0})^* \otimes |\varphi_k\rangle\langle\varphi_\ell|$$

$$\equiv \sum_{k \in I} U_{k0}\, \rho\, (U_{k0})^* \otimes |\varphi_k\rangle\langle\varphi_k|. \tag{8.6}$$

The equivalence in (8.6) means that a macroscopic apparatus cannot operationally determine the correlations between the quantum states, which means that we cannot distinguish between superpositions and mixtures of states. An alternative point of view is that off-diagonal terms lead, due to external perturbations, to unpredictable phases which are not observable. As the macro states of the apparatus $\{\varphi_k \mid k \in I_j\}$ are assumed to be distinguishable, we can say that the observable assumes the value μ_j if the apparatus is in the subspace spanned by the states $\{\varphi_k \mid k \in I_j\}$. The state of the object after the measurement is then given by

$$\rho' = \frac{\Lambda_j^{\mathsf{T}}(\rho)}{\operatorname{Tr} \Lambda_j^{\mathsf{T}}(\rho)}$$

with

$$\Lambda_j^{\mathsf{T}}(\rho) := \sum_{k \in I_j} U_{k0}\, \rho\, (U_{k0})^*, \tag{8.7}$$

and the transpose reminds of the fact that the maps are defined on the Banach space $\mathcal{T}(\mathfrak{H})$ of states of the system. The probability of obtaining the value μ_j is equal to $\operatorname{Tr} \Lambda_j^{\mathsf{T}}(\rho)$ and the mean value of the measured observable is

$$\langle E \rangle_\rho = \sum_j \mu_j \operatorname{Tr} \Lambda_j^{\mathsf{T}}(\rho).$$

Finally, the state of the object, after performing the measurement but without recording the result, is given by the mapping

$$\rho \mapsto \Lambda^{\mathsf{T}}(\rho), \quad \Lambda^{\mathsf{T}} := \sum_j \Lambda_j^{\mathsf{T}}. \tag{8.8}$$

This example introduces several mathematical notions which, after generalization, form the foundation of the operational approach to quantum mechanics. The transformation-valued measure

$$\{\mu_j\} \mapsto \Lambda_j^{\mathsf{T}}$$

on the set of measurement outcomes is called a *state transformer* with associated *positive operator-valued measure* (POVM)

$$\{\mu_j\} \mapsto \Lambda_j(\mathbf{1}),$$

with

$$\operatorname{Tr} \rho \Lambda_j(X) := \operatorname{Tr} X \Lambda_j^{\mathsf{T}}(\rho), \quad \rho \in \mathcal{T}(\mathfrak{H}), \ X \in \mathcal{B}(\mathfrak{H}).$$

This POVM is identified with an unsharp observable E.

One can easily check that sharp observables, corresponding to self-adjoint operators with spectral decomposition (8.1) and their related maps (8.2) and (8.3), can be treated as special cases setting

$$\mu_j := \lambda_j, \quad I_j := \{j\} \text{ and } U_{j0} = U_{0j} := P_j.$$

The maps on the Banach space $\mathcal{T}(\mathfrak{H})$ which appeared in our example possess a very particular structure given by the *sandwich formula*

$$\rho \mapsto \Phi^{\mathsf{T}}(\rho) := \sum_j W_j \rho W_j^*, \quad \rho \in \mathcal{T}(\mathfrak{H}) \tag{8.9}$$

with

$$\sum_j W_j^* W_j \leq \mathbf{1}. \tag{8.10}$$

The map (8.9) transforms positive trace-class operators among themselves. If there is equality in (8.10), then normalization is preserved and states are transformed into states. Positivity preservation is an obvious consequence of the explicit form (8.9) but it is an interesting fact that the maps satisfy the stronger property of *complete positivity*. This is an important mathematical condition with various physical consequences and will be discussed in details in the next section. The discrete-valued measurement process that was presented here can be extended to a much wider setting, allowing for outcomes in an arbitrary measurable space.

This model of a measurement introduces us also to the theory of open systems. Formally, one can treat the interaction of a quantum system with a quantum environment as a measurement process without recording of an outcome and the map (8.8) with the structure (8.7) is an example of a *reduced dynamical map* for an open quantum system. In Section 8.2, we study this topic in detail and show that the most general quantum dynamical maps are indeed of the form (8.8).

Example 8.1 shows how positive operator-valued measures can describe joint measurements of non-commuting observables. We know that it is impossible to make simultaneous sharp measurements of non-commuting observables; see Example 2.44. The canonical example of such a situation is a simultaneous approximative measurement of the position and the momentum of a particle. Experimentally, this is relevant, for example, in a bubble chamber where trajectories of elementary particles moving in external electric and magnetic fields are visualized and provide information about both position and momentum of the particle. This example goes beyond the restricted framework of above and shows how to generalize the structures that we have introduced.

Example 8.1 For the construction of the state transformer and the operator-valued measure corresponding to a joint measurement of position and momentum, we use canonical coherent states for the Weyl operators $W(q,p)$; see Sections 2.4.2 and 3.2. Choosing an arbitrary normalized reference vector $\varphi_0 \in \mathcal{L}^2(\mathbb{R})$, we define a family

$$|q\,p\rangle := W(q,p)\,|\varphi_0\rangle$$

of normalized pure states labelled by the points of the phase space. They satisfy the overcompleteness relation

$$\frac{1}{2\pi}\int_{\mathbb{R}^2} dq dp\, |q\,p\rangle\langle q\,p| = \mathbb{1}.$$

The state transformer given by

$$F \subset \mathbb{R}^2 \mapsto \Lambda_F^\mathsf{T}, \quad \Lambda_F^\mathsf{T}(\rho) := \frac{1}{2\pi}\int_F dq dp\, P_{q\,p}\,\rho\,P_{q\,p}$$

with

$$P_{q\,p} := |q\,p\rangle\langle q\,p|$$

produces the positive operator-valued measure

$$F \subset \mathbb{R}^2 \mapsto \mathcal{O}(F) := \frac{1}{2\pi}\int_F dq dp\, P_{q\,p}.$$

We leave it as an easy exercise to show that the marginal positive operator-valued measures

$$G \subset \mathbb{R} \mapsto \mathcal{O}^q(G) := \mathcal{O}(G \times \mathbb{R})$$

and

$$J \subset \mathbb{R} \mapsto \mathcal{O}^p(J) := \mathcal{O}(\mathbb{R} \times J)$$

are smeared spectral measures for the sharp position and momentum observables, with the squared modulus of the reference vector φ_0 in position and momentum representation as smoothening functions.

8.2 Open quantum systems

In the theory of open systems, we divide the universe into two parts: our open system and the rest, which is called environment or reservoir or, under additional thermodynamic equilibrium conditions, heat bath. The open system is typically rather small or possesses at least a well-defined structure and our aim is to study its evolution in detail. On the contrary, the influence of the environment should be described in a coarse-grained fashion: only a small number of global parameters matter. In the physical literature, this last condition is achieved by introducing a very simple model for the environment: a system of independent harmonic oscillators. This successful Ansatz can be justified in the following way. The environment is a large complicated system composed of weakly interacting particles in a stationary or equilibrium state. The open system is essentially coupled to the fluctuations of certain reservoir observables which are sums of a large number of small, almost independent, contributions averaged over a macroscopic time scale. Such situations are mathematically modelled by limit theorems, like the central limit theorems, which lead, in the classical case to Wiener or Poisson processes (noises). Quantum analogues exist also as operator-valued noises on the Fock Hilbert space of a free bosonic field which is equivalent to a system of harmonic oscillators; see Section 7.2.3.

In the classical theory, the associated evolution equations for mixed states of the open system yield the Fokker–Planck or the linear Boltzmann equation and, in the quantum domain, the completely positive dynamical semigroups.

Our brief survey of the quantum theory of open systems begins with the presentation of the general setting. The Hilbert space of the open system is denoted by $\mathfrak{H}$ and that of the reservoir by $\mathfrak{H}_R$. For the case of an infinite system, $\mathfrak{H}_R$ is the GNS-representation space associated with a given reference state ω_R. The state ω_R is given by a density matrix on $\mathfrak{H}_R$. A single time-step of the global dynamics is governed by a unitary operator $U : \mathfrak{H} \otimes \mathfrak{H}_R \mapsto \mathfrak{H} \otimes \mathfrak{H}_R$ and the initial state of the open system is denoted by ρ. We restrict ourselves to weakly coupled systems where we may assume that the initial state of the total system is simply given by the product of the initial states of reservoir and open system, i.e. $\rho \otimes \omega_R$. The *reduced dynamics* is then the map

$$\rho \mapsto \Lambda^{\mathsf{T}}(\rho) := \mathrm{Tr}_{\mathfrak{H}_R} \, U \, \rho \otimes \omega_R \, U^*. \tag{8.11}$$

We now compute the partial trace in (8.11). Choose, as in (8.5), an orthonormal basis $\{\varphi_k\}$ in $\mathfrak{H}_R$ diagonalizing ω_R

$$\omega_R = \sum_k \lambda_k \, |\varphi_k\rangle\langle\varphi_k|.$$

Next, we write the action of U with respect to this basis on the tensor space:

$$U = \sum_{ij} U_{ij} \otimes |\varphi_i\rangle\langle\varphi_j|,$$

where the U_{ij} are operators acting on the open system and satisfy

$$\sum_m U_{im}(U_{jm})^* = \sum_m (U_{mi})^* U_{mj} = \delta_{ij}\, \mathbf{1}.$$

We now obtain

$$\Lambda^{\mathsf{T}}(\rho) = \sum_{ij} \lambda_j\, U_{ij}\rho(U_{ij})^* = \sum_\alpha W_\alpha \rho (W_\alpha)^*, \qquad (8.12)$$

with

$$W_\alpha := W_{ij} := \lambda_j\, U_{ij} \quad \text{and} \quad \sum_\alpha (W_\alpha)^* W_\alpha = \mathbf{1}.$$

In Heisenberg picture

$$\Lambda(X) = \sum_\alpha W_\alpha^* X W_\alpha. \qquad (8.13)$$

The dynamical maps (8.12) and (8.13) can be defined in the more general context of C*-algebras. These maps enjoy the property of complete positivity, which is the subject of Section 8.3.

In most applications, we need a one-parameter family of dynamical maps parametrized by a time variable $t \in \mathbb{R}^+$:

$$\Lambda_t^{\mathsf{T}}(\rho) = \mathrm{Tr}_{\mathfrak{H}_{\mathrm{R}}}\, U(t)\rho \otimes \omega_{\mathrm{R}} U(t)^* \qquad (8.14)$$

where $U(t) := e^{-itH}$ is the unitary Hamiltonian dynamics of the open system plus the reservoir.

Such a general dynamics is intractable except for a few simple models such as linearly coupled harmonic oscillators or the Friedrichs model. Under suitable assumptions on the dynamical behaviour of the reservoir, strong enough to guarantee central limit type results for rescaled temporal fluctuations of the reservoir similar to those studied in Section 7.2.3, one can show that the memory effects in the dynamics become negligible and one can replace the exact dynamics (8.14) by an approximate Markovian Schrödinger dynamics $\{\Gamma_t^{\mathsf{T}} \mid t \in \mathbb{R}^+\}$ which satisfies the semigroup law

$$\Gamma_{t+s}^{\mathsf{T}}(\rho) = \Gamma_t^{\mathsf{T}}(\Gamma_s^{\mathsf{T}}(\rho)).$$

Example 8.2 In many examples of derivations of reduced dynamics, one obtains the *Markovian master equation*

$$\frac{d\rho_t}{dt} = L^{\mathsf{T}}(\rho_t)$$

with

$$L^{\mathsf{T}}(\rho) := -i[H, \rho] + \frac{1}{2}\sum_j \left([V_j\rho, V_j^*] + [V_j, \rho V_j^*]\right).\tag{8.15}$$

H is self-adjoint and the V_j are bounded operators. Then

$$\rho_t = \Gamma_t^{\mathsf{T}}(\rho_0) = e^{tL^{\mathsf{T}}}(\rho_0).$$

From the structure of L^{T}, it can be seen that Γ^{T} maps density matrices into density matrices. Preservation of normalization is obvious from (8.15) by the cyclicity of the trace. Preservation of positivity follows from Trotter's formula

$$\Gamma_t^{\mathsf{T}}(\rho_0) = \lim_{n\to\infty} \left(e^{t\Phi^{\mathsf{T}}/n}\, e^{-tK^{\mathsf{T}}/n}\right)^n(\rho_0).$$

The operations Φ^{T} and K^{T} act on operators as

$$\Phi^{\mathsf{T}}(\rho) := \sum_j V_j\,\rho\,V_j^*,$$

$$K^{\mathsf{T}}(\rho) := W\rho + \rho W^* \quad \text{with } W := iH + \frac{1}{2}\sum_j V_j V_j^*.$$

8.3   Complete positivity

The purpose of this section is to give a mathematically precise description of the notion of complete positivity and of some of its consequences. In the previous section, we considered in the context of measurement theory some linear transformations of the states on the observables. There is a dual Heisenberg picture to this on the level of the algebra of observables. These maps are compatible with the probabilistic interpretation of quantum mechanics as they preserve positivity and total probability, i.e. normalization. It turns out that they have the much stronger property of complete positivity, which has no commutative counterpart. Complete positivity essentially means that a transformation is robust for forming tensor products: a tensor product of completely positive maps remains completely positive. This is essential for the description of open systems, where we wish to couple a small system to a larger heat bath and study the relaxation properties of the small system. Complete positivity imposes more restrictive conditions than mere positivity on the parameters that define a linear transformation of the observables. In this way, one can, e.g. in the case of a semigroup of completely positive maps, obtain inequalities between several decay rates that enter in the map.

Definition 8.3 *A linear map* Γ *from a C*-algebra* $\mathfrak{A}$ *into a C*-algebra* $\mathfrak{B}$ *is positive if* $\Gamma(x)$ *is positive whenever x is positive.* Γ *is unity-preserving if* $\Gamma(\mathbf{1}) = \mathbf{1}$.*

A positive map Γ is automatically self-adjoint: $\Gamma(a^*) = \Gamma(a)^*$. Moreover, $\|\Gamma\| = \|\Gamma(\mathbf{1})\|$. Therefore, it has norm 1 if it is unity-preserving. Positive multiples, sums and compositions of positive maps remain positive, but there is a serious problem with respect to tensor products. Suppose that we couple both $\mathfrak{A}$ and $\mathfrak{B}$ to a same third (finite-dimensional) algebra $\mathfrak{C}$ and that we extend Γ trivially to $\mathfrak{A} \otimes \mathfrak{C}$ by

$$\Gamma \otimes \mathrm{id} : a \otimes c \mapsto \Gamma(a) \otimes c,$$

then $\Gamma \otimes \mathrm{id}$ might fail to be positive (id denotes the identity map). This can already happen in the case $\mathfrak{C} = \mathcal{M}_2$, as the following example shows.

Example 8.4 Consider on $\mathcal{M}_2$ the map

$$\Gamma : \begin{pmatrix} a_{11} & a_{12} \\ a_{21} & a_{22} \end{pmatrix} \mapsto \begin{pmatrix} \delta a_{11} + (1 - \delta)a_{22} & \mu a_{12} \\ \bar{\mu} a_{21} & (1 - \delta')a_{11} + \delta' a_{22} \end{pmatrix},$$

where $\delta, \delta' \in \mathbb{R}$ and $\mu \in \mathbb{C}$ are three parameters. Clearly, Γ is unity-preserving. We claim that Γ is positive on $\mathcal{M}_2$ iff

$$0 \leq \delta \leq 1, \ 0 \leq \delta' \leq 1 \ \text{and} \ |\mu| \leq \sqrt{\delta\delta'} + \sqrt{(1 - \delta)(1 - \delta')}. \tag{8.16}$$

We then show that the conditions (8.16) are not strong enough to ensure that $\Gamma \otimes \mathrm{id}$ is positive on $\mathcal{M}_2 \otimes \mathcal{M}_2$.

Γ is positive iff it maps positive matrices into positive matrices. A 2×2 matrix

$$A = \begin{pmatrix} a_{11} & a_{12} \\ a_{21} & a_{22} \end{pmatrix}$$

is positive iff

$$a_{11} \geq 0, \ a_{21} = \overline{a_{12}}, \ \text{and} \ |a_{12}|^2 \leq a_{11}a_{22}; \tag{8.17}$$

see Example 2.24. Using (8.17), it is a simple exercise to show that positivity of Γ is equivalent to (8.16).

Applying $\Gamma \otimes \mathrm{id}$ to the positive matrix

$$\begin{pmatrix} 1 & 0 & 0 & 1 \\ 0 & 0 & 0 & 0 \\ 0 & 0 & 0 & 0 \\ 1 & 0 & 0 & 1 \end{pmatrix} = \begin{pmatrix} \begin{pmatrix} 1 & 0 \\ 0 & 0 \end{pmatrix} & \begin{pmatrix} 0 & 1 \\ 0 & 0 \end{pmatrix} \\ \begin{pmatrix} 0 & 0 \\ 1 & 0 \end{pmatrix} & \begin{pmatrix} 0 & 0 \\ 0 & 1 \end{pmatrix} \end{pmatrix} \in \mathcal{M}_2 \otimes \mathcal{M}_2$$

we obtain

$$\begin{pmatrix} \delta & 0 & 0 & \mu \\ 0 & 1 - \delta' & 0 & 0 \\ 0 & 0 & 1 - \delta & 0 \\ \bar{\mu} & 0 & 0 & \delta' \end{pmatrix}, \tag{8.18}$$

which is positive iff

$$0 \leq \delta \leq 1, \ \ 0 \leq \delta' \leq 1 \ \text{and} \ |\mu| \leq \sqrt{\delta \delta'}. \tag{8.19}$$

This condition, necessary for positivity of $\Gamma \otimes \text{id}$, is more restrictive than (8.16).

Definition 8.5 *A map* Γ *from a C*-algebra* $\mathfrak{A}$ *into a C*-algebra* $\mathfrak{B}$ *is d-positive if* $\Gamma \otimes \text{id}$ *from* $\mathfrak{A} \otimes \mathcal{M}_d$ *to* $\mathfrak{B} \otimes \mathcal{M}_d$ *is positive.* Γ *is completely positive if it is d-positive for any* $d \in \mathbb{N}$.

Theorem 8.6 *Let* Γ *be a unity-preserving completely positive map from* $\mathfrak{A}$ *to* $\mathfrak{B}$; *then it satisfies the 2-positivity inequality*

$$\Gamma(a)\Gamma(a^*) \leq \Gamma(aa^*). \tag{8.20}$$

Furthermore, let $\Gamma(a)\Gamma(a^*) = \Gamma(aa^*)$ *for some* $a \in \mathfrak{A}$; *then*

$$\Gamma(ab) = \Gamma(a)\Gamma(b), \quad b \in \mathfrak{A}.$$

Proof Let a be an arbitrary element in $\mathfrak{A}$; then

$$0 \leq \begin{pmatrix} \mathbf{1} & 0 \\ a^* & 0 \end{pmatrix} \begin{pmatrix} \mathbf{1} & a \\ 0 & 0 \end{pmatrix} = \begin{pmatrix} \mathbf{1} & a \\ a^* & aa^* \end{pmatrix}.$$

As Γ is 2-positive, we must also have that

$$0 \leq \begin{pmatrix} \mathbf{1} & \Gamma(a) \\ \Gamma(a^*) & \Gamma(aa^*) \end{pmatrix},$$

which is equivalent to the inequality (8.20).

The second statement follows by applying inequality (8.20) to a linear combination $a + \alpha b$ and letting α vary in $\mathbb{C}$. □

Theorem 8.7 *A $d \times d$ matrix $A = [a_{ij}]$ over a C*-algebra $\mathfrak{A}$ is positive iff one of the following two conditions is satisfied:*

(a) *A is a sum of matrices of the form $[(x_i)^* x_j]$ with $x_\ell \in \mathfrak{A}$, $\ell = 1, 2, \ldots d$.*

(b) *For any choice $(x_1, x_2, \ldots, x_d)$ of d elements in $\mathfrak{A}$*

$$\sum_{i,j=1}^{d} x_i^* a_{ij} x_j \ \text{is positive in} \ \mathfrak{A}.$$

Proof In order to prove (a) we use that $A \in \mathcal{M}_d(\mathfrak{A})$ is positive iff it is of the form B^*B with $B = [b_{ij}] \in \mathcal{M}_d(\mathfrak{A})$. Writing $A = B^*B$, we obtain

$$A = [(B^*B)_{ij}] = \left[\sum_{k=1}^{d}(B^*)_{ik}B_{kj}\right] = \sum_{k=1}^{d}[(b_{ki})^*b_{kj}].$$

It is also obvious that $[(x_i)^*x_j]$ is positive in $\mathcal{M}_d(\mathfrak{A})$ as it is of the form B^*B with

$$B = \begin{pmatrix} x_1 & x_2 & \cdots & x_d \\ 0 & 0 & \cdots & 0 \\ \vdots & \vdots & \ddots & \vdots \\ 0 & 0 & \cdots & 0 \end{pmatrix}.$$

Consider an arbitrary representation π of $\mathfrak{A}$ on a Hilbert space $\mathfrak{H}$ and the representation $\mathrm{id}_d \otimes \pi$ of $\mathcal{M}_d(\mathfrak{A})$ on $\mathbb{C}^d \otimes \mathfrak{H}$

$$\mathrm{id}_d \otimes \pi([a_{ij}]) \, [\varphi_k] := \left[\sum_{\ell=1}^{d}\pi(a_{k\ell})\varphi_\ell\right]. \tag{8.21}$$

Any representation of $\mathcal{M}_d(\mathfrak{A})$ on a Hilbert space is of the form (8.21). Now $\mathrm{id}_d \otimes \pi(A)$ is positive iff for all vectors $[\varphi_k]$

$$0 \le \Big\langle [\varphi_k], \, \mathrm{id}_d \otimes \pi(A)\,[\varphi_\ell]\Big\rangle = \sum_{k,\ell=1}^{d}\langle \varphi_k, \pi(a_{k\ell})\,\varphi_\ell\rangle.$$

Choosing for the φ_k vectors of the form $\pi(x_k)\zeta$ with $x_k \in \mathfrak{A}$ and $\zeta \in \mathfrak{H}$, we obtain

$$0 \le \sum_{k,\ell=1}^{d}\langle \pi(x_k)\zeta, \pi(a_{k\ell})\,\pi(x_\ell)\zeta\rangle = \Big\langle \zeta, \sum_{k,\ell=1}^{d}\pi\big((x_k)^*a_{k\ell}\,x_\ell\big)\zeta\Big\rangle,$$

and, therefore, $\pi\Big(\sum_{k,\ell=1}^{d}(x_k)^*a_{k\ell}x_\ell\Big) \ge 0$. As π is arbitrary, $\sum_{k,\ell=1}^{d}(x_k)^*a_{k\ell}x_\ell$ is positive.   □

Theorem 8.8 (Choi)   Let $\{e_{ij} \mid i, j = 1, 2, \ldots, d\}$ be a system of matrix units in $\mathcal{M}_d$. A linear map $\Gamma : \mathcal{M}_d \to \mathfrak{A}$ is completely positive iff

$$[\Gamma(e_{ij})] \text{ is positive in } \mathcal{M}_d(\mathfrak{A}). \tag{8.22}$$

Proof We have to show that

$$\mathrm{id}_k \otimes \Gamma : \mathcal{M}_k(\mathcal{M}_d) \to \mathcal{M}_k(\mathfrak{A}) : [A_{ij}] \mapsto [\Gamma(A_{ij})]$$

positive for each $k = 1, 2, \ldots$ is equivalent to condition (8.22). Notice that each of the A_{ij} appearing in (8.22) is itself a $d \times d$ scalar matrix.

Suppose first that Γ is completely positive. As $[e_{ij}]$ is a positive matrix in $\mathcal{M}_d(\mathcal{M}_d)$, condition (8.22) has to be satisfied.

Conversely, by Theorem 8.7(a) it is sufficient to check whether, for all possible choices of $\{Y_1, Y_2, \ldots, Y_k\}$ in $\mathcal{M}_d$, $\mathrm{id}_k \otimes \Gamma([(Y_i)^* Y_j])$ is positive in $\mathcal{M}_k(\mathfrak{A})$. By Theorem 8.7(b) this is the case iff for all choices of $\{x_1, x_2, \ldots, x_k\}$ in $\mathfrak{A}$

$$\sum_{i,j=1}^{k} x_i^* \Gamma\big((Y_i)^* Y_j\big) x_j \quad \text{is positive in } \mathfrak{A}. \tag{8.23}$$

We now write each of the $Y_i \in \mathcal{M}_d$ as a linear combination of matrix units

$$Y_i = \sum_{\ell,m=1}^{d} y_i^{\ell m} e_{\ell m}, \quad y_i^{\ell m} \in \mathbb{C}$$

and rewrite (8.23) as

$$\begin{aligned}
\sum_{i,j=1}^{k} x_i^* & \Gamma\bigg(\Big(\textstyle\sum_{\ell m} y_i^{\ell m} e_{\ell m}\Big)^* \textstyle\sum_{\ell' m'} y_j^{\ell' m'} e_{\ell' m'}\bigg) x_j \\
&= \sum_{i,j,\ell,\ell',m,m'} \overline{y_i^{\ell m}} y_j^{\ell' m'} x_i^* \Gamma\big(e_{m\ell} e_{\ell' m'}\big) x_j \\
&= \sum_{i,j,\ell,m,m'} \overline{y_i^{\ell m}} y_j^{\ell m'} x_i^* \Gamma\big(e_{mm'}\big) x_j \\
&= \sum_{\ell} \sum_{m,m'} \Big(\textstyle\sum_i y_i^{\ell m} x_i\Big)^* \Gamma\big(e_{mm'}\big) \textstyle\sum_j y_j^{\ell m'} x_j.
\end{aligned} \tag{8.24}$$

If condition (8.22) is satisfied, then (8.24) is positive and, therefore, Γ is completely positive. $\qquad\square$

Corollary 8.9 *A completely positive map Γ from the $d \times d$ matrices into a C^*-algebra $\mathfrak{A}$ is completely positive iff it is of the form*

$$\Gamma([a_{ij}]) = \sum_{\ell=1}^{d} \sum_{i,j=1}^{d} a_{ij} (x_{i\ell})^* x_{j\ell}$$

with $x_{i\ell} \in \mathfrak{A}$ for $i, \ell = 1, 2, \ldots, d$.

It follows from Theorems 8.8 and 8.7 that a map Γ from $\mathcal{M}_d$ to a C^*-algebra $\mathfrak{A}$ is completely positive iff it is d-positive.

Example 8.10 In order to check under which conditions the mapping Γ of Example 8.4 is completely positive, we use Theorem 8.8. Γ is completely positive iff it is 2-positive and this is the case iff the matrix (8.22), which is exactly equal to the matrix (8.18), is positive. Therefore, the necessary and sufficient conditions for Γ to be completely positive are exactly these of eqn (8.19).

Theorem 8.11 (Stinespring) *Let $\Gamma : \mathfrak{A} \to \mathcal{B}(\mathfrak{H})$ be a unity-preserving completely positive mapping from a C^*-algebra $\mathfrak{A}$ into the bounded linear operators on a Hilbert space $\mathfrak{H}$. There exist a Hilbert space $\mathfrak{K}$, a representation π of $\mathfrak{A}$ into $\mathcal{B}(\mathfrak{K})$ and an isometry $V : \mathfrak{H} \to \mathfrak{K}$ such that*

$$\Gamma(a) = V^*\pi(a)V, \quad a \in \mathfrak{A},$$
$$\pi(\mathfrak{A})V\mathfrak{H} \text{ is dense in } \mathfrak{K}. \tag{8.25}$$

$\mathfrak{K}$, π and V are uniquely determined, up to unitary equivalence.

Proof Stinespring's theorem generalizes the GNS representation theorem, Theorem 5.14 and its proof is similar. We first define an inner product on the algebraic tensor product of $\mathfrak{A}$ and $\mathfrak{H}$:

$$\langle a \otimes \varphi, b \otimes \psi \rangle_\Gamma := \langle \varphi, \Gamma(a^*b)\,\psi \rangle.$$

$\langle \cdot, \cdot \rangle_\Gamma$ is positive because Γ is completely positive but it could be degenerate. As in the proof of the GNS theorem, we divide $\mathfrak{A} \otimes \mathfrak{H}$ by the subspace spanned by the elements in the kernel of $\langle \cdot, \cdot \rangle_\Gamma$. Let $\mathfrak{K}$ be the Hilbert space obtained by completing this quotient. The representation π and the isometry V are now defined on the equivalence class of $a \otimes \varphi$ by

$$\pi(x)\,[a \otimes \varphi] := [xa \otimes \varphi] \quad \text{and} \quad V \varphi := [\mathbf{1} \otimes \varphi]$$

The properties stated in the theorem can now readily be checked. □

A state ω on $\mathfrak{A}$ is a particular case of a completely positive map with range $\mathbb{C}$. If we put in the Stinespring representation $\Omega := V\,1$, then we recover the GNS representation theorem. Complete positivity is only a new notion within the context of non-commutative systems. Even if only the domain or the range of the map is Abelian, complete positivity coincides with ordinary positivity.

Theorem 8.12 *A positive map $\Gamma : \mathfrak{A} \to \mathfrak{B}$ is completely positive if either $\mathfrak{A}$ or $\mathfrak{B}$ is Abelian.*

The following theorem shows that the irreversible dynamical maps discussed in Section 8.1 and 8.2 are indeed the most general completely positive unity-preserving maps on $\mathcal{B}(\mathfrak{H})$.

Theorem 8.13 (Kraus) *Any completely positive map $\Gamma : \mathcal{B}(\mathfrak{H}) \to \mathcal{B}(\mathfrak{K})$ is of the form*

$$\Gamma(A) = \sum_j W_j^* A W_j, \tag{8.26}$$

where the W_j are bounded linear operators from $\mathfrak{K}$ to $\mathfrak{H}$ and the sum converges in strong operator sense. When $\mathfrak{H}$ is finite-dimensional, one can restrict the sum in (8.26) to at most $\dim(\mathfrak{H})^2$ terms.

Proof Let $(\mathfrak{L}, \pi, V)$ be the canonical objects associated with Γ by the Stinespring theorem. As π is a unitary representation of an algebra of the type $B(\mathfrak{H})$ on $\mathfrak{L}$, we know that it is of the form

$$\pi(A) = A \otimes \mathbf{1}$$

for a decomposition $\mathfrak{L} = \mathfrak{H} \otimes \mathfrak{M}$ of the representation space. Let $\{f_j\}$ be an orthonormal basis of $\mathfrak{M}$ and define the bounded linear operators W_j on $\mathfrak{K}$ by

$$V\varphi = \sum_j W_j \varphi \otimes f_j.$$

Then (8.25) is precisely the expression (8.26).

If $\mathfrak{H}$ is finite-dimensional, then we can use Theorem 8.8 to restrict the summation in (8.26) to at most $\dim(\mathfrak{H})^2$ terms.   □

Example 8.14 Returning to the completely positive map Γ of Example 8.4, we can explicitly write the action of Γ as in (8.26). As $|\mu| \leq \sqrt{\delta\delta'}$, we can introduce the parameterization

$$\mu = \sqrt{\delta\delta'}\sin(\phi)e^{i\psi}, \quad 0 \leq \phi \leq \pi/2, \ 0 \leq \psi < 2\pi.$$

Introducing the matrices

$$W_1 := \begin{pmatrix} \overline{\alpha} & 0 \\ 0 & \overline{\beta} \end{pmatrix}, \qquad W_2 := \begin{pmatrix} \beta' & 0 \\ 0 & \alpha' \end{pmatrix},$$

$$W_3 := \begin{pmatrix} 0 & 0 \\ \sqrt{1-\delta} & 0 \end{pmatrix}, \qquad W_4 = \begin{pmatrix} 0 & \sqrt{1-\delta'} \\ 0 & 0 \end{pmatrix},$$

with

$$\alpha := \sqrt{\delta}\cos(\phi/2)e^{i\psi/2}, \qquad \alpha' := \sqrt{\delta'}\cos(\phi/2)e^{i\psi/2},$$
$$\beta := \sqrt{\delta'}\sin(\phi/2)e^{-i\psi/2}, \qquad \beta' := \sqrt{\delta}\sin(\phi/2)e^{-i\psi/2},$$

one verifies that

$$\Gamma(A) = \sum_{j=1}^{4} W_j^* A W_j.$$

Linear transformations of observables that arise in situations where a system evolves under a dynamical map, or where it is coupled to another system or where observables of a coupled system are projected on a subsystem are all completely positive. As compositions of completely positive maps are again completely positive, any reasonable physical operation enjoys this property. Basic physical operations are the following:

- Any homomorphism from a C*-algebra $\mathfrak{A}$ to a C*-algebra $\mathfrak{B}$ is completely positive. This is an immediate consequence of the characterization of positive elements in Theorem 8.7. This applies in particular to identifications of a system with a subsystem of a larger one such as

$$A \mapsto A \otimes \mathbf{I}, \quad A \in \mathfrak{A},$$

or its dual version

$$\omega \in \mathcal{S}(\mathfrak{A} \otimes \mathfrak{B}) \mapsto \omega_{\mathfrak{A}} \quad \text{with} \quad \omega_{\mathfrak{A}}(A) := \omega(A \otimes \mathbf{I}).$$

In particular, for algebras of the type $\mathcal{B}(\mathfrak{H})$ and states given by density matrices we recover the partial trace operation (2.67).

- Let $\omega_{\mathfrak{B}}$ be a state on $\mathfrak{B}$, then we can obtain *marginal operators* of the composed system $\mathfrak{A} \otimes \mathfrak{B}$ by extending the map

$$A \otimes B \mapsto \omega_{\mathfrak{B}}(B)\, A.$$

The dual map is given by

$$\omega_{\mathfrak{A}} \mapsto \omega_{\mathfrak{A}} \otimes \omega_{\mathfrak{B}}.$$

- The tensor product and composition of two completely positive maps are again completely positive.

8.4 Quantum dynamical semigroups

We now come to a more abstract version of the reduced Markovian dynamics that was introduced in Section 8.1, formulated in an algebraic language: a *strongly continuous quantum dynamical semigroup* in Heisenberg picture is a one-parameter family $\{\Gamma_t \mid t \in \mathbb{R}^+\}$ of maps on a C*-algebra $\mathfrak{A}$ satisfying

$$\Gamma_t \text{ is completely positive,}$$
$$\Gamma_{t+s} = \Gamma_t \Gamma_s, \quad t, s \in \mathbb{R}^+,$$
$$t \mapsto \Gamma_t \text{ is strongly continuous.}$$

The transposed semigroup $\{\Gamma_t^{\mathsf{T}} \mid t \in \mathbb{R}^+\}$ is a quantum dynamical semigroup in Schrödinger picture

$$\Gamma_t^{\mathsf{T}}(\omega) := \omega \circ \Gamma_t, \quad \omega \in \mathcal{S}(\mathfrak{A}).$$

Quantum dynamical semigroups are special instances of strongly continuous one-parameter semigroups of *contractions*, i.e. mappings on Banach spaces with norm not larger than 1 and therefore the classical theory can be applied. In particular, standard results ensure the existence of a densely defined generator L of the semigroup. $\text{Dom}(L)$ consists of all elements x such that $t \mapsto \Gamma_t(x)$ is differentiable in the strong sense in $t = 0$ and

$$L(x) := \lim_{t \downarrow 0} \frac{\Gamma_t(x) - x}{t}, \quad x \in \text{Dom}(L).$$

The *Hille–Yoshida* theorem characterizes the generators of contraction semigroups.

Theorem 8.15 (Hille–Yoshida) *An operator L defined on a dense subspace $\text{Dom}(L)$ of a Banach space $\mathfrak{B}$ is a generator of a strongly continuous semigroup $\{T(t) \mid t \geq 0\}$ of contractions iff it satisfies the following conditions:*

$$\|(\mathbf{1} - \lambda L)x\| \geq \|x\|, \quad x \in \text{Dom}(L), \quad \lambda \geq 0,$$
$$\overline{(\mathbf{1} - \lambda L)\,\text{Dom}(L)} = \mathfrak{B}, \quad \text{for some } \lambda > 0. \tag{8.27}$$

If L satisfies (8.27), then T_t can be found by the strong operator limits:

$$T(t)\,x = \lim_{n \to \infty} \left(\mathbf{1} - tL/n\right)^{-n} x$$
$$= \lim_{\epsilon \downarrow 0} \exp\left(tL(\mathbf{1} - \epsilon L)^{-1}\right) x.$$

The natural question then arises: What is the structure of a generator of a quantum dynamical semigroup? The answer is not known in general but important partial results are available. We present such a structural theorem for the finite-dimensional case.

Theorem 8.16 (Gorini–Kossakowski–Lindblad–Sudarshan) *A map L generates a dynamical semigroup on matrix algebra $\mathcal{M}_d$ iff*

$$L(A) = i[H, A] + \Phi(A) - \tfrac{1}{2}(\Phi(\mathbf{1})A + A\Phi(\mathbf{1})) \tag{8.28}$$

where Φ is a completely positive map on $\mathcal{M}_d$ and $H = H^* \in \mathcal{M}_d$.

Proof The 'if' part is simple, noticing that

$$L(A) = \Phi(A) - K(A)$$

with

$$K(A) := \tfrac{1}{2}(VA + AV^*) \quad \text{and} \quad V := \Phi(\mathbf{1}) + iH$$

and that $e^{t\Phi}$ and e^{-tK} are completely positive semigroups. Then, by Trotter's product formula

$$e^{tL} = \lim_{n \to \infty} \left[e^{t\Phi/n} e^{-tK/n}\right]^n$$

is also completely positive; moreover, $L(\mathbf{1}) = 0$ implies unity preservation.

To prove the 'only if' part we use the isomorphism between two finite-dimensional linear spaces,

(1) $\mathcal{B}(\mathcal{M}_d)$, the space of linear maps on the algebra $\mathcal{M}_d$,
(2) $\mathcal{M}_d \otimes \mathcal{M}_d$, which can be viewed as the space $d \times d$ matrices with entries consisting of $d \times d$ matrices.

Let $\{e_{ij}\}$ be the standard system of matrix units in $\mathcal{M}_d$, then the isomorphism is given by

$$\Gamma \in \mathcal{B}(\mathcal{M}_d) \equiv \hat{\Gamma} \in \mathcal{M}_d \otimes \mathcal{M}_d \quad \text{with} \quad \hat{\Gamma}_{ij} := \Gamma(e_{ij}). \tag{8.29}$$

One should notice that this is just an isomorphism of linear spaces, which does not turn composition of maps into multiplication of matrices. Due to Theorem 8.8, there is one-to-one correspondence between completely positive maps Γ on $\mathcal{M}_d$ and positive matrices $\hat{\Gamma}$ in $\mathcal{M}_d \otimes \mathcal{M}_d$. Under the isomorphism (8.29), the map $J := (1/d)\,\text{id}$ on $\mathcal{M}_d$ becomes a projection operator $\hat{J}$ in $\mathcal{M}_d \otimes \mathcal{M}_d$ with entries $\hat{J}_{ij} = (1/d)e_{ij}$. We use this map $\hat{J}$ to decompose $\hat{\Gamma}$ as follows:

$$\hat{\Gamma} = (\mathbf{1} - \hat{J})\hat{\Gamma}(\mathbf{1} - \hat{J}) + \hat{J}(\hat{\Gamma} - \tfrac{1}{2}\hat{\Gamma}\hat{J}) + (\hat{\Gamma} - \tfrac{1}{2}\hat{J}\hat{\Gamma})\hat{J}.$$

The first term in this decomposition is a positive operator and therefore the image of a completely positive map on $\mathcal{M}_d$. The second and third terms correspond, under the isomorphism (8.29), to right- and left-multiplication of $d \times d$ matrices. This can be seen as follows. The left-multiplication L_S by a matrix $S = [s_{ij}]$ is mapped into

$$\widehat{L_S}_{ij} = L_S(e_{ij}) = Se_{ij} = \sum_k s_{ki} e_{kj}.$$

Now such matrices $\widehat{L_S}$ are precisely characterized by the property $\widehat{L_S}\hat{J} = \widehat{L_S}$. The analogous property holds for right-multiplication.

The corresponding decomposition of the completely positive map Γ reads

$$\Gamma(A) = \Gamma^{\|}(A) + SA + AS^*,$$

where $\Gamma^{\|}$ is also a completely positive map. Replacing the single map Γ by a one-parameter semigroup $\{\Gamma_t = e^{tL} \mid t \in \mathbb{R}^+\}$, we obtain

$$L(A) = \lim_{t \downarrow 0} \frac{1}{t}(\Gamma_t(A) - A)$$

$$= \lim_{t \downarrow 0} \frac{1}{t}\Gamma_t^{\|}(A) + \lim_{t \downarrow 0} \frac{1}{t}(S_t - \frac{1}{2})A + \lim_{t \downarrow 0} \frac{1}{t}A(S_t^* - \frac{1}{2})$$

$$= \Phi(A) - \tfrac{1}{2}(B^*A + AB).$$

Φ is completely positive as a limit of completely positive maps. We then put $H := (1/4i)(B - B^*)$ and $\frac{1}{2}(B + B^*) = \Phi(\mathbf{1})$ in order to satisfy the unity preservation condition $L(\mathbf{1}) = 0$. $\square$

Corollary 8.17 *A generator L of a completely positive unity-preserving semi-group on $\mathcal{M}_d$ can be written as*

$$L(A) = i[H, A] + \frac{1}{2} \sum_j \{V_j^*[A, V_j] + [V_j^*, A]V_j\} \tag{8.30}$$

with finitely many terms in the sum. Its Schrödinger picture version reads

$$L^{\mathsf{T}}(\rho) = -i[H, \rho] + \frac{1}{2} \sum_j \{[V_j\rho, V_j^*] + [V_j, \rho V_j^*]\}. \tag{8.31}$$

The form of L is not uniquely determined by (8.30).

Proof The corollary follows immediately from Theorem 8.16 and the general form of a completely positive transformation of $\mathcal{M}_d$ as obtained in Theorem 8.13. □

Example 8.18 The simplest example of an open system is a two-level system, i.e. a spin-$\frac{1}{2}$, coupled to a reservoir. In Heisenberg picture, we consider the following generator L of a completely positive semigroup on $\mathcal{M}_2$:

$$L(A) = \frac{i\omega}{2}[\sigma^z, A] + \frac{\gamma}{2}\left(\sigma^+[A, \sigma^-] + [\sigma^+, A]\sigma^-\right)$$
$$+ \frac{\kappa}{2}\left(\sigma^-[A, \sigma^+] + [\sigma^-, A]\sigma^+\right) - \frac{\eta}{2}[\sigma^z, [\sigma^z, A]],$$

where the Pauli matrices σ were introduced in Example 4.6 and $\gamma, \kappa, \eta \geq 0$. Exponentiating this generator, we obtain the semigroup

$$\begin{pmatrix} a_{11} & a_{12} \\ a_{21} & a_{22} \end{pmatrix} \mapsto \begin{pmatrix} \delta(t)a_{11} + (1 - \delta(t))a_{22} & \mu(t)a_{12} \\ \overline{\mu(t)}a_{21} & (1 - \delta'(t))a_{11} + \delta'(t)a_{22} \end{pmatrix}$$

of Example 8.4, with

$$\mu(t) = e^{-i\omega t}e^{-(1/2)(\gamma+\kappa+\eta)t},$$
$$\delta(t) = \frac{\kappa + \gamma e^{-(\gamma+\kappa)t}}{\kappa + \gamma}, \tag{8.32}$$
$$\delta'(t) = \frac{\gamma + \kappa e^{-(\gamma+\kappa)t}}{\kappa + \gamma}.$$

Clearly, there are two time scales involved in the decay of the matrix elements in (8.32) and the decay rate of the off-diagonal matrix elements is at least one-half of that of the diagonal ones.

Theorem 8.16 has been generalized to the case of $\mathfrak{A} = \mathcal{B}(\mathfrak{H})$ with infinite-dimensional $\mathfrak{H}$ but under the assumption that the generator L is a bounded operator on $\mathcal{B}(\mathfrak{H})$. The structure of L is exactly the same as in (8.28) or as the more explicit versions given in (8.30) and (8.31). The operators appearing in these formulae are bounded but possibly infinite sums have to be considered. Convergence can be shown in an appropriate topology (ultra-weak operator convergence). This result can be generalized further to bounded generators of dynamical semigroups acting on some C*-algebras. There is no general characterization of unbounded generators of quantum dynamical semigroup. Nevertheless, large classes of Markovian master equations of the form (8.30) and (8.31) with unbounded operators H and V_j or even with more singular objects like quantum fields and sums over j replaced by integrals possess solutions (in a precisely defined sense) given in terms of quantum dynamical semigroups.

8.5 Quasi-free completely positive maps

The class of unital completely positive quasi-free maps between CAR-algebras will be important for future applications. Such maps $\Gamma : \mathfrak{A}(\mathfrak{H}) \to \mathfrak{A}(\mathfrak{K})$ are determined by a linear operator $V : \mathfrak{H} \to \mathfrak{K}$ and a symbol $R \in B(\mathfrak{H})$. The combinatorial rule to evaluate Γ on a monomial in the fields $c^\#$ is

$$\Gamma\big(c^*(\varphi_1)\cdots c^*(\varphi_m)c(\psi_n)\cdots c(\psi_1)\big)$$
$$= \sum \epsilon\, \omega_R\big(c^*(\varphi_{j_1})\cdots c^*(\varphi_{j_{m-k}})c(\psi_{s_{n-\ell}})\cdots c(\psi_{s_1})\big)$$
$$\times\, c^*(V\varphi_{i_1})\cdots c^*(V\varphi_{i_k})c(V\psi_{r_\ell})\cdots c(V\psi_{r_1}). \qquad (8.33)$$

The summation in (8.33) extends over all ordered partitions

$$\{1,\dots,m\} = \{i_1,\dots,i_k\} \cup \{j_1,\dots,j_{m-k}\}, \quad i_1 < \cdots < i_k, \; j_1 < \cdots < j_{m-k}$$
$$\{1,\dots,n\} = \{r_1,\dots,r_\ell\} \cup \{s_1,\dots,s_{n-\ell}\}, \quad r_1 < \cdots < r_\ell, \; s_1 < \cdots < s_{n-\ell}$$

and ϵ takes the values ± 1 according to whether

$$(j_1,\dots,j_{m-k},s_{n-\ell},\dots,s_1,i_1,\dots,i_k,\dots,r_\ell,\dots,r_1)$$

arises through an even or an odd permutation of $(1,\dots,m,n,\dots,1)$. The functional ω_R is quasi-free and computed according to (6.25). In particular,

$$\Gamma(\mathbf{1}) = \mathbf{1},$$
$$\Gamma(c^*(\varphi)) = c^*(V\varphi),$$
$$\Gamma(c^*(\varphi)c(\psi)) = c^*(V\varphi)c(V\psi) + \langle\psi, R\varphi\rangle,$$
$$\Gamma(c^*(\varphi_1)c^*(\varphi_2)c(\psi_2)c(\psi_1)) = c^*(V\varphi_1)c^*(V\varphi_2)c(V\psi_2)c(V\psi_1)$$
$$+\langle\psi_1, R\varphi_2\rangle c^*(V\varphi_1)c(V\psi_2) - \langle\psi_1, R\varphi_1\rangle c^*(V\varphi_1)c(V\psi_1)$$
$$+\langle\psi_2, R\varphi_2\rangle c^*(V\varphi_1)c(V\psi_1) - \langle\psi_2, R\varphi_1\rangle c^*(V\varphi_2)c(V\psi_1)$$
$$-\langle\psi_2, R\varphi_1\rangle\langle\psi_1, R\varphi_2\rangle + \langle\psi_1, R\varphi_1\rangle\langle\psi_2, R\varphi_2\rangle.$$

If Γ is completely positive, then it transforms quasi-free states on $\mathfrak{A}(\mathfrak{K})$ into quasi-free states on $\mathfrak{A}(\mathfrak{H})$:

$$\omega_{Q_2} \circ \Gamma = \omega_{Q_1} \quad \text{with} \quad Q_1 = V^* Q_2 V + R.$$

Therefore, $0 \le Q_2 \le \mathbf{1}$ should imply $0 \le V^* Q_2 V + R \le \mathbf{1}$ or, equivalently,

$$\|V\| \le 1 \quad \text{and} \quad 0 \le R \le \mathbf{1} - V^* V. \tag{8.34}$$

In fact, the necessary conditions (8.34) are also sufficient to guarantee that Γ is completely positive. This can be seen by purifying Γ to a quasi-free automorphism on a larger algebra and then recovering Γ by projecting onto the original algebra. We begin by describing the last operation.

Let ω be an even state on a CAR-algebra $\mathfrak{A}(\mathfrak{K})$, then the map

$$\mathrm{id} \wedge \omega : \, c^{\#}(\varphi_1 \oplus \psi_1) \cdots c^{\#}(\varphi_m \oplus \psi_n)$$
$$\mapsto \sum \epsilon \, \omega\big(c^{\#}(\psi_{j_1}) \cdots c^{\#}(\psi_{j_{m-k}})\big) \, c^{\#}(\varphi_{i_1}) \cdots c^{\#}(\varphi_{i_k}) \tag{8.35}$$

extends to a completely positive unital map from $\mathfrak{A}(\mathfrak{H} \oplus \mathfrak{K})$ to $\mathfrak{A}(\mathfrak{H})$. In (8.35), the summation is taken over the ordered partitions $\{i_1, \dots, i_k\} \cup \{j_1, \dots, j_{n-k}\}$ of $\{1, \dots, n\}$ and ϵ is the signature of the permutation $(i_1, \dots, i_k, j_1, \dots, j_{n-k})$ of $\{1, \dots, n\}$. This can directly be verified using Theorem 6.12 and the fact that an element in a C*-algebra is positive iff its expectation in an arbitrary state is positive.

Consider now a pair of operators V, R which satisfy condition (8.34). The operator

$$\tilde{V} := \begin{pmatrix} V & -\sqrt{\mathbf{1} - V V^*} \\ \sqrt{\mathbf{1} - V^* V} & V^* \end{pmatrix}$$

is a unitary operator from $\mathfrak{H} \oplus \mathfrak{K} \to \mathfrak{K} \oplus \mathfrak{H}$. This follows from the identity

$$V^* \sqrt{\mathbf{1} - V V^*} = \sqrt{\mathbf{1} - V^* V} \, V^*.$$

Therefore, $\alpha_{\tilde{V}} : c(\eta) \mapsto c(\tilde{V} \eta)$ extends to a quasi-free automorphism of $\mathfrak{A}(\mathfrak{H} \oplus \mathfrak{K})$. Moreover, there exists a symbol Q on $\mathfrak{K}$ with $0 \le Q \le \mathbf{1}$ and such that

$$\langle \varphi_1, R \varphi_2 \rangle = \langle \sqrt{\mathbf{1} - V^* V} \, \varphi_1, Q \sqrt{\mathbf{1} - V^* V} \, \varphi_2 \rangle,$$

for $\varphi_1, \varphi_2 \in \mathfrak{H}$. But

$$\Gamma = (\omega_Q \wedge \mathrm{id}) \circ \alpha_{\tilde{V}} \circ \jmath_{\mathfrak{H}} \tag{8.36}$$

with $\jmath_{\mathfrak{H}}$ the natural injection of $\mathfrak{A}(\mathfrak{H})$ into $\mathfrak{A}(\mathfrak{H} \oplus \mathfrak{K})$; see (6.20). As all three maps on the right-hand side of (8.36) are completely positive, Γ inherits this property.

The graded product construction extends to completely positive maps between CAR-algebras. If $\Gamma_1 : \mathfrak{A}(\mathfrak{H}_1) \to \mathfrak{A}(\mathfrak{K}_1)$ and $\Gamma_2 : \mathfrak{A}(\mathfrak{H}_2) \to \mathfrak{A}(\mathfrak{K}_2)$ are completely positive maps and Γ_1 is even in the sense that it intertwines the parity automorphisms (6.23) of $\mathfrak{A}(\mathfrak{H}_1)$ and $\mathfrak{A}(\mathfrak{K}_2)$, then $\Gamma_1 \wedge \Gamma_2$, defined in the obvious way, extends to a completely positive map from $\mathfrak{A}(\mathfrak{H}_1 \oplus \mathfrak{H}_2)$ to $\mathfrak{A}(\mathfrak{K}_1 \oplus \mathfrak{K}_2)$. A special instance of this is given by quasi-free completely positive maps determined by operators of the form

$$V = W \oplus \mathbf{1} \quad \text{and} \quad R = T \oplus 0$$

with $\|W\| \leq 1$ and $0 \leq T \leq \mathbf{1} - W^*W$ in an orthogonal decomposition $\mathfrak{H} = \mathfrak{K} \oplus \mathfrak{L}$. The map on $\mathfrak{A}(\mathfrak{H})$ is then the product of the quasi-free map defined by W and T and the identity map on $\mathfrak{A}(\mathfrak{L})$.

The structure of strongly continuous one-parameter semigroups of quasi-free unity-preserving completely positive semigroup $\{\Gamma_t \mid t \in \mathbb{R}^+\}$ on a CAR-algebra $\mathfrak{A}(\mathfrak{H})$ is now readily described. Each Γ_t is constructed in terms of an operator $V(t)$ and a symbol $R(t)$. These families of operators have to satisfy the following requirements:

(1) $\{V(t) \mid t \in \mathbb{R}^+\}$ is a strongly continuous semigroup of contractions on $\mathfrak{H}$.
(2) $\{R(t) \mid t \in \mathbb{R}^+\}$ is a strongly continuous family of operators on $\mathfrak{H}$ subject to a positivity and a cocycle condition:

$$0 \leq R(t) \leq \mathbf{1} - V(t)^*V(t),$$
$$R(t_1 + t_2) = V(t_2)^* R(t_1) V(t_2) + R(t_2)$$

It is then possible to characterize Γ in terms of generators;

$$V(t) = e^{tD} \quad \text{and} \quad R(t) = \int_0^t ds\, e^{sD^*} A e^{sD}, \tag{8.37}$$

where D is the generator of a contraction semigroup on $\mathfrak{H}$ and A is a linear operator on $\mathfrak{H}$ such that

$$0 \leq A \leq -(D + D^*).$$

Example 8.19 Consider the canonical orthonormal basis $\{e_1, e_2, \dots, e_d\}$ in $\mathbb{C}^d$ and the Fock representation of the CAR-algebra $\mathfrak{A}(\mathbb{C}^d)$ with the short-hand notations c_j^* and c_j for the Fock creation and annihilation operators $c^*(e_j)$ and $c(e_j)$. The most general quasi-free dynamical semigroup $\{\Gamma_t \mid t \in \mathbb{R}^+\}$ is governed by the generator

$$L(X) = i \sum_{k,\ell=1}^d h_{k\ell}\, [c_k^* c_\ell, X] + \frac{1}{2} \sum_{k,\ell=1}^d \Big(a_{k\ell}\, (c_\ell[X, c_k^*] + [c_\ell, X]c_k^*)$$
$$+ b_{k\ell}\, (c_\ell^*[X, c_k] + [c_\ell^*, X]c_k) \Big).$$

$H = [h_{k\ell}]$ is a Hermitian matrix and the matrices $A = [a_{k\ell}]$ and $B = [b_{k\ell}]$ are positive. The link with (8.37) is given by

$$D = iH - \tfrac{1}{2}(A + B).$$

In particular, for $d = 1$ we recover the semigroup of Example 8.18 with $\eta = 0$, $a = \kappa$, $b = \gamma$, $\sigma^- = c$, $\sigma^+ = c^*$, $\sigma^z = 2c^*c - \mathbf{1}$ and $h = \omega$.

8.6   Notes

The operational meaning of a measurement as an interaction between the observed system and the apparatus was already recognized by von Neumann (1932) for the case of his idealized measurement. Later on, Davies and Lewis (1970) and Ludwig (1974) developed a more general approach which included unsharp measurements. The statistical analysis of quantum measurements and the role of symmetry groups for the so-called covariant measurements is discussed in the monograph of Holevo (1982). For a recent exposition of quantum measurement theory, we refer to the book by Busch *et al.* (1995).

The description of open quantum systems by means of the projection technique and the generalized master equation was initiated by Nakajima, Prigogine, Résibois and Zwanzig, see the text by Haake (1973). A mathematically rigorous formalism, including the earlier results on Markovian master equations, was given in Davies (1976). Rigorous derivations of Markovian master equations (see Example 8.2) based on physically different limiting procedures, can be found in the papers (Davies 1974; Gorini and Kossakowski 1976; Dümcke 1985). One may refer to the text by Alicki and Lendi (1987) for a review on these later results.

Fundamental mathematical results on completely positive maps are presented in (Stinespring 1955; Størmer 1963; Choi 1972, 1975), and in the book by Kraus (1983). The importance of complete positivity for the theory of quantum measurements and quantum open systems has been recognized in (Kraus 1971; Gorini *et al.* 1976). The book by Paulsen (1986) collects a number of results on complete positivity.

The theory of one-parameter semigroups on Banach spaces, including Theorem 8.15, can be found in the books of Yosida (1965) and Kato (1984). The general form of the generator of a completely positive quantum dynamical semigroup was established in (Gorini *et al.* 1976) in the finite-dimensional case and for bounded generators on $\mathcal{B}(\mathfrak{H})$ in (Lindblad 1976). Our proof of Theorem 8.16 is similar to that of Streater (1995). For more general operator algebras, see (Christensen and Evans 1979). Quasi-free dynamical semigroups generalize the physical examples of Markovian master equations for a damped harmonic oscillator and a two-level atom. They were presented in (Weidlich and Haake 1965) and for unstable particles in the paper (Alicki 1978); see for more information the book by Agarwal (1971). Quasi-free maps for fermions appeared in (Evans 1979; Fannes and Rocca 1980) and for bosons in the paper (Demoen *et al.* 1977). Quasi-free dynamical semigroups have been considered in (Vanheuverzwijn 1977) for the CCR and in (Evans 1980) for the CAR. Large classes of quantum dynamical semigroups with unbounded generators have been studied in (Davies 1977; Sinha 1994; Holevo 1996; Chebotarev and Fagnola 1998).

Further developments were related to the equivalence of quantum dynamical semigroups and stochastic differential equations either with the non-commutative noise introduced in (Hudson and Parthasarathy 1984) or with commutative noises as in (Alicki and Fannes 1987). The former initiated the field of quantum stochastic calculus reviewed in the book by Parthasarathy (1992), while the later found applications to numerical computations in quantum optics under the name of state diffusion model (see the book by Carmichael (1993)).

9

ENTROPY

The term entropy appears in many different places and was given a number of meanings in various contexts.

Thermodynamical entropy   The name entropy was introduced by Clausius for the quantity

$$\Delta S = S_{\text{final}} - S_{\text{initial}} = \int \frac{dQ}{T},$$

where Q is the heat energy absorbed by a system in evolving in a quasi-static and reversible way from an initial to a final state and T is the instantaneous absolute temperature. The entropy S is assumed to depend only on the actual state of the system and not on its history, with zero value at $T = 0$, which is the third law of thermodynamics. For an isolated system the entropy does not decrease in time. This is the second law of thermodynamics and the increase of entropy is a measure of the irreversibility of the processes that drive isolated systems towards equilibrium.

Entropy in classical statistical mechanics   The goal of statistical mechanics is to connect the microscopic mechanical description of physical systems to thermodynamical behaviour. It deals with systems composed of very many elementary particles and, although such a system supposedly obeys the laws of classical mechanics, it is totally pointless to describe them in mechanical terms. This would indeed suppose a perfect knowledge of an enormous amount of initial conditions and the slightest error would immediately blow up, this is called molecular chaos. Boltzmann and Gibbs introduced the statistical approach, replacing a single system by an ensemble of systems whose elements satisfy a limited number of global constraints, such as fixed average particle density, fixed energy density, ... Let W be the volume in phase space occupied by the ensemble; then the entropy is given by

$$S = k \log W.$$

A more refined treatment connects the entropy to the velocity distribution $f(\mathbf{v})$ of a gas

$$S = -k \int f(\mathbf{v}) \log f(\mathbf{v}) \, d\mathbf{v}. \tag{9.1}$$

Information entropy   The information entropy is designed to measure the information contained in long messages that are generated by fixing the probabilities of the occurrences of the different letters that form the message. It is equivalent to measuring how many strings of letters satisfy global statistical constraints: if there are *a priori* many possible messages, then the knowledge of a particular one contains a large amount of information. Moreover, this measure should be invariant with respect to the different codings and encodings of the message.

Example 9.1 Suppose that we are given an alphabet $\{\epsilon_1, \epsilon_2, \ldots, \epsilon_d\}$ of d letters and that we are to write a long message of length n in this alphabet, with the only constraint that the letter ϵ_j should appear with relative frequency μ_j, $j = 1, 2, \ldots, d$. Obviously, $\mu = (\mu_1, \mu_2, \ldots, \mu_d)$ is a probability measure on the alphabet. The number $N(n, \mu)$ of possible messages is

$$N(n, \mu) = \frac{n!}{(n\mu_1)!(n\mu_2)! \cdots (n\mu_d)!} \sim \exp n \left(\sum_{j=1}^{d} -\mu_j \log \mu_j \right).$$

The quantity

$$S(\mu) := -\sum_{j=1}^{d} \mu_j \log \mu_j \tag{9.2}$$

is the *Shannon entropy* of the measure μ. The number of messages that satisfy our constraint grows exponentially with their length at an exponential rate determined by $S(\mu)$. This is precisely Boltzmann's point of view, where one determines the number of possible realizations of a large system given some global constraints

$$nS(\mu) = \log N(n, \mu).$$

S is a measure for the information contents of our demand that messages should asymptotically contain the letter ϵ_j with relative frequency μ_j. If the single letter ϵ_{j_0} appears with probability 1, then only one message can be written, the uncertainty is zero and indeed the entropy vanishes. On the contrary, if all letters are equidistributed, which corresponds to the measure $\mu_0 = (1/d, 1/d, \ldots, 1/d)$, then $S(\mu_0) = \log d$ and this is the maximal possible value for S.

Example 9.2 Let $\mu = (\mu_1, \mu_2, \ldots, \mu_d)$ be a probability measure on a discrete system of d points. Unless μ is concentrated at a single point, there is an uncertainty about the state of our system encoded in μ. The *entropy* aims at measuring this uncertainty by associating the number

$$S(\mu) := \sum_{j=1}^{d} -\mu_j \log \mu_j$$

with μ. Another useful point of view is to look at $S(\mu)$ as the information gained about the state of a large deterministic system by performing a measurement on the system with only d possible outcomes. This particular function $\mu \mapsto S(\mu)$ enjoys a number of natural properties that one would wish for a measure of uncertainty. Among them,

$S(\mu) \geq 0$ and $S(\mu) = 0$ iff μ is degenerated;

S reaches its maximal value $\log d$ on a system with d states;

S is additive on product systems: $S(\mu_1 \times \mu_2) = S(\mu_1) + S(\mu_2)$;

S depends continuously on μ.

In fact, the form of S can be derived on the basis of these axiomatic demands.

Entropy in quantum statistical mechanics   For quantum systems, von Neumann proposed

$$S(\rho) := \text{Tr}(-\rho \log \rho) = \sum_j -p_j \log p_j \qquad (9.3)$$

as a definition of entropy. Here, ρ is a density matrix with eigenvalues $\{p_1, p_2, \dots\}$ on the Hilbert space of the system. The expression (9.3) generalizes in a natural way both the expression (9.1) of the Boltzmann entropy and that of the Shannon entropy (9.2).

9.1   von Neumann entropy

Some basic properties of von Neumann's entropy are collected in this section. These properties are not only fundamental in statistical mechanics but they also play an essential role in studying dynamical systems. Some properties, such as the strong subadditivity, are quite hard to prove and we refer to the literature for details.

9.1.1   *Technical preliminaries*

A real function f defined on a closed convex subset C of $\mathbb{R}^d$ with non-empty interior is *convex* if it satisfies the inequality

$$f(\lambda \mathbf{x} + (1 - \lambda)\mathbf{y}) \leq \lambda f(\mathbf{x}) + (1 - \lambda)f(\mathbf{y}) \qquad (9.4)$$

whenever $0 \leq \lambda \leq 1$ and $\mathbf{x}, \mathbf{y} \in C$. The function f is *strictly convex* if the equality in (9.4) is only attained for trivial cases, i.e. $\mathbf{x} = \mathbf{y}$ or $\lambda = 0$ or 1. By iterating inequality (9.4), we see that for a convex function f

$$f\left(\sum_j \lambda_j \mathbf{x}_j\right) \leq \sum_j \lambda_j f(\mathbf{x}_j) \qquad (9.5)$$

whenever $0 \leq \lambda_j$, $\sum_j \lambda_j = 1$ and $\mathbf{x}_j \in C$. If μ is a probability measure and f is convex, then *Jensen's inequality* holds:

$$\int_C f(\mathbf{x})\, \mu(d\mathbf{x}) \leq f\left(\int_C \mathbf{x}\, \mu(d\mathbf{x})\right). \qquad (9.6)$$

This is a continuous version of (9.5). For the case that f is twice continuously differentiable in the interior of C and continuous on its boundary, convexity is equivalent to

$$0 \leq \left[\frac{\partial^2 f}{\partial x_i \partial x_j}\right]$$

in the interior of C. If f is once continuously differentiable, then we can also express convexity by saying that f remains above every tangent plane:

$$f(\mathbf{y}) \geq f(\mathbf{x}) + (\mathbf{y} - \mathbf{x}) \cdot \nabla f(\mathbf{x}). \tag{9.7}$$

A function f is *concave* if $-f$ is convex.

We can estimate matrix elements of a convex function of a self-adjoint operator with Jensen's inequality. Let φ be a normalized vector in a Hilbert space $\mathfrak{H}$, A a self-adjoint operator on $\mathfrak{H}$ with spectral measure $E^A(d\lambda)$, such that $\varphi \in \mathrm{Dom}(A)$ and f a convex function defined on a domain that contains the support of the measure $\|E^A(d\lambda)\varphi\|^2$; then

$$f(\langle \varphi, A\,\varphi \rangle) \leq \langle \varphi, f(A)\,\varphi \rangle.$$

It suffices to apply Jensen's inequality to the probability measure $\|E^A(d\lambda)\varphi\|^2$.

Example 9.3 The function

$$\eta : [0, 1] \to \mathbb{R} : x \mapsto \begin{cases} -x \log x & \text{for } x > 0, \\ 0 & \text{for } x = 0 \end{cases} \tag{9.8}$$

is strictly concave and vanishes only at $x = 0$ and $x = 1$ (Fig. 9.1).

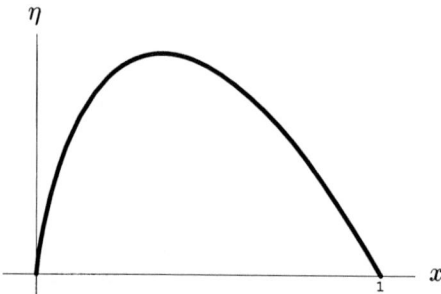

FIG. 9.1. The function $x \mapsto \eta(x)$.

The next lemma shows that $A \mapsto \mathrm{Tr}\, f(A)$ is convex whenever f is.

Lemma 9.4 *Let $f : [a, b] \to \mathbb{R}$ be a convex function, let A and B be Hermitian $d \times d$ matrices such that $[a, b]$ contains the eigenvalues of both A and B and let $0 \leq \lambda \leq 1$ then*

$$\mathrm{Tr}\, f(\lambda A + (1 - \lambda)B) \leq \lambda \,\mathrm{Tr}\, f(A) + (1 - \lambda)\,\mathrm{Tr}\, f(B).$$

Proof First remark that the spectrum of $\lambda A + (1 - \lambda)B$ is contained in the convex hull of the spectra of A and B, so that the statement of the lemma makes sense. Consider now an orthonormal basis $\{\varphi_i \mid i = 1, 2, \ldots, d\}$ of eigenvectors of $\lambda A + (1 - \lambda B)$ and write

$$\operatorname{Tr} f(\lambda A + (1 - \lambda)B) = \sum_{i=1}^{d} f(\langle \varphi_i, (\lambda A + (1 - \lambda)B)\, \varphi_i \rangle)$$

$$\leq \sum_{i=1}^{d} \lambda f(\langle \varphi_i, A\varphi_i \rangle) + (1 - \lambda) f(\langle \varphi_i, B\varphi_i \rangle)$$

$$\leq \sum_{i=1}^{d} \lambda \langle \varphi_i, f(A)\, \varphi_i \rangle + (1 - \lambda)\langle \varphi_i, f(B)\, \varphi_i \rangle$$

$$= \lambda \operatorname{Tr} f(A) + (1 - \lambda) \operatorname{Tr} f(B).$$

In the second estimate we have used Jensen's inequality (9.6). □

Lemma 9.5 (Klein) *Let $f :]a, b[\to \mathbb{R}$ be a once continuously differentiable convex function and let A and B be two Hermitian $d \times d$ matrices such that the eigenvalues of both A and B are contained in $]a, b[$; then*

$$\operatorname{Tr} f(B) \geq \operatorname{Tr}\big(f(A) + (B - A)\, f'(A)\big). \tag{9.9}$$

Moreover, if f is strictly convex on $]a, b[$, equality in (9.9) holds iff $A = B$.

Proof Consider the orthonormal bases $\{\varphi_i \mid i = 1, 2, \ldots, d\}$ and $\{\psi_i \mid i = 1, 2, \ldots, d\}$ of eigenvectors of A and B and their corresponding eigenvalues $\{\alpha_i \mid i = 1, 2, \ldots, d\}$ and $\{\beta_i \mid i = 1, 2, \ldots, d\}$ and use the one-dimensional version of (9.7) to obtain

$$\operatorname{Tr} f(B) = \sum_{i=1}^{d} \langle \varphi_i, f(B)\, \varphi_i \rangle$$

$$= \sum_{i=1}^{d} \sum_{j=1}^{d} f(\beta_j)\, |\langle \varphi_i, \psi_j \rangle|^2 \tag{9.10}$$

$$\geq \sum_{i=1}^{d} \sum_{j=1}^{d} \big(f(\alpha_i) + (\beta_j - \alpha_i)\, f'(\alpha_i)\big)\, |\langle \varphi_i, \psi_j \rangle|^2 \tag{9.11}$$

$$\geq \sum_{i=1}^{d} \sum_{j=1}^{d} \langle \varphi_i, \big(f(A) + (B - A)\, f'(A)\big)\, \psi_j \rangle \langle \psi_j, \varphi_i \rangle$$

$$= \operatorname{Tr}\big(f(A) + (B - A) f'(A)\big).$$

In order to prove that the equality implies $A = B$, we remark that because of the strict convexity of f we can only have equality between (9.10) and (9.11) if $(\beta_j - \alpha_i)\, |\langle \varphi_i, \psi_j \rangle|^2 = 0$ for every pair i, j, but then $A = B$. □

A real function f on $\mathbb{R}$ which satisfies the requirement that $f(A) \leq f(B)$ whenever A and B are bounded self-adjoint operators on a Hilbert space $\mathfrak{H}$ with $A \leq B$ is called *(increasing) operator monotone.*

Lemma 9.6 *Let* $0 < A \leq B$ *be bounded operators on a Hilbert space* $\mathfrak{H}$ *with bounded inverses* A^{-1} *and* B^{-1}*; then* $B^{-1} \leq A^{-1}$.

Proof That $A \leq B$ can be written as

$$\|A^{1/2}\psi\| \leq \|B^{1/2}\psi\|.$$

Next we use that

$$\|B^{-1/2}\varphi\|^2 = \sup_{\{\psi|\,\|B^{1/2}\psi\|\leq\|\varphi\|\}} |\langle\varphi,\psi\rangle|^2.$$

As

$$\{\psi \mid \|B^{1/2}\psi\| \leq \|\varphi\|\} \subset \{\psi \mid \|A^{1/2}\psi\| \leq \|\varphi\|\},$$

we conclude that

$$\|B^{-1/2}\varphi\|^2 \leq \|A^{-1/2}\varphi\|^2 \text{ or } B^{-1} \leq A^{-1}.$$

$\square$

9.1.2 Properties of von Neumann's entropy

The results in this section are formulated for matrices but they remain true in the infinite-dimensional case whenever their formulation makes sense.

Theorem 9.7 (a) $S(\rho) = S(U\rho U^*)$ *for any density matrix* ρ *on* $\mathbb{C}^d$ *and any unitary* U *on* $\mathbb{C}^d$.

(b) *For any density matrix* ρ *on* $\mathbb{C}^d$

$$0 \leq S(\rho) \leq \log d. \tag{9.12}$$

Proof The first statement is obvious.

In order to prove (9.12), let $\{p_1, p_2, \ldots, p_d\}$ be the eigenvalue list of ρ and assume that all $p_j > 0$; then, by the concavity of the logarithm on $\mathbb{R}^+$,

$$S(\rho) = \sum_{j=1}^{d} \eta(p_j) = \sum_{j=1}^{d} p_j \log \frac{1}{p_j} \leq \log\left(\sum_{j=1}^{d} 1\right) = \log d.$$

The general case, with some of the p_j possibly vanishing, follows by continuity.
$\square$

Theorem 9.8 *Let $\{\rho_1, \rho_2, \ldots, \rho_n\}$ be a collection of density matrices on $\mathbb{C}^d$ and let $\{\lambda_1, \lambda_2, \ldots, \lambda_n\}$ a probability measure on $\{1, 2, \ldots, n\}$; then*

$$\sum_{j=1}^{n} \lambda_j S(\rho_j) \leq S\left(\sum_{j=1}^{n} \lambda_j \rho_j\right), \tag{9.13}$$

$$S\left(\sum_{j=1}^{n} \lambda_j \rho_j\right) \leq \sum_{j=1}^{n} \lambda_j S(\rho_j) + \sum_{j=1}^{n} \eta(\lambda_j). \tag{9.14}$$

Moreover, the concavity in (9.13) is strict.

Proof The strict concavity of the entropy is an immediate consequence of Lemma 9.4. For a strictly positive operator A, i.e. an operator such that $A \geq \epsilon \mathbf{1}$, we may define $\log A$ through the integral

$$\log A = \int_1^\infty dt \left(t^{-1} \mathbf{1} - (t\mathbf{1} + A)^{-1}\right).$$

Because $x \mapsto (t + x)^{-1}$ is decreasing operator monotone (see Lemma 9.6) $\log$ is increasing operator monotone. We then have the following implications:

$$\lambda_j \rho_j \leq \sum_{k=1}^{n} \lambda_k \rho_k$$

$$\Downarrow$$

$$\log(\lambda_j \rho_j) \leq \log\left(\sum_{k=1}^{n} \lambda_k \rho_k\right)$$

$$\Downarrow$$

$$\lambda_j \rho_j^{1/2} \log(\lambda_j \rho_j) \rho_j^{1/2} \leq \rho_j^{1/2} \log\left(\sum_{k=1}^{n} \lambda_k \rho_k\right) \rho_j^{1/2}.$$

Summing over j and taking the trace yields the inequality (9.14). $\square$

A continuity property of entropy is given by the following

Theorem 9.9 *Let ρ and σ be two density matrices on $\mathbb{C}^d$ such that*

$$\|\rho - \sigma\|_1 = \mathrm{Tr}\,|\rho - \sigma| \leq \tfrac{1}{3};$$

then

$$|S(\rho) - S(\sigma)| \leq \|\rho - \sigma\|_1 \log d + \eta(\|\rho - \sigma\|_1).$$

The next results are concerned with relations between the entropy of a density matrix on a composed system with various combinations of entropies of subsystems. We denote by ρ_{12} a density matrix on $\mathbb{C}^{d_1} \otimes \mathbb{C}^{d_2}$ and by ρ_1 and ρ_2 the partial traces of ρ_{12} with respect to $\mathbb{C}^{d_2}$ and $\mathbb{C}^{d_1}$. In terms of the states defined by ρ_{12}, ρ_1 and ρ_2 this means that we consider the restrictions of a state of a composed system to its components.

Theorem 9.10 *Let ρ_1 and ρ_2 be two density matrices on $\mathbb{C}^{d_1}$ and $\mathbb{C}^{d_2}$; then*

$$S(\rho_1 \otimes \rho_2) = S(\rho_1) + S(\rho_2),$$
$$S(\lambda \rho_1 \oplus (1-\lambda)\rho_2) = \lambda S(\rho_1) + (1-\lambda)S(\rho_2) + \eta(\lambda) + \eta(1-\lambda), \quad 0 \le \lambda \le 1.$$

Proof This is an immediate consequence of the fact that the eigenvalues of $\rho_1 \otimes \rho_2$ are exactly the products of the eigenvalues of ρ_1 and ρ_2.

Similarly, the eigenvalues of $\lambda \rho_1 \oplus (1-\lambda)\rho_2$ are obtained by joining the eigenvalue list of $\lambda \rho_1$ with that of $(1-\lambda)\rho_2$. $\square$

Theorem 9.11 *The entropy is* subadditive, *i.e. with ρ_{12}, ρ_1 and ρ_2 as above*

$$S(\rho_{12}) \le S(\rho_1) + S(\rho_2). \tag{9.15}$$

Moreover, equality holds iff $\rho_{12} = \rho_1 \otimes \rho_2$, i.e. iff the two subsystems are independent.

Proof We use Klein's inequality (9.9) with the choice $f = -\eta$, $A = \rho_1 \otimes \rho_2$ and $B = \rho_{12}$. As above, we may first restrict attention to strictly positive ρ_{12} and then pass to the general case by continuity. It is then sufficient to observe that $\log \rho_1 \otimes \rho_2 = \log \rho_1 \otimes \mathbb{I}_2 + \mathbb{I}_1 \otimes \log \rho_2$. $\square$

Theorem 9.12 (Lieb–Ruskai)   *The entropy is* strongly subadditive, *that is, using a notation similar to the one above,*

$$S(\rho_{123}) + S(\rho_2) \le S(\rho_{12}) + S(\rho_{23}). \tag{9.16}$$

The next corollary is a consequence of the preceding theorem.

Corollary 9.13 *The entropy satisfies the* triangle inequality

$$|S(\rho_1) - S(\rho_2)| \le S(\rho_{12}). \tag{9.17}$$

9.1.3   Mean entropy

Given a shift-invariant state ω on a spin chain as in Example 4.4 which is specified by a consistent set of local density matrices $\{\rho_\Lambda\}$, an entropy $S(\rho_\Lambda)$ can be computed for each finite subset Λ of $\mathbb{Z}$. Due to shift-invariance, this entropy depends only on the shape of Λ and not on its location in $\mathbb{Z}$. It turns out that, for reasonable subsets Λ, $S(\rho_\Lambda)$ typically grows linearly with the size of Λ. The strong subadditivity property of the entropy ensures the existence of the *mean entropy* $\sigma(\omega)$ *of* ω:

$$\sigma(\omega) := \lim_{\Lambda \to \mathbb{Z}} \frac{1}{|\Lambda|} S(\rho_\Lambda). \tag{9.18}$$

In (9.18), the sets Λ tend to $\mathbb{Z}$ in the *sense of van Hove*. A sequence Λ_j of finite subsets tends to $\mathbb{Z}$ in the sense of van Hove if the number of points in Λ_j tends to infinity and if for any $n \in \mathbb{N}$ the ratio $N(+,j)/N(-,j)$ approaches 1 for $j \to \infty$. $N(+,j)$ is the minimal number of disjoint intervals of length n needed to cover Λ_j and $N(-,j)$ is the maximal number of disjoint intervals of length n contained in Λ_j. $|\Lambda|$ is the number of sites in Λ. The limit in (9.18) is actually obtained as an infimum over intervals

$$\sigma(\omega) = \inf_n \frac{1}{n} S(\rho_{[1,2,\dots,n]}) \tag{9.19}$$

The next result slightly strengthens (9.19):

Theorem 9.14 *Let ω be a shift-invariant state on a quantum spin chain $\mathfrak{A}^{\mathbb{Z}}$. Denote by S_n the entropy of the segment $\{1,2,\dots,n\}$ and by σ the mean entropy of ω; then*

$$S_{n+1} - S_n \le S_n - S_{n-1}, \quad n = 2,3,\dots, \tag{9.20}$$

$$\frac{S_{n+1}}{n+1} \le \frac{S_n}{n}, \quad n = 1,2,\dots, \tag{9.21}$$

$$\sigma = \lim_{n\to\infty} \frac{S_n}{n} = \lim_{n\to\infty} \left(S_{n+1} - S_n \right). \tag{9.22}$$

Proof Let ρ_Λ denote the reduced density matrix of ω corresponding to $\Lambda \subset \mathbb{Z}$, finite. We first apply the strong subadditivity inequality to the sets $\{1,2,\dots,n\}$ and $\{2,3,\dots,n+1\}$ and write by translation invariance that

$$S_n = S(\rho_{\{k,k+1,\dots,k+n\}}), \quad k \in \mathbb{Z}.$$

This leads to

$$S_{n+1} + S_{n-1} \le 2S_n, \tag{9.23}$$

proving (9.20).

The result (9.21) can be shown by induction. Suppose indeed that $kS_{k+1} \le (k+1)S_k$ for $k < n$. We use (9.23) and write

$$nS_{n+1} \le 2nS_n - nS_{n-1} = (n+1)S_n + (n-1)S_n - nS_{n-1}$$
$$\le (n+1)S_n$$

by the induction hypothesis. We initiate the induction by using subadditivity on an interval of length 2: $S_2 \le 2S_1$.

It remains to show that (9.22) holds. Remark that, by the triangle inequality, $S_{n+1} \ge S_n - S_1$. Therefore, $n \mapsto S_{n+1} - S_n$ is a monotonically decreasing sequence that is bounded from below, which allows us to define

$$\alpha := \lim_{n\to\infty} \left(S_{n+1} - S_n \right).$$

Fix now $\epsilon > 0$ and $n_0 \in \mathbb{N}$ such that $|S_{n+1} - S_n - \alpha| \leq \epsilon$ whenever $n \geq n_0$. We then write

$$|S_n - S_{n_0} - (n - n_0)\alpha|$$
$$\leq |S_n - S_{n-1} - \alpha| + |S_{n-1} - S_{n-2} - \alpha| + \cdots + |S_{n_0+1} - S_{n_0} - \alpha|$$
$$\leq (n - n_0)\epsilon.$$

This shows that $\sigma := \lim_{n \to \infty} S_n/n$ exists and is equal to α, i.e.

$$\sigma = \lim_{n \to \infty} \frac{S_n}{n} = \lim_{n \to \infty} \left(S_{n+1} - S_n \right).$$

$\square$

9.1.4 *Entropy of quasi-free states*

We compute here the mean entropy of a quasi-free state for a system of fermions living on a chain. The algebra is the CAR-algebra $\mathfrak{A}(\ell^2(\mathbb{Z}))$ and the state is the gauge- and shift-invariant quasi-free state ω_Q determined by the symbol Q (see Section 6.2.3)

$$Q \in \mathcal{B}(\ell^2(\mathbb{Z})), \quad 0 \leq Q \leq \mathbf{1}. \tag{9.24}$$

Shifting the fermions along the chain is described by the quasi-free automorphism

$$\gamma(c^*(\varphi)) := c^*(T\,\varphi),$$

with

$$T\,e_i := e_{i+1}, \quad i \in \mathbb{Z}$$

on the canonical basis of $\ell^2(\mathbb{Z})$. The necessary and sufficient condition for ω_Q to be shift-invariant is

$$Q = T Q T^* \tag{9.25}$$

Conditions (9.24) and (9.25) are equivalent to

$$(Q\,\varphi)^\wedge(\theta) = q(\theta)\,\varphi^\wedge(\theta),$$

where q is a measurable real function on the unit circle $\mathbb{T}$ such that

$$0 \leq q(\theta) \leq 1, \quad \theta \in \mathbb{T}$$

and

$$\varphi^\wedge(\theta) := \sum_{j \in \mathbb{Z}} \varphi(j)\,e^{ij\theta}.$$

The matrix elements q_{jk} of Q in the canonical basis of $\ell^2(\mathbb{Z})$ are given by

$$q_{jk} := \langle e_j, Q\,e_k \rangle = \frac{1}{2\pi} \int_{\mathbb{T}} d\theta\, e^{-i(j-k)\theta}\, q(\theta) = q^\wedge(j - k).$$

They are constant along parallels to the main diagonal. An operator A on $\mathcal{B}(\ell^2(\mathbb{N}))$ which is given by a matrix $[a_{jk}]$ with $a_{j+\ell\,k+\ell} = a_{jk}$ for $j, k, \ell \in \mathbb{N}$ is a *Toeplitz operator*.

Let P_n be the projector from $\ell^2(\mathbb{Z})$ onto the finite-dimensional subspace spanned by $\{e_1, e_2, \ldots, e_n\}$. The restriction of ω_Q to the subalgebra $\mathfrak{A}(P_n\ell^2(\mathbb{Z}))$ is again quasi-free with symbol P_nQP_n. Due to the shift-invariance of ω_Q, the matrices P_nQP_n are the upper left $n \times n$ corners of the Toeplitz operator

$$Q^+ := [q_{jk}]_{j,k \geq 0} = \frac{1}{2\pi} \int_{\mathbb{T}} d\theta \, e^{-i(j-k)\theta} \, q(\theta).$$

These states are given by density matrices ρ_n on the Fock space $\Gamma_{\text{a.s.}}(\ell^2(\mathbb{Z}))$ of finite-rank. We show that

$$\lim_{n \to \infty} \frac{1}{n} S(\rho_n) = \frac{1}{2\pi} \int_{\mathbb{T}} d\theta \, \Big(\eta(q(\theta)) + \eta(1 - q(\theta)) \Big), \tag{9.26}$$

η being the standard entropy function (9.8).

Lemma 9.15 *Let ω_Q be the gauge-invariant quasi-free state of $\mathfrak{A}(\mathbb{C}^n)$ determined by the symbol Q with spectrum $\{q_1, q_2, \ldots, q_n\}$ and corresponding eigenvectors $\{|1\rangle, |2\rangle, \ldots, |n\rangle\}$. The state ω_Q is given by a density matrix ρ_Q on Fock space $\Gamma_{\text{a.s.}}(\mathbb{C}^n)$:*

$$\omega_Q(x) = \operatorname{Tr} \rho_Q \, \pi_F(x),$$

which factorizes into

$$\rho_Q = \bigotimes_{j=1}^{n} \begin{pmatrix} q_j & 0 \\ 0 & 1 - q_j \end{pmatrix} \tag{9.27}$$

with respect to the decomposition

$$\Gamma_{\text{a.s.}}(\mathbb{C}^n) = \bigotimes_{j=1}^{n} \Gamma_{\text{a.s.}}(\mathbb{C}|j\rangle) \equiv \bigotimes^{n} \mathbb{C}^2 \equiv \mathbb{C}^{2^n}.$$

Its spectrum is given by

$$\operatorname{Sp}(\rho_Q) = \{r_\Lambda \mid \Lambda \subset \{1, 2, \ldots, n\}\}$$

with

$$r_\Lambda := \prod_{j \in \Lambda} q_j \prod_{\ell \in \{1,2,\ldots,n\} \backslash \Lambda} (1 - q_\ell).$$

In particular,

$$S(\rho_Q) = \operatorname{Tr}\Big(-Q \log Q - (\mathbf{1} - Q) \log(\mathbf{1} - Q)\Big).$$

The structure of the density matrix ρ_Q of a quasi-free state is precisely that which was found in Example 6.5. It is also obvious that the lemma can be extended to include the case of infinite-dimensional one-particle spaces. In order to do so, the infinite tensor product (9.27) has to be meaningful. A detailed analysis shows that this is the case iff

$$\|Q\|_1 := \operatorname{Tr} Q = \sum_j q_j < \infty. \tag{9.28}$$

Quasi-free states which satisfy (9.28) are *normal with respect to the Fock representation*. In order to compute the mean entropy of a shift-invariant state, we use a general result by Szegö on the spectrum of Toeplitz operators.

Let A be a Hermitian Toeplitz operator on $\ell^2(\mathbb{N})$ corresponding to the symbol a, i.e. a is a real measurable function on $\mathbb{T}$ which we suppose to be bounded, $\|a\|_\infty < \infty$. Denote by $A(n)$ the $n \times n$ matrix $[a_{jk}]$ with

$$a_{jk} = \frac{1}{2\pi} \int_{\mathbb{T}} d\theta \, e^{i(k-j)\theta} \, a(\theta)$$

and let its eigenvalues be given by $\{\lambda_1(n), \lambda_2(n), \dots, \lambda_n(n)\}$. The *empirical eigenvalue distribution* of $A(n)$ is the probability measure

$$\mu_n(dx) := \frac{1}{n} \sum_{j=1}^n \delta(x - \lambda_j(n)) \, dx.$$

Theorem 9.16 (Szegö) *Let A be a Toeplitz operator as above; then the empirical eigenvalue distributions μ_n converges $*$-weakly to μ when $n \to \infty$. The probability measure μ is determined by*

$$\mu(\Delta) = \frac{1}{2\pi} \int_{a^{-1}(\Delta)} d\theta$$

for any Borel subset Δ of $\mathbb{R}$. Thus, for any continuous complex function f defined on the essential support of a, one has

$$\lim_{n \to \infty} \frac{1}{n} \sum_{j=1}^n f(\lambda_j(n)) = \frac{1}{2\pi} \int_0^{2\pi} d\theta \, f(a(\theta)). \tag{9.29}$$

One can obtain a proof of Theorem 9.16 by computing the moments of the measure μ, i.e. by plugging in for f in (9.29) the functions $x \mapsto x^d$. For such a choice, the left-hand side of (9.29) is the limit of the normalized trace of the dth power of $A(n)$ which can explicitly be evaluated. Formula (9.26) for the mean entropy of a shift-invariant quasi-free state is obtained by choosing in Theorem 9.16

$$f(x) := \eta(x) + \eta(1 - x).$$

9.2 Relative entropy

9.2.1 Finite-dimensional case

For two $d \times d$ density matrices ρ and σ, the *relative entropy* $S(\rho, \sigma)$ is the quantity

$$S(\rho, \sigma) := \begin{cases} \text{Tr}\left(\rho \log \rho - \rho \log \sigma\right) & \text{if } \text{Ker}(\sigma) \subset \text{Ker}(\rho), \\ +\infty & \text{otherwise.} \end{cases} \qquad (9.30)$$

For a canonical Gibbs state $\sigma = \exp{-\beta H}/\mathcal{Z}$ at inverse temperature β, we find that

$$S(\rho, \sigma) = \text{Tr}\, \rho\, \beta H - S(\rho) + \log \mathcal{Z} = \beta F(\rho) - \beta F(\sigma).$$

In these expressions, $S(\rho)$ is the entropy of the state ρ and $F(\rho)$ is its *non-equilibrium free energy*

$$F(\rho) := \text{Tr}\, \rho\, H - \frac{1}{\beta} S(\rho).$$

Therefore, the relative entropy $S(\rho, \sigma)$ measures the increase in free energy in passing from the equilibrium state σ to ρ. We group in the next theorem some of its basic properties

Theorem 9.17 *The relative entropy of finite-dimensional density matrices enjoys the following properties:*

(a) *Positivity:* $S(\rho, \sigma) \geq 0$ *and* $(\rho, \sigma) = 0$ *iff* $\rho = \sigma$.

(b) *Additivity:* $S(\rho_1 \otimes \rho_2, \sigma_1 \otimes \sigma_2) = S(\rho_1, \sigma_1) + S(\rho_2, \sigma_2)$.

(c) *Joint convexity: For* $0 \leq \lambda \leq 1$

$$S(\lambda \rho_1 + (1 - \lambda)\rho_2, \lambda \sigma_1 + (1 - \lambda)\sigma_2) \leq \lambda S(\rho_1, \sigma_1) + (1 - \lambda) S(\rho_2, \sigma_2).$$

(d) *Behaviour with respect to completely positive maps: For any unity-preserving completely positive* Γ *from* $\mathcal{M}_d$ *to* $\mathcal{M}_k$

$$S(\Gamma^{\mathsf{T}}(\rho), \Gamma^{\mathsf{T}}(\sigma)) \leq S(\rho, \sigma). \qquad (9.31)$$

(e) *Lower bound:* $S(\rho, \sigma) \geq \frac{1}{2} \|\rho - \sigma\|_1^2$.

Proof (a) follows from Klein's inequality (9.5) choosing $f(x) = x \log x$, $A = \sigma$ and $B = \rho$.

(b) is the consequence of a simple computation.

The proofs of the other properties are based on strong subadditivity. $\square$

Example 9.18 A completely positive unity-preserving transformation Γ of $\mathcal{M}_d$ is *bistochastic* if it preserves the trace on $\mathcal{M}_d$:

$$\operatorname{Tr}\Gamma(X) = \operatorname{Tr}X, \quad X \in \mathcal{M}_d.$$

Writing out (9.31)(b) with ρ the tracial state on $\mathcal{M}_d$, we obtain

$$S(\Gamma(\sigma)) \geq S(\sigma).$$

Bistochastic completely positive maps are entropy increasing.

The properties of the relative entropy in Theorem 9.17 extend to the infinite-dimensional case whenever their statement makes sense.

9.2.2   Maximum entropy principle

Suppose that we are given a number of constraints on the state of a finite system specifying the expectation values of a number of observables. We may then ask for the least biased state within the convex set of states $\mathcal{C}$ that satisfy our constraints. The *maximum entropy principle* specifies that we should look in $\mathcal{C}$ for the state with maximal entropy. Due to the strict concavity of the entropy, the maximal entropy principle selects a unique state in $\mathcal{C}$. Often, as in the case of a spin chain, we have a local structure at hand and we are rather interested in shift-invariant states with a given energy density ... In such a case, we should rather look for states which maximize the mean entropy. This problem is mostly intractable unless we specify some further properties of the states that we consider such as certain decoupling rules for the correlation functions of the state.

Example 9.19 *(Gibbs states of finite systems)*   On $\mathcal{B}(\mathfrak{H})$, the Gibbs state

$$\rho_\beta := \frac{e^{-\beta H}}{\operatorname{Tr}\left(e^{-\beta H}\right)}$$

is the state ρ which has the largest possible von Neumann entropy given that its average energy $\operatorname{Tr}\rho H$ is fixed. The self-adjoint operator H is the Hamiltonian of the system. This statement follows directly from the positivity of the relative entropy:

$$0 \leq S(\rho, \rho_\beta) = -S(\rho) - \operatorname{Tr}\rho\log\rho_\beta = -S(\rho) - \operatorname{Tr}\rho_\beta\log\rho_\beta.$$

Example 9.20 *(Hartree–Fock approximation, with gauge and shift invariance)*
The *variational principle in statistical mechanics* characterizes thermal equilibrium states as these translation-invariant states which minimize the non-equilibrium mean free energy. Returning to the standard situation of Example 6.4, we restrict our attention in the computation of the mean equilibrium free energy of an interacting fermion system to the closed convex set $\mathcal{C}$ generated by the gauge- and shift-invariant quasi-free states. This is the *Hartree–Fock*

approximation and it yields by construction an upper bound for the equilibrium free energy density. Because of the affinity of the mean non-equilibrium free energy and particle density, we may restrict our attention to the shift- and gauge-invariant quasi-free states and forget about their convex mixtures. Let Q be a symbol of such a state, i.e.

$$(Q\,\varphi)^\wedge(\mathbf{k}) = q(\mathbf{k})\,\varphi^\wedge(\mathbf{k}), \quad \varphi \in \mathcal{L}^2(\mathbb{R}^3) \tag{9.32}$$

with

$$0 \le q(\mathbf{k}) \le 1 \quad \text{and} \quad \varphi^\wedge(\mathbf{k}) := \int_{\mathbb{R}^3} d\mathbf{x}\, \varphi(\mathbf{x})\, e^{-2\pi i \mathbf{k}\cdot\mathbf{x}}.$$

The restricted variational principle consists in computing the infimum of

$$f_q := e_q - \frac{1}{\beta}\sigma_q \tag{9.33}$$

over the set of symbols (9.32) subject to the condition that the average particle density n_q is fixed. The mean energy e_q is the sum of a kinetic part t_q and a potential part v_q, and σ_q is the entropy density. Explicit computation of all these quantities leads to

$$t_q = \frac{2\pi^2\hbar^2}{m}\int_{\mathbb{R}^3} d\mathbf{k}\,\|\mathbf{k}\|^2\, q(\mathbf{k}),$$

$$v_q = \frac{1}{2}\int_{\mathbb{R}^3} d\mathbf{x}\, V(\mathbf{x})\left(q^\wedge(\mathbf{0})^2 - q^\wedge(\mathbf{x})\, q^\wedge(-\mathbf{x})\right),$$

$$n_q = q^\wedge(\mathbf{0}),$$

$$\sigma_q = \int_{\mathbb{R}^3} d\mathbf{k}\,\left(\eta(q(\mathbf{k})) + \eta(1 - q(\mathbf{k}))\right).$$

A minimizer q_0 of (9.33) with the constraints that the particle density equals ρ, satisfies the non-linear Hartree–Fock equation

$$q_0 = \frac{1}{1 + e^{\beta(H_{\mathrm{eff}} - \zeta)}} \quad \text{with} \quad (q_0)^\wedge(\mathbf{0}) = \rho$$

and

$$H_{\mathrm{eff}}(\mathbf{k}) := \frac{2\pi^2\hbar^2}{m}\|\mathbf{k}\|^2$$
$$+ \frac{1}{2}\int_{\mathbb{R}^3} d\mathbf{x}\, V(\mathbf{x})\left(2q^\wedge(\mathbf{0}) - e^{-2\pi i\mathbf{k}\cdot\mathbf{x}}\, q^\wedge(-\mathbf{x}) - e^{2\pi i\mathbf{k}\cdot\mathbf{x}}\, q^\wedge(\mathbf{x})\right).$$

Example 9.21 *(Quantum particle on curved space)* Suppose that Δ is a self-adjoint generalized Laplacian on a compact Riemannian manifold $\mathfrak{X}$ of dimension *d.* Consider it as minus the Hamiltonian of a free particle; then, due to Theorem 3.1,

$$\rho_\beta := \mathcal{Z}^{-1}(\beta) \exp[-\beta\Delta] \quad \text{with} \quad \mathcal{Z}(\beta) := \text{Tr} \, e^{-\beta\Delta}$$

is a well-defined density matrix satisfying the maximum entropy principle. The asymptotic formulas for the entropy and the mean energy in the limit of large temperatures or large volumes can be derived from Weyl's formula (see Theorem 3.9):

$$S(\rho_\beta) \sim -\frac{d}{2} \log\beta + \log \text{Vol}(\mathfrak{X}) \quad \text{and} \quad \text{Tr}\,\rho_\beta\Delta \sim \frac{d}{2\beta}.$$

9.2.3   Algebraic setting

The construction of the relative entropy can also be performed in the general framework of algebraic statistical mechanics. It still carries the meaning of increase in total free energy due to passing from a state σ to a state ρ.

Consider a state σ on a C*-algebra $\mathfrak{A}$, with GNS-representation $(\mathfrak{H}, \pi, \Omega)$. We still use the notation σ for the state $X \mapsto \langle \Omega, X\,\Omega \rangle$ on the von Neumann algebra $\mathfrak{R} := \pi(\mathfrak{A})''$ generated by $\pi(\mathfrak{A})$. In general, Ω is not separating for $\mathfrak{R}$ and therefore the projector P on $\{\mathfrak{R}'\Omega\}$ differs from $\mathbf{1}$. The mapping

$$S : X\,\Omega \mapsto PX^*\Omega, \quad X \in \mathfrak{R}$$

extends to a closable, antilinear operator, still denoted by S, with polar decomposition $S = J\Delta^{1/2}$. The *natural positive cone* $\mathcal{P}$ is given by

$$\mathcal{P} := \overline{\{XJXJ\Omega \mid X \in \mathfrak{R}\}}.$$

Any normal state ρ on $\mathfrak{R}$ with support contained in the support of σ can uniquely be written as a vector state

$$\rho(X) = \langle \Phi, X\,\Phi \rangle, \quad X \in \mathfrak{R}$$

with $\Phi \in \mathcal{P}$. The *relative modular operator* $\Delta_{\Phi,\Omega}$ is then $S^*_{\Phi,\Omega}S_{\Phi,\Omega}$, where $S_{\Phi,\Omega}$ is the closure of the antilinear mapping

$$X\,\Omega \mapsto PX^*\Phi, \quad X \in \mathfrak{R}.$$

We can therefore write that

$$\rho(X^*X) = \langle X^*\Phi, X^*\Phi \rangle = \langle X^*\Phi, PX^*\Phi \rangle$$
$$= \langle S_{\Phi,\Omega}X\,\Omega, S_{\Phi,\Omega}X\,\Omega \rangle = \langle X\,\Omega, \Delta_{\Phi,\Omega}X\,\Omega \rangle.$$

Definition 9.22 *With the notations of above, the* relative entropy $S(\rho,\sigma)$ *of two states* ρ *and* σ *on a C*-algebra* $\mathfrak{A}$ *is given by*

$$S(\rho,\sigma) := \begin{cases} \langle \Phi, \log\Delta_{\Phi,\Omega}\Phi \rangle & \textit{if } \rho \textit{ and } \sigma \textit{ are as above and} \\ & \quad\quad \Phi \in \text{Dom}(|\log\Delta_{\Phi,\Omega}|), \\ +\infty & \textit{otherwise.} \end{cases}$$

For the case of density matrices, this construction returns (9.30). Let σ be a density matrix on $\mathfrak{H}$ supported by the closed subspace $\mathfrak{K}$. The GNS-space $\mathfrak{H}$ is then naturally identified with $\mathfrak{H} \otimes \mathfrak{K}$, $\pi(x) = x \otimes \mathbf{1}$ and $\Omega = \sum_j s_j^{1/2} \varphi_j \otimes \varphi_j$, where the s_j are the eigenvalues of σ arranged in decreasing order and repeated according to their multiplicity, and where $\{\varphi_j \mid j\}$ is an orthonormal basis of eigenvectors of σ with $\sigma \varphi_j = s_j \varphi_j$. The operator P is the projector on $\mathfrak{K} \otimes \mathfrak{K}$ and the positive cone $\mathcal{P}$ consists of all vectors of the type $\sum_{k,\ell} X_{k\ell} \varphi_k \otimes \varphi_\ell$, where $X = [X_{k\ell}]$ is a non-negative Hilbert–Schmidt operator on $\mathfrak{K}$. If ρ is another density matrix on $\mathfrak{H}$ with support in $\mathfrak{K}$, then $\Delta_{\rho,\sigma} = \rho \otimes \sigma^{-1}$ and the relative entropy $S(\rho, \sigma)$ in Definition 9.22 reduces to the usual formula (9.30) for density matrices.

9.3 Notes

von Neumann (1932) proposed the expression (9.3) for the entropy of a density matrix. The basic properties of entropy, which are summarized in Theorems 9.7, 9.8 and 9.11, can be found in the review article (Wehrl 1978). The continuity property of Theorem 9.9 is in (Fannes 1973b) and the proof of the strong subadditivity, Theorem 9.12, in (Lieb and Ruskai 1973). Theorem 9.14 is a well-known folklore result. The expression (9.26) for the mean entropy of a quasi-free fermion state was proven in a slightly different way in (Fannes 1973a). For a proof of Szegö's theorem see the book by Grenander and Szegö (1958).

The form (9.30) of the quantum relative entropy is due to (Umegaki 1962) and its behaviour with respect to dynamical maps was studied in (Lindblad 1975). The lower bound in Theorem 9.17 can be found in (Hiai et al. 1983).

The maximum entropy principle was formulated in (Jaynes 1957) and independently by Ingarden and Urbanik; see Ingarden et al. (1997) for a modern exposition with applications. The development of the relative entropy in an algebraic setting is the subject of (Araki 1976, 1977). A modern comprehensive presentation of quantum entropies and their applications can be found in the monograph of Ohya and Petz (1993).

10

DYNAMICAL ENTROPY

Dynamical instability is believed to be omnipresent in many-particle systems and to be the basic mechanism that is responsible for irreversible behaviour, either driving systems towards equilibrium or causing transport phenomena when they are subject to external gradients. This dynamical instability of large systems seems to be generic as opposed to integrability.

Classical deterministic systems typically exhibit an exponential divergence of phase space trajectories corresponding to slightly modified initial conditions. This leads to a very complicated phase portrait which could be modelled by a succession of stretching and folding operations. A long-time analytical treatment of the motion is infeasible and statistical methods become relevant. As an overall characteristic of the motion, one could try to measure at which rate repeated observations of a system reveal information about its initial state or, equivalently, at which rate the system wipes out the details of its initial condition. Dynamical entropy is a tool that is designed to obtain this rate for chaotic systems with few degrees of freedom. For systems with many degrees of freedom, one should rather look for a mean dynamical entropy. More detailed descriptions such as Lyapunov exponents have been developed which measure the exponential instability in a given direction. Again, for complex systems one expects rather a continuum of exponents.

Figures 10.1 and 10.2 show an entropy computation for two quite different classical dynamical systems: a rotation of the circle by an angle which is an irrational multiple of 2π and the angle-doubling map. The Lebesgue measure is invariant under both dynamics, it is mixing for the second but only ergodic for the first. The figures show the temporal behaviour of the entropy $S(n)$ produced by a repeated coarse-grained observation of the dynamics up to time n. This coarse-grained picture is obtained by dividing the circle in three more or less equal arcs. It clearly appears from the figures that the entropy growth is sublinear, in fact logarithmic, for the irrational rotation (Fig. 10.1), while it becomes linear for the doubling map (Fig. 10.2). The dynamical entropy, which is essentially the growth rate of the entropy, can be shown to be exactly zero for the first case and $\log 2$ for the second and it coincides with the sum of the positive Lyapunov exponents for sufficiently smooth systems.

For quantum systems, the situation is different. Due to the uncertainty relations, the quantization of a classical dynamical system with a compact phase space typically belongs to a finite-dimensional Hilbert space and the evolution is almost periodic at the Heisenberg time scale. Nevertheless, in a coarse-grained

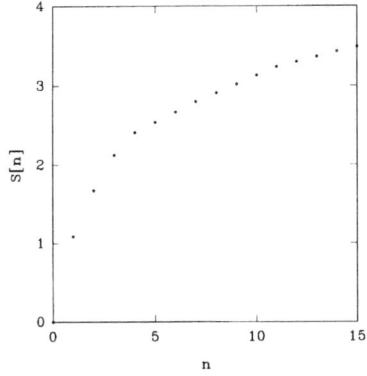

FIG. 10.1. Entropy of the irrational rotation.

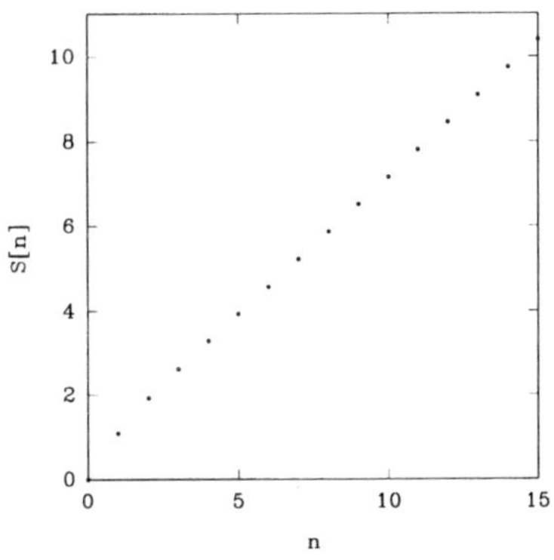

FIG. 10.2. Entropy of the angle-doubling map.

picture and for very short times one can see the onset of chaotic behaviour of the
corresponding classical limit. This is signalled by a linear growth of the entropy
$S(n)$. Figure 10.3 shows the behaviour of the entropy of the quantum kicked
top in a chaotic regime for a system in a space of dimension 41. The Lyapunov
exponent of the corresponding classical system is about 1. One can observe a
linear growth of $S(n)$ up to times of the order $\log 41 \approx 4$ and saturation becomes
apparent at times comparable to the Heisenberg time $n_H = 41$. One should also
observe that $S(n)$ does not tend to infinity with n.

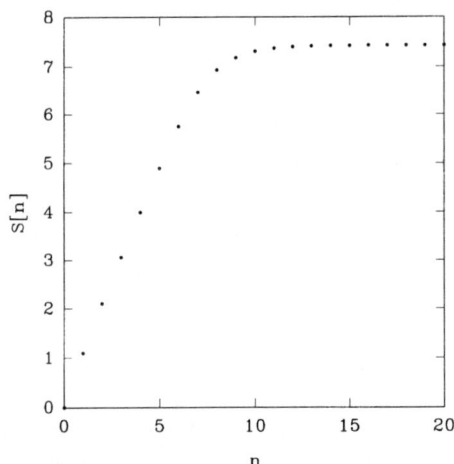

FIG. 10.3. Entropy of quantum kicked
top.

The leading dynamical behaviour of a quantum dynamical system should be
extracted by considering a coupled limit between the time and the quantization

parameter that measures the non-commutativity of the system. The discrete nature of the system makes it difficult to introduce good notions of Lyapunov exponents.

Infinite-particle systems, such as those studied in statistical mechanics, have a continuous dynamical spectrum, but hardly any result has been obtained on their dynamical behaviour. A number of tractable model systems have been developed to exhibit dynamical relaxation. For such models, the initial linear growth in information can persist forever.

There are several possible approaches to measuring the information revealed by a system. Repeated observations of the system is a common ingredient to most of them. One could use a commuting set of orthogonal projections as basic measuring device, that is to say a single self-adjoint observable or a sharp measurement. This works for a single measurement, but composing two or more such sharp measurements does not lead to a sharp joint measurement. This is generally the case for repeated observations at different times.

Another possibility is to use completely positive maps from a classical system to the dynamical system. This approach was finalized in the work by Connes, Narnhofer and Thirring and goes under the name of CNT-approach. Many results have been obtained in this framework. We shall, however, not follow this line here and refer for further details to the literature.

In this chapter, we present a notion of entropy based on the use of coarse-grained observations obtained in terms of finite quantum probability spaces. This leads to a shift model on a quantum spin half-chain and we measure the asymptotic growth of information by the von Neumann mean entropy. A number of models are considered in more detail.

10.1   Operational partitions

We briefly recall here the unsharp measurement process presented in Section 8.1 and argue that such measurements implicitly use the notion of operational partition. An unsharp measurement of a quantum system belonging to a Hilbert space $\mathfrak{H}$ with possible outcomes labelled by $j = 1, 2, \ldots, k$ is determined by specifying k operators $W_1, W_2, \ldots, W_k$ on $\mathfrak{H}$ such that $\sum_j W_j^* W_j = \mathbf{1}$. If a state of a quantum system is described by a density matrix ρ, then the probability p_j of observing the outcome j is given by

$$p_j = \operatorname{Tr} \rho\, W_j^* W_j = \operatorname{Tr} W_j\, \rho\, W_j^*.$$

As the measurements of the different j's need not be orthogonal in the sense that $W_i^* W_j \neq 0$ for $i \neq j$, there is more information contained in the measurement, namely, the correlations

$$\operatorname{Tr} \rho\, W_i^* W_j, \quad i, j = 1, 2, \ldots, k.$$

A useful way to encode this information is to consider the matrix $[\operatorname{Tr} \rho\, W_i^* W_j]$ which is itself a $k \times k$ density matrix. This suggests a description of the unsharp measurement rather in terms of a state on the $k \times k$ matrices than a probability measure on the set $\{1, 2, \ldots, k\}$. Such extended measurements produce in fact a coarse-grained probabilistic picture of the observed quantum system in terms of $k \times k$ matrices which are the basic discrete quantum probability spaces.

In general, coarse-graining aims at providing a reduced description of a non-commutative probability space $(\mathfrak{A}, \omega)$ in terms of the basic discrete model of $k \times k$ matrices $\mathcal{M}_k$ equipped with a density matrix ρ. It amounts to lumping the states on $\mathfrak{A}$ together, preserving the probabilistic interpretation. Moreover, the procedure should be robust for trivial compositions with non-interacting systems, i.e. we do not only require simple positivity but rather complete positivity of a coarse-graining map.

Coarse-graining is a completely positive and unity-preserving map Γ from the $k \times k$ matrices $\mathcal{M}_k$ to $\mathfrak{A}$. This map pulls back the states ω on $\mathfrak{A}$ in the $k \times k$ density matrix $\rho[\Gamma]$; more explicitly

$$\operatorname{Tr} \rho[\Gamma]\, A := \omega\big(\Gamma(A)\big),$$

or, in terms of matrix elements,

$$\rho[\Gamma]_{ij} := \omega\big(\Gamma(e_{ji})\big),$$

where $\{e_{ij} \mid i, j = 1, 2, \ldots, k\}$ is the canonical system of matrix units in $\mathcal{M}_k$. The general structure of a completely positive map from $\mathcal{M}_k$ into $\mathfrak{A}$ was found in Corollary 8.9. We consider here extreme maps Γ, i.e. maps that admit no non-trivial decompositions in sums of, possibly non-unital, completely positive maps. Such maps can be expressed in terms of *operational partitions of unity*. A k-tuple $\mathbf{X} = (x_1, x_2, \ldots, x_k)$ of elements of $\mathfrak{A}$ is an *operational partition of unity*, in short a *partition*, if

$$\sum_{i=1}^{k} x_i^* x_i = \mathbf{1}.$$

We call k the *size of* $\mathbf{X}$ and denote it by $|\mathbf{X}|$. The corresponding *coarse-grained map* reads

$$\Gamma_{\mathbf{X}}([a_{ij}]) = \sum_{ij} a_{ij}\, x_i^* x_j$$

and the density matrix built with the partition $\mathbf{X}$ is given by

$$\rho[\mathbf{X}]_{ij} \equiv \rho[\Gamma_{\mathbf{X}}]_{ij} = \big[\omega(x_j^* x_i)\big]. \tag{10.1}$$

It is denoted by $\rho[\mathbf{X}]$ and called the *correlation matrix of the partition* $\mathbf{X}$. The partition $\mathbf{X}$ is *bistochastic* (see Example (9.18)) if it satisfies the additional requirement

$$\sum_{i=1}^{k} x_i\, x_i^* = \mathbf{1}. \tag{10.2}$$

Example 10.1 *(Classical partitions)*   To a partition $\mathbf{C} = (C_1, C_2, \dots, C_k)$ of a probability space $(\mathbf{X}, \mu)$ in non-overlapping measurable subsets we may associate the operational partition $\mathbf{C} = (I_{C_1}, I_{C_2}, \dots, I_{C_k})$ of indicator functions of the sets C_j. In this case, the correlation matrix $\rho[\mathbf{C}]$ is diagonal with $\mu(C_j)$ at the site jj. A general operational partition $\mathbf{F}$ of $\mathcal{L}^\infty(\mathbf{X}, \mu)$ consists of a k-tuple $(f_1, f_2, \dots, f_k)$ of measurable functions which generally overlap and satisfy

$$\sum_{j=1}^{k} |f_j|^2 = 1.$$

The corresponding correlation matrix has matrix elements

$$\rho[\mathbf{F}]_{ij} = \int_{\mathbf{X}} \mu(dx)\, \overline{f_j(x)}\, f_i(x)$$

which are all typically non-zero.

Example 10.2 Suppose that we observe a quantum spin with total spin j which is in the pure state $\varphi \in \mathbb{C}^{2j+1}$. The basic observables for this system are the three components of the spin-j representation $\mathbf{J}$ of SU(2). They are characterized by

$$(\mathbf{J})^* = \mathbf{J}, \quad \mathbf{J} \times \mathbf{J} = i\,\mathbf{J} \quad \text{and} \quad \mathbf{J} \cdot \mathbf{J} = j(j+1). \tag{10.3}$$

We could model this system in a natural way by considering the partition

$$\mathbf{X} := \frac{1}{\sqrt{j(j+1)}}\, \mathbf{J}$$

of unity and this produces a three-dimensional, i.e. spin-1, matrix model of the original system. We decompose the correlation matrix $\rho[\mathbf{X}]$ into real and imaginary part

$$\rho[\mathbf{X}] = \sigma - i\,\eta.$$

As $\rho[\mathbf{X}]$ is Hermitian, η is a skew-adjoint 3×3 matrix with real matrix elements and can therefore be written in terms of a vector $\mathbf{s} \in \mathbb{R}^3$ as

$$\eta = \frac{1}{j(j+1)} \begin{pmatrix} 0 & s_z & s_y \\ -s_z & 0 & s_x \\ -s_y & -s_x & 0 \end{pmatrix}.$$

Using the commutation relations (10.3), we find that

$$\mathbf{s} = \langle \varphi, \mathbf{J}\, \varphi \rangle.$$

We could then enquire about the expected value of the spin in the direction of an arbitrary unit vector $\mathbf{e}$ in $\mathbb{R}^3$ and its variance. Both quantities can be expressed in terms of $\rho[\mathbf{X}]$ or, equivalently, of σ and $\mathbf{s}$:

$$\langle J_{\mathbf{e}} \rangle_\varphi = \langle \varphi, \mathbf{e} \cdot \mathbf{J}\varphi \rangle = \mathbf{e} \cdot \mathbf{s}$$

and

$$\Delta_\varphi (J_{\mathbf{e}})^2 = j(j+1)\langle \mathbf{e}, \sigma\, \mathbf{e} \rangle - (\mathbf{e} \cdot \mathbf{s})^2.$$

Example 10.3 *(Partitions in Fermi fields)* In the one-particle space $\mathfrak{H}$ of a fermion system, we fix k vectors $(\varphi_1, \varphi_2, \ldots, \varphi_k)$ subject to the condition that

$$\sum_{j=1}^{k} \|\varphi_j\|^2 = 1.$$

For such a choice, $\mathbf{X} := (a^*(\varphi_1), a^*(\varphi_2), \ldots, a^*(\varphi_k), a(\varphi_k), \ldots, a(\varphi_1))$ is an operational partition of unity. If the reference state ω of $\mathfrak{A}(\mathfrak{H})$ is gauge-invariant, then $\rho[\mathbf{X}]$ is of the form

$$\rho[\mathbf{X}] = \begin{pmatrix} R & 0 \\ 0 & D - R \end{pmatrix},$$

where R is the $k \times k$ one-particle correlation matrix

$$R_{ij} := \omega(a^*(\varphi_j)a(\varphi_i))$$

and D is the diagonal matrix with $\|\varphi_j\|^2$ at jth site. In terms of measurements, $\mathbf{X}$ describes the detection of single fermions with different sensitivities according to the mode of the particle.

Any unital C*-algebra $\mathfrak{A}$ admits a multitude of partitions of arbitrary length. Indeed, fixing $k-1$ observables x_j in $\mathfrak{A}$ in such that $\sum_j x_j^* x_j \leq \mathbf{1}$, it suffices to choose for x_k the element $\sqrt{\mathbf{1} - \sum_j x_j^* x_j}$ in order to obtain a partition $(x_1, x_2, \ldots, x_k)$.

Our next aim is to find an equivalent representation of the correlation matrix (10.1) on the Hilbert space version of the probability space.

Theorem 10.4 *Let $\mathbf{X} = (x_1, x_2, \ldots, x_k)$ be a size k partition of unity in a C*-algebra $\mathfrak{A}$ and consider the GNS-representation $(\mathfrak{H}, \pi, \Omega)$ of the state ω on $\mathfrak{A}$. Consider on $\mathcal{B}(\mathfrak{H})$ the unity-preserving completely positive map*

$$\Pi_{\mathbf{X}} : Y \mapsto \sum_{i=1}^{k} \pi(x_i)^* Y\, \pi(x_i).$$

It is the dual map of $\rho \mapsto \Pi_{\mathbf{X}}^{\mathsf{T}}(\rho)$ with ρ density matrix on $\mathfrak{H}$ and

$$\operatorname{Tr} \Pi_{\mathbf{X}}^{\mathsf{T}}(\rho)\, Y := \operatorname{Tr} \rho\, \Pi_{\mathbf{X}}(Y), \quad Y \in \mathcal{B}(\mathfrak{H}).$$

Then $\rho[\mathbf{X}]$ *and*

$$\hat{\rho}[\mathbf{X}] := \Pi_{\mathbf{X}}^{\mathsf{T}}(|\Omega\rangle\langle\Omega|) = \sum_{j=1}^{k} |\pi(x_j)\,\Omega\rangle\langle\pi(x_j)\,\Omega| \tag{10.4}$$

have, up to multiplicities of zero, the same spectra.

Proof Let ξ be the normalized vector

$$\xi := \sum_{i=1}^{k} e_i \otimes \pi(x_i)\,\Omega$$

in $\mathbb{C}^k \otimes \mathfrak{H}$, where $\{e_1, e_2, \dots, e_k\}$ is an orthonormal basis in $\mathbb{C}^k$. By Theorem 10.6, which will be proved shortly, the restrictions of the vector state $|\xi\rangle\langle\xi|$ to the $k \times k$ matrices $\mathcal{M}_k$ and to $\mathcal{B}(\mathfrak{H})$ have the same eigenvalues up to multiplicities of zero. We now compute these restrictions. By $\operatorname{Tr}_{\mathfrak{H}}$ and $\operatorname{Tr}_{\mathbb{C}^k}$, we denote the partial traces on the spaces $\mathfrak{H}$ and $\mathbb{C}^k$, then

$$\operatorname{Tr}_{\mathfrak{H}} |\xi\rangle\langle\xi| = \sum_{i,j=1}^{k} \operatorname{Tr}_{\mathfrak{H}} |e_i \otimes \pi(x_i)\,\Omega\rangle\langle e_i \otimes \pi(x_i)\,\Omega|$$

$$= \sum_{i,j=1}^{k} \langle\Omega, \pi(x_j^* x_i)\,\Omega\rangle\, |e_i\rangle\langle e_j|$$

$$= \sum_{i,j=1}^{k} \omega(x_j^* x_i)\, |e_i\rangle\langle e_j|$$

$$= \rho[\mathbf{X}]$$

and

$$\operatorname{Tr}_{\mathbb{C}^k} |\xi\rangle\langle\xi| = \sum_{i,j=1}^{k} \operatorname{Tr}_{\mathbb{C}^k} |e_i \otimes \pi(x_i)\,\Omega\rangle\langle e_i \otimes \pi(x_i)\,\Omega|$$

$$= \sum_{i=1}^{k} |\pi(x_i)\,\Omega\rangle\langle\pi(x_i)\,\Omega|$$

$$= \Pi_{\mathbf{X}}^{\mathsf{T}}(|\Omega\rangle\langle\Omega|).$$

This proves the theorem. □

In the proof of Theorem 10.4 we used a few technical results, which we shall now explain.

Definition 10.5 *A complex conjugation on a Hilbert space $\mathfrak{H}$ is a complex antilinear map $\varphi \in \mathfrak{H} \mapsto \overline{\varphi} \in \mathfrak{H}$ such that*

$$\overline{\overline{\varphi}} = \varphi \quad and \quad \langle \overline{\varphi}, \overline{\psi} \rangle = \langle \psi, \varphi \rangle.$$

Theorem 10.6 *Let $\varphi \mapsto \overline{\varphi}$ be a complex conjugation on an Hilbert space $\mathfrak{H}$, then*

$$\varphi \otimes \psi \in \mathfrak{H} \otimes \mathfrak{K} \mapsto |\psi\rangle\langle\overline{\varphi}| \in \mathcal{B}(\mathfrak{H}, \mathfrak{K}) \tag{10.5}$$

extends to an isomorphism between $\mathfrak{H} \otimes \mathfrak{K}$ and the Hilbert–Schmidt operators $\mathcal{T}_2(\mathfrak{H}, \mathfrak{K})$ from $\mathfrak{H}$ to $\mathfrak{K}$ equipped with the trace scalar product

$$\langle A, B \rangle := \mathrm{Tr}_{\mathfrak{H}} A^* B = \mathrm{Tr}_{\mathfrak{K}} B A^*.$$

In particular, for any vector $\varphi \in \mathfrak{H} \otimes \mathfrak{K}$, there exist orthonormal bases $\{e_1, e_2, \dots\}$ in $\mathfrak{H}$ and $\{f_1, f_2, \dots\}$ in $\mathfrak{K}$ and non-negative numbers $\kappa_1 \geq \kappa_2 \geq \cdots$ such that

$$\sum_j \kappa_j^2 < \infty \quad and \quad \varphi = \sum_j \kappa_j\, e_j \otimes f_j. \tag{10.6}$$

Proof One checks immediately that

$$\mathrm{Tr}\left(|\psi\rangle\langle\overline{\varphi}|\right)^* |\psi'\rangle\langle\overline{\varphi'}| = \langle \varphi \otimes \psi, \varphi' \otimes \psi' \rangle.$$

Therefore, the map (10.5) is certainly an isometry from $\mathfrak{H} \otimes \mathfrak{K}$ into $\mathcal{T}_2(\mathfrak{H}, \mathfrak{K})$.

Conversely, let K be a Hilbert–Schmidt operator from $\mathfrak{H}$ to $\mathfrak{K}$. As K is compact, there are orthonormal bases $\{e_1, e_2, \dots\}$ in $\mathfrak{H}$ and $\{f_1, f_2, \dots\}$ in $\mathfrak{K}$ and non-negative numbers κ_j such that

$$K = \sum_j \kappa_j\, |f_j\rangle\langle e_j| \quad and \quad \sum_j \kappa_j^2 < \infty.$$

But this means that K is the image under the map (10.5) of the vector $\sum_j \kappa_j\, \overline{e_j} \otimes f_j$ in $\mathfrak{H} \otimes \mathfrak{K}$. □

Theorem 10.7 *Let $\mathfrak{H}_1$ and $\mathfrak{H}_2$ be Hilbert spaces and φ_{12} be a normalized vector in $\mathfrak{H}_1 \otimes \mathfrak{H}_2$. The reduced density matrices ρ_1 and ρ_2,*

$$\mathrm{Tr}\,\rho_1 A_1 := \langle \varphi_{12}, A_1 \otimes \mathbf{1}_2 \varphi_{12} \rangle \quad and \quad \mathrm{Tr}\,\rho_2 A_2 := \langle \varphi_{12}, \mathbf{1}_1 \otimes A_2 \varphi_{12} \rangle,$$

have, up to multiplicities of zero, the same spectrum. Moreover, non-zero eigenvalues of ρ_1 and ρ_2 have the same multiplicity.

Proof We write φ_{12}, as in (10.6),

$$\varphi_{12} = \sum_j \kappa_j \, e_j \otimes f_j$$

and compute the reduced density matrix ρ_1

$$\begin{aligned}
\mathrm{Tr}\,\rho_1 \, A_1 &:= \langle \varphi_{12}, A_1 \otimes \mathbf{I}_2 \, \varphi_{12} \rangle \\
&= \Big\langle \sum_j \kappa_j \, e_j \otimes f_j, A_1 \otimes \mathbf{I}_2 \sum_j \kappa_\ell \, e_\ell \otimes f_\ell \Big\rangle \\
&= \sum_{j,\ell} \kappa_j^2 \, \delta_{j,\ell} \, \langle e_j, A_1 \, e_\ell \rangle \\
&= \sum_j \kappa_j^2 \, \mathrm{Tr}\, |e_j\rangle\langle e_j| \, A_1.
\end{aligned}$$

Therefore,

$$\rho_1 = \sum_j \kappa_j^2 \, |e_j\rangle\langle e_j| \tag{10.7}$$

and, as $\{e_1, e_2, \dots\}$ is orthonormal, (10.7) is the spectral decomposition of ρ_1. Similarly,

$$\rho_2 = \sum_j \kappa_j^2 \, |f_j\rangle\langle f_j|$$

and the theorem follows. $\qquad\square$

Two operational partitions $\mathbf{X} = (x_1, x_2, \dots, x_k)$ and $\mathbf{Y} = (y_1, y_2, \dots, y_\ell)$ of sizes k and ℓ can be composed to yield a partition $\mathbf{X} \circ \mathbf{Y}$ of size $k\ell$, just by multiplying their elements

$$\mathbf{X} \circ \mathbf{Y} := (x_1 y_1, x_1 y_2, \dots, x_k y_\ell). \tag{10.8}$$

Indeed,

$$\begin{aligned}
\sum_{i=1}^k \sum_{j=1}^\ell (x_i y_j)^* x_i y_j &= \sum_{i=1}^k \sum_{j=1}^\ell y_j^* x_i^* x_i y_j \\
&= \sum_{j=1}^\ell y_j^* \Big(\sum_{i=1}^k x_i^* x_i \Big) y_j \\
&= \sum_{j=1}^\ell y_j^* y_j = \mathbf{1}. \tag{10.9}
\end{aligned}$$

On the level of the GNS-space of the state ω, composing partitions amounts to composing the associated completely positive maps in reversed order

$$\Pi_{\mathbf{X} \circ \mathbf{Y}} = \Pi_{\mathbf{Y}} \circ \Pi_{\mathbf{X}} \quad \text{and} \quad (\Pi_{\mathbf{X} \circ \mathbf{Y}})^{\mathsf{T}} = (\Pi_{\mathbf{X}})^{\mathsf{T}} \circ (\Pi_{\mathbf{Y}})^{\mathsf{T}}.$$

It is important to note that composing partitions is generally a non-commutative operation: $\mathbf{X} \circ \mathbf{Y} \neq \mathbf{Y} \circ \mathbf{X}$. In the sequel, we shall need to consider the set $\mathcal{P}(\mathfrak{A}_0)$ of partitions with elements belonging to some unital $*$-subalgebra $\mathfrak{A}_0$ of $\mathfrak{A}$.

This set, equipped with the composition law (10.9), carries a natural semi-group structure. There are two important subsemigroups of $\mathcal{P}(\mathfrak{A}_0)$: the *bistochastic partitions* $\mathcal{P}_{\mathrm{b}}(\mathfrak{A}_0)$ (see (10.2)) and the partitions $\mathcal{P}_{\mathrm{u}}(\mathfrak{A}_0)$ in *scalar multiples of unitaries*, i.e.

$$\mathbf{X} = (\lambda_1 u_1, \lambda_2 u_2, \dots, \lambda_n u_n)$$

with u_j unitary and $\lambda_j \in \mathbb{C}$ satisfying $\sum_j |\lambda_j|^2 = 1$. Note that

$$\mathcal{P}_{\mathrm{u}}(\mathfrak{A}_0) \subset \mathcal{P}_{\mathrm{b}}(\mathfrak{A}_0) \subset \mathcal{P}(\mathfrak{A}_0).$$

10.2   Dynamical entropy

For computational purposes one may introduce dynamical entropy directly in terms of correlation matrices of partitions; see Section 10.2.2. It is, however, also possible to introduce a symbolic dynamics which is a shift on a quantum spin chain and then consider the mean entropy of the reference state on the chain.

10.2.1   *Symbolic dynamics*

Let $(\mathfrak{X}, T, \mu)$ be a stationary classical dynamical system with a not necessarily invertible dynamical map T. Stationarity means that $\mu(T^{-1}(A)) = \mu(A)$ for any measurable subset A of $\mathfrak{X}$. A *symbolic dynamics* mimics the evolution by a shift dynamics on a space of symbolic sequences $\Omega := (\mathcal{S}_k)^{\mathbb{N}_0}$ written with letters belonging to an alphabet $\mathcal{S}_k = \{1, 2, \dots, k\}$ of length k. To each letter j there corresponds a measurable subset C_j of $\mathfrak{X}$ such that $\mathbf{C} := \{C_j \mid j\}$ is a partition of $\mathfrak{X}$, i.e. C_j's corresponding to different j's are disjoint and $\mathfrak{X}$ is, up to a set of μ-measure 0, equal to the union of the C_j. To each finite configuration ω_Λ with the letter s_j at the site $j \in \Lambda$ one associates the measurable subset

$$C_{\omega_\Lambda} := \bigcap_{j \in \Lambda} T^{-j}(C_{s_j}).$$

By construction, the action of the left-shift γ on Ω intertwines with the evolution on $\mathfrak{X}$ in the sense that on local configurations

$$C_{\gamma(\omega)} = T^{-1}(C_\omega).$$

The symbolic dynamics produces a coarse-grained picture of $(\mathfrak{X}, T, \mu)$ as it lumps all phase space points together that cannot be distinguished by any of the refined partitions C_ω. The construction provides us also with a probability measure $\mu^\mathbf{C}$ on Ω which is the pull-back of μ. On the cylinder set $[\omega_\Lambda]$ of all configurations in Ω that coincide with a given configuration ω_Λ on Λ:

$$\mu^\mathbf{C}([\omega_\Lambda]) := \mu(C_{\omega_\Lambda}).$$

The stationarity of μ implies that $\mu^\mathbf{C}$ is shift-invariant.

On an algebraic level we have obtained a mapping

$$\Gamma^\mathbf{C} : \mathcal{C}(\Omega) \to \mathcal{L}^\infty(\mathfrak{X}, \mu)$$

determined by the requirement

$$\Gamma^\mathbf{C}(f) = \sum_{\omega \in \Omega^\Lambda} f(\omega)\, I_{C_\omega}$$

for local functions with dependence region Λ. Γ enjoys the following properties:

(a) $\Gamma^\mathbf{C}$ is a unity-preserving homomorphism;
(b) $\Gamma^\mathbf{C}$ intertwines the shift and the dynamics, i.e. $\Gamma^\mathbf{C} \circ \gamma = \Theta \circ \Gamma^\mathbf{C}$ with $\Theta(g)(x) := g(T\,x)$.

These properties mean that $\Gamma^\mathbf{C}$ is a homomorphism between the symbolic shift dynamics $(\mathcal{C}(\Omega), \gamma, \mu^\mathbf{C})$ and $(\mathcal{L}^\infty(\mathfrak{X}, \mu), \Theta, \mu)$. The partition $\mathbf{C}$ is *generating* if $\Gamma^\mathbf{C}$ extends to a measure space isomorphism between $(\mathcal{L}^\infty(\Omega, \mu^\mathbf{C}), \gamma, \mu^\mathbf{C})$ and $(\mathcal{L}^\infty(\mathfrak{X}, \mu), \Theta, \mu)$. The isomorphism identifies almost everywhere $x \in \mathfrak{X}$ with the symbolic sequence

$$\omega = (s_0, s_1, s_2, \dots) \quad \text{with} \quad T^k(x) \in C_{s_k}, \quad k = 0, 1, 2, \dots .$$

We now consider a C*-dynamical system $(\mathfrak{A}, \Theta, \omega)$ in discrete time with single-step evolution given by a unital homomorphism Θ, e.g. Θ preserves the algebraic operations and $\mathbf{1}$ but is possibly non-invertible. We do not necessarily assume that the reference state ω is invariant under Θ although in many application this will be the case. We call the dynamical system *stationary* if $\omega \circ \Theta = \omega$. The basic idea is to observe an evolving system repeatedly by making joint measurements at successive discrete times. For a single coarse-grained observation, described by $\mathbf{X} = (x_1, x_2, \dots, x_k)$, the correlation matrix $\rho[\mathbf{X}]$ provides us with the maximal information that is contained in that observation; see Section 8.1. The information contained in observations at times $0, 1, \dots, m-1$ is collected in a density matrix on $\mathcal{M}_k^{\otimes m}$. Unlike the classical case, we cannot hope to construct a homomorphism from a shift on a spin half-chain to $(\mathfrak{A}, \Theta, \omega)$ that intertwines the shift and the dynamics, simply because usually $[x_i, \Theta(x_j)] \neq 0$ for elements x_i, x_j of a partition $\mathbf{X}$. Therefore, we have to settle with completely positive maps instead of homomorphisms. It is, however, possible to save some aspects of the intertwining and of the locality in time.

Starting with a partition $\mathbf{X} = (x_1, x_2, \ldots, x_k)$, we may consider the completely positive unity-preserving map

$$\tilde{\Gamma} : \mathcal{M}_k \otimes \mathfrak{A} \to \mathfrak{A} : \tilde{\Gamma}([y_{ij}]) := \sum_{i,j=1}^{k} x_i^* y_{ij}\, x_j.$$

Each matrix $A = [a_{ij}] \in \mathcal{M}_k$ generates a linear transformation $\tilde{\Gamma}_A$ of $\mathfrak{A}$ by putting

$$\tilde{\Gamma}_A(y) := \tilde{\Gamma}(A \otimes y) = \sum_{ij} a_{ij}\, x_i^* x_j.$$

The map $\tilde{\Gamma}_{\mathbf{1}}$ is unity-preserving and completely positive. We may now construct a map $\Gamma^{\mathbf{X}}$ from $\mathcal{M}_k^{\otimes \mathbb{N}_0}$ to $\mathfrak{A}$ that plays a role similar to $\Gamma^{\mathbf{C}}$. On elementary tensors with $A_j \in \mathcal{M}_k$

$$\Gamma^{\mathbf{X}}(A_0 \otimes A_1 \otimes \cdots \otimes A_{m-1}) := \tilde{\Gamma}_{A_0}\big(\Theta\big(\tilde{\Gamma}_{A_1}\big(\cdots \Theta\big(\tilde{\Gamma}_{A_{m-1}}(\mathbf{1})\big) \cdots\big)\big)\big). \quad (10.10)$$

We now have that

(a) $\Gamma^{\mathbf{X}}$ is a unity-preserving and completely positive,

(b) $\Gamma^{\mathbf{X}} \circ \gamma = \tilde{\Gamma}_{\mathbf{1}} \circ \Theta \circ \Gamma^{\mathbf{X}}$.

The reference state ω defines a state $\omega^{\mathbf{X}}$ on $\mathcal{M}_k^{\otimes \mathbb{N}_0}$ by putting

$$\omega^{\mathbf{X}}(a) := \omega\big(\Gamma^{\mathbf{X}}(a)\big), \quad a \in \mathcal{M}_k^{\otimes \mathbb{N}_0}.$$

Here too we may observe a difference with the classical systems: the state $\omega^{\mathbf{X}}$ is not necessarily shift-invariant. A sufficient condition to have $\omega^{\mathbf{X}}$ translation-invariant is that

$$\omega \text{ is } \Theta\text{-invariant} \quad \text{and} \quad \omega\Big(\sum_{j=1}^{k} x_j^* y\, x_j\Big) = \omega(y), \quad y \in \mathfrak{A}.$$

An instance of such a situation is given by a Θ-invariant tracial state ω on $\mathfrak{A}$ and a bistochastic partition $\mathbf{X}$. In concrete examples, it happens that $\omega^{\mathbf{X}} \circ \gamma^n$ converges when $n \to \infty$. We may then use the limit to define a stationary quantum shift on a full spin chain.

10.2.2 The entropy

Consider the correlation matrices $\rho[\mathbf{X}^m]$ associated with the operational partition

$$\mathbf{X}^m := \Theta^{m-1}(\mathbf{X}) \circ \Theta^{m-2}(\mathbf{X}) \circ \ldots \circ \Theta(\mathbf{X}) \circ \mathbf{X}, \quad (10.11)$$

where the time evolution $\Theta(\mathbf{X})$ of an operational partition $\mathbf{X} = (x_1, x_2, \ldots, x_k)$ is defined as

$$\Theta(\mathbf{X}) := (\Theta(x_1), \Theta(x_2), \ldots, \Theta(x_k))$$

and the composition of partitions is that given in (10.8).

The explicit matrix elements of $\rho[\mathbf{X}^m]$ are computed as

$$\rho[\mathbf{X}^m]_{\mathbf{i},\mathbf{j}} = \omega\big(x_{j_1}^* \,\Theta(x_{j_2}^*) \cdots \Theta^{m-1}(x_{j_m}^*)\, \Theta^{m-1}(x_{i_m}) \cdots \Theta(x_{i_2}) x_{i_1}\big). \qquad (10.12)$$

The multi-indices $\mathbf{i}$ and $\mathbf{j}$ in (10.12) are $(i_1, i_2, \dots, i_m)$ and $(j_1, j_2, \dots, j_m)$. The set $\{\rho[\mathbf{X}^m] \mid m \in \mathbb{N}_0\}$ is a consistent family of density matrices on $(\mathcal{M}_k)^{\otimes \mathbb{N}_0}$. Consistency can explicitly be checked by taking the partial trace of $\rho[\mathbf{X}^{m+1}]$ over the rightmost factor in $\otimes_{j=1}^{m+1} \mathbb{C}^k$. For m-tuples $\mathbf{i}$ and $\mathbf{j}$ and putting $\mathbf{i} \times (\ell) := (i_1, i_2, \dots, i_m, \ell)$

$$\sum_{\ell=1}^{k} \rho[\mathbf{X}^{m+1}]_{\mathbf{i}\times(\ell), \mathbf{j}\times(\ell)}$$

$$= \sum_{\ell=1}^{k} \omega\big(x_{j_1}^* \,\Theta(x_{j_2}^*) \cdots \Theta^{m-1}(x_{j_m}^*)\, \Theta^m(x_\ell^*)\, \Theta^m(x_\ell)\, \Theta^{m-1}(x_{i_m}) \cdots x_{i_1}\big)$$

$$= \omega\big(x_{j_1}^* \,\Theta(x_{j_2}^*) \cdots \Theta^{m-1}(x_{j_m}^*)\, \Theta^{m-1}(x_{i_m}) \cdots x_{i_1}\big)$$

$$= \rho[\mathbf{X}^m]_{\mathbf{i},\mathbf{j}}.$$

In fact, from (10.10) it is easily seen that the $\rho[\mathbf{X}^m]$ are the reduced density matrices of the state $\omega^{\mathbf{X}}$ of the previous section.

In the GNS-representation of ω, this construction looks as follows. To $\rho[\mathbf{X}]$ corresponds the density matrix

$$\hat{\rho}[\mathbf{X}] = \Pi_{\mathbf{X}}^{\mathsf{T}}(|\Omega\rangle\langle\Omega|) = \sum_{j=1}^{k} |\pi(x_j)\,\Omega\rangle\langle\pi(x_j)\,\Omega|.$$

For m time steps

$$\hat{\rho}[\mathbf{X}^m] = \Pi_{\Theta^{m-1}(\mathbf{X})}^{\mathsf{T}} \circ \cdots \circ \Pi_{\Theta(\mathbf{X})}^{\mathsf{T}} \circ \Pi_{(\mathbf{X})}^{\mathsf{T}}(|\Omega\rangle\langle\Omega|). \qquad (10.13)$$

Assume now that the dynamics is implemented by a unitary operator u:

$$\pi(\Theta(x)) = u^* \pi(x)\, u.$$

This is certainly the case for a stationary system (see Theorem 5.22) but it holds in a more general situation. We use the notation Θ^{T} to denote the dual action of Θ in the representation. It acts on a density matrix ρ on the GNS-space as

$$\Theta^{\mathsf{T}}(\rho) := u\,\rho\,u^*.$$

Repeating the same coarse-grained observation a time step later we obtain for any two vectors φ and ψ in the representation space

$$\left(\Theta^{\mathsf{T}} \circ \Pi_{\mathbf{X}}^{\mathsf{T}}\right)^2 (|\varphi\rangle\langle\psi|)$$

$$= \Theta^{\mathsf{T}} \circ \Pi_{\mathbf{X}}^{\mathsf{T}} \circ \Theta^{\mathsf{T}} \circ \Pi_{\mathbf{X}}^{\mathsf{T}}(|\varphi\rangle\langle\psi|)$$

$$= \sum_{j_2=1}^{k} \sum_{j_1=1}^{k} |u\,\pi(x_{j_2})\,u\,\pi(x_{j_1})\,\varphi\rangle\langle u\,\pi(x_{j_2})\,u\,\pi(x_{j_1})\,\psi|$$

$$= u^2 \left(\sum_{j_2=1}^{k} \sum_{j_1=1}^{k} |\pi(\Theta(x_{j_2}))\,\pi(x_{j_1})\,\varphi\rangle\langle u^*\pi(\Theta(x_{j_2}))\,\pi(x_{j_1})\,\psi|\right)(u^*)^2$$

$$= u^2\,\Pi_{\Theta(\mathbf{X})\circ\mathbf{X}}^{\mathsf{T}}(|\varphi\rangle\langle\psi|)\,(u^*)^2.$$

This computation extends to an arbitrary number of time steps. We have therefore shown that for any density matrix ρ on the GNS-space and any $m \in \mathbb{N}$

$$\left(\Theta^{\mathsf{T}} \circ \Pi_{\mathbf{X}}^{\mathsf{T}}\right)^m (\rho) = u^m\,\Pi_{\mathbf{X}^m}^{\mathsf{T}}(\rho)\,(u^*)^m. \tag{10.14}$$

Formula (10.14) expresses the equivalence of two kinds of repeated measurements. The first one is similar to the Schrödinger picture and is given in terms of a discrete time semigroup while the second is rather similar to the interaction picture; see (3.1.2).

Our next goal is to obtain a quantitative measure for the rate at which we can extract information from a dynamical system by repeated measurements. Basically, we use the largest von Neumann entropy density of symbolic models of the system. Technically, this is done in several stages. We first consider the entropy of a single partition. Next we introduce the rate at which this entropy increases, when we refine the coarse-grained picture by letting it evolve in time. Finally, we consider the supremum over a suitable class of models of the system.

Definition 10.8 *The entropy* $\mathsf{H}[\omega, \mathbf{X}]$ *of a partition* $\mathbf{X}$ *is the von Neumann entropy* $S(\rho[\mathbf{X}])$ *of the correlation matrix* $\rho[\mathbf{X}]$ *or, equivalently, by Theorem 10.4 of the density matrix* $\hat{\rho}[\mathbf{X}]$

$$\mathsf{H}[\omega, \mathbf{X}] = S(\rho[\mathbf{X}]) = S(\hat{\rho}[\mathbf{X}]).$$

Definition 10.9 *The dynamical entropy* $\mathsf{h}[\omega, \Theta, \mathbf{X}]$ *of an operational partition* $\mathbf{X}$ *of a dynamical system* $(\mathfrak{A}, \Theta, \omega)$ *is defined to be*

$$\mathsf{h}[\omega, \Theta, \mathbf{X}] := \limsup_{m\to\infty} \frac{1}{m}\,\mathsf{H}[\omega, \mathbf{X}^m]$$

$$= \limsup_{m\to\infty} \frac{1}{m}\,\mathsf{H}[\omega, \Theta^{m-1}(\mathbf{X}) \circ \Theta^{m-2}(\mathbf{X}) \circ \ldots \circ \Theta(\mathbf{X}) \circ \mathbf{X}].$$

The fact that we have to take a limsup instead of a limit is due to the lack of translation invariance of the state $\omega^{\mathbf{X}}$ on the half-chain. In specific models, however, the state on the half-chain is often asymptotically shift-invariant and its mean entropy exists as a limit. Finally, we arrive at the definition of the dynamical entropy by taking the supremum of the $h[\omega, \Theta, \mathbf{X}]$ over a suitable class of partitions. In order to specify this class in a given model we require that the elements of the partition $\mathbf{X}$ belong to a distinguished $*$-subalgebra $\mathfrak{A}_0$ of $\mathfrak{A}$ which is globally invariant under Θ. Moreover, we may put a restriction on the type of partition that we consider, such as limiting the sup over bistochastic or even unitary partitions. In many situations it appears that it suffices already to consider partitions belonging to the semigroups $\mathcal{P}_b(\mathfrak{A}_0)$ or $\mathcal{P}_u(\mathfrak{A}_0)$ introduced at the end of Section 10.1.

Definition 10.10 *The dynamical entropy* $h[\omega, \Theta, \mathfrak{A}_0]$ *of the C*-dynamical system* $(\mathfrak{A}, \Theta, \omega)$ *with respect to the Θ-invariant $*$-subalgebra $\mathfrak{A}_0$ of $\mathfrak{A}$ is given by*

$$h[\omega, \Theta, \mathfrak{A}_0] := \sup_{\mathbf{X} \subset \mathcal{P}(\mathfrak{A}_0)} h[\omega, \Theta, \mathbf{X}]. \tag{10.15}$$

Definitions similar to (10.15) can be given for the cases of bistochastic or unitary partitions with the obvious relations

$$h_u[\omega, \Theta, \mathfrak{A}_0] \leq h_b[\omega, \Theta, \mathfrak{A}_0] \leq h[\omega, \Theta, \mathfrak{A}_0].$$

10.3   Some technical results

We now prove some technical properties of the entropy of a partition and of the dynamical entropy which will prove useful in treating concrete models in the next section.

Lemma 10.11 *Let ρ be a density matrix on a Hilbert space $\mathfrak{H}$ and consider a size k partition $\mathbf{X} = (x_1, x_2, \ldots, x_k)$ in $\mathcal{B}(\mathfrak{H})$. If $\rho[\mathbf{X}]$ is the $k \times k$ correlation matrix $[\rho(x_j^* x_i)]$, then*

(a) $S(\rho[\mathbf{X}]) \leq S(\rho) + \log d$, *where d is the dimension of $\mathfrak{H}$;*
(b) $S(\rho[\mathbf{X}]) \leq S(\rho) + S(\sigma)$ *with* $\sigma = \sum_{i=1}^{k} x_i \rho x_i^*$;
(c) $S(\sigma) - S(\rho) \leq S(\rho[\mathbf{X}])$.

Proof   As σ is a density matrix on $\mathfrak{H}$, we always have the bound $S(\sigma) \leq \log d$; see Theorem 10.4. Therefore the first inequality is an immediate consequence of the second.

In order to prove (b), we consider an orthonormal basis $\{\varphi_1, \varphi_2, \ldots\}$ of eigenvectors of ρ:

$$\rho \varphi_j = p_j \varphi_j.$$

Next, let $\{e_1, e_2, \ldots, e_k\}$ be the canonical orthonormal basis in $\mathbb{C}^k$ and consider for each j a vector $\eta_j \in \mathbb{C}^k \otimes \mathfrak{H}$,

$$\eta_j := \sum_{\ell=1}^{k} e_\ell \otimes x_\ell \varphi_j.$$

A simple calculation yields

$$\langle \eta_i, \eta_j \rangle = \sum_{\ell_1=1}^{k} \sum_{\ell_2=1}^{k} \langle e_{\ell_1} \otimes x_{\ell_1} \varphi_i, e_{\ell_2} \otimes x_{\ell_2} \varphi_j \rangle$$

$$= \sum_{\ell=1}^{k} \langle x_\ell \varphi_i, x_\ell \varphi_j \rangle$$

$$= \sum_{\ell=1}^{k} \langle \varphi_i, x_\ell^* x_\ell \varphi_j \rangle$$

$$= \langle \varphi_i, \varphi_j \rangle = \delta_{ij}.$$

But then the density matrix

$$\tau := \sum_j p_j |\eta_j\rangle\langle\eta_j|$$

on $\mathbb{C}^k \otimes \mathfrak{H}$ has, up to multiplicities of zero, the same eigenvalues, and therefore the same entropy, as ρ. We now complete the proof by computing the reduced density matrices of τ and using the triangle inequality (9.17),

$$\mathrm{Tr}_{\mathbb{C}^k} \sum_j p_j |\eta_j\rangle\langle\eta_j| = \mathrm{Tr}_{\mathbb{C}^k} \sum_j \sum_{\ell_1=1}^{k} \sum_{\ell_2=1}^{k} p_j |e_{\ell_1} \otimes x_{\ell_1} \varphi_j\rangle\langle e_{\ell_2} \otimes x_{\ell_2} \varphi_j|$$

$$= \sum_j \sum_{\ell=1}^{k} p_j |x_\ell \varphi_j\rangle\langle x_\ell \varphi_j|$$

$$= \sum_{\ell=1}^{k} x_\ell \rho x_\ell^* = \sigma.$$

Tracing over $\mathfrak{H}$ yields

$$\mathrm{Tr}_{\mathfrak{H}} \sum_j p_j |\eta_j\rangle\langle\eta_j| = \mathrm{Tr}_{\mathfrak{H}} \sum_j \sum_{\ell_1=1}^{k} \sum_{\ell_2=1}^{k} p_j |e_{\ell_1} \otimes x_{\ell_1} \varphi_j\rangle\langle e_{\ell_2} \otimes x_{\ell_2} \varphi_j|$$

$$= \sum_{\ell_1=1}^{k} \sum_{\ell_2=1}^{k} \left(\mathrm{Tr}_{\mathfrak{H}} \rho x_{\ell_2}^* x_{\ell_1} \right) |e_{\ell_1}\rangle\langle e_{\ell_2}| = \rho[\mathbf{X}].$$

Finally, (c) is a consequence of the same triangle inequality (9.17). □

Remark that the inequality (c) remains meaningful for infinite systems after passing to the thermodynamical limit, i.e. even if $S(\rho)$ and $S(\sigma)$ tend to infinity, still $S(\rho[\mathbf{X}])$ may furnish a useful bound on the entropy production caused by a local perturbation.

As a direct consequence of the previous lemma, we have

Theorem 10.12 *Let* $(\mathfrak{A}, \Theta, \omega)$ *be a quantum dynamical system and suppose that* $\mathfrak{A}$ *is finite-dimensional; then*

$$\mathsf{h}[\omega, \Theta, \mathfrak{A}] = 0. \qquad (10.16)$$

Proof We can always assume that the algebra $\mathfrak{A}$ is a subalgebra of the $d \times d$ matrices $\mathcal{M}_d$. The state ω can be written in terms of a density matrix ρ. Then, using Lemma 10.11, we have

$$S\big(\rho[\mathbf{X}^n]\big) \leq S(\rho) + \log d \leq 2 \log d. \qquad (10.17)$$

Dividing both sides by n and taking the limit yields

$$\mathsf{h}[\omega, \Theta, \mathbf{X}] = 0.$$

Since $\mathbf{X}$ is arbitrary $\mathsf{h}[\omega, \Theta, \mathfrak{A}] = 0$. $\square$

Hence, finite quantum systems are too small to display a non-zero asymptotic entropy production. Nevertheless, there is entropy production in finite systems, as discussed in Section 12.4.

Lemma 10.13 *Let* $\mathbf{X}$ *and* $\mathbf{Y}$ *be two partitions of a quantum probability space* $(\mathfrak{A}, \omega)$ *and suppose that* $\mathbf{X}$ *is bistochastic (see (10.2)); then*

$$S(\rho[\mathbf{X} \circ \mathbf{Y}]) \geq S(\rho[\mathbf{Y}]).$$

Proof We use the GNS-form $\hat{\rho}[\mathbf{X} \circ \mathbf{Y}]$ of $\rho[\mathbf{X} \circ \mathbf{Y}]$ as in (10.4). This density matrix can be decomposed using, in (10.9),

$$\hat{\rho}[\mathbf{X} \circ \mathbf{Y}] = (\Pi_{\mathbf{X} \circ \mathbf{Y}})^{\mathsf{T}} \big(|\Omega\rangle\langle\Omega|\big) = (\Pi_{\mathbf{X}})^{\mathsf{T}} \circ (\Pi_{\mathbf{Y}})^{\mathsf{T}} \big(|\Omega\rangle\langle\Omega|\big).$$

As $\Pi_{\mathbf{X}}$ is bistochastic and as such maps are entropy-increasing (see Example 9.18) the statement follows. $\square$

In the next lemma, we prove a continuity property.

Lemma 10.14 *Let* $\mathbf{X}^{(r)} = \big(x_1^{(r)}, x_2^{(r)}, \ldots, x_k^{(r)}\big)$ *and* $\mathbf{Y}^{(r)} = \big(y_1^{(r)}, y_2^{(r)}, \ldots, y_k^{(r)}\big)$ *with* $r = 1, 2, \ldots, n$ *be two families of partitions in a* C^*-algebra $\mathfrak{A}$ *and let* ω *be a state on* $\mathfrak{A}$. *Assume that these partitions satisfy*

$$\big\|x_j^{(r)} - y_j^{(r)}\big\| \leq \epsilon_r, \ \ j = 1, 2, \ldots, k \ \ and \ \ 2k \sum_r \epsilon_r \leq \frac{1}{3};$$

then

$$\left| \mathsf{H}\big[\omega, \mathbf{X}^{(n)} \circ \mathbf{X}^{(n-1)} \circ \cdots \circ \mathbf{X}^{(1)}\big] - \mathsf{H}\big[\omega, \mathbf{Y}^{(n)} \circ \mathbf{Y}^{(n-1)} \circ \cdots \circ \mathbf{Y}^{(1)}\big] \right|$$

$$\leq 2k \log(2k^n) \bigg(\sum_{r=1}^{n} \epsilon_r\bigg) - 2k \bigg(\sum_{r=1}^{n} \epsilon_r\bigg) \log\bigg(2k \sum_{r=1}^{n} \epsilon_r\bigg).$$

Proof We use the estimate of Theorem 10.7 for entropies

$$|S(\rho) - S(\sigma)| \le \|\rho - \sigma\|_1 \log d + \eta(\|\rho - \sigma\|_1)$$

for $\|\rho - \sigma\|_1 \le \frac{1}{3}$. For notational simplicity, we consider from now on two pairs of partitions; generalization to n is obvious. We express the entropies $H[\omega, \mathbf{X}^{(2)} \circ \mathbf{X}^{(1)}]$ and $H[\omega, \mathbf{Y}^{(2)} \circ \mathbf{Y}^{(1)}]$ using Definition 10.8 and (10.13):

$$H[\omega, \mathbf{X}^{(2)} \circ \mathbf{X}^{(1)}] = S(\hat{\rho}[\mathbf{X}^{(2)} \circ \mathbf{X}^{(1)}]),$$

$$\hat{\rho}[\mathbf{X}^{(2)} \circ \mathbf{X}^{(1)}] = \Pi^{\mathsf{T}}_{\mathbf{X}^{(2)}} \circ \Pi^{\mathsf{T}}_{\mathbf{X}^{(1)}}(|\Omega\rangle\langle\Omega|).$$

Both density matrices $\hat{\rho}[\mathbf{X}^{(2)} \circ \mathbf{X}^{(1)}]$ and $\hat{\rho}[\mathbf{Y}^{(2)} \circ \mathbf{Y}^{(1)}]$ are supported by a space of dimension at most k^2 and in order to estimate their norm difference, we use

$$\left\| \Pi^{\mathsf{T}}_{\mathbf{X}^{(2)}} \circ \Pi^{\mathsf{T}}_{\mathbf{X}^{(1)}} - \Pi^{\mathsf{T}}_{\mathbf{Y}^{(2)}} \circ \Pi^{\mathsf{T}}_{\mathbf{Y}^{(1)}} \right\|$$
$$= \left\| (\Pi^{\mathsf{T}}_{\mathbf{X}^{(2)}} - \Pi^{\mathsf{T}}_{\mathbf{Y}^{(2)}}) \circ \Pi^{\mathsf{T}}_{\mathbf{X}^{(1)}} + \Pi^{\mathsf{T}}_{\mathbf{Y}^{(2)}} \circ (\Pi^{\mathsf{T}}_{\mathbf{X}^{(1)}} - \Pi^{\mathsf{T}}_{\mathbf{Y}^{(1)}}) \right\|$$
$$\le \left\| \Pi^{\mathsf{T}}_{\mathbf{X}^{(2)}} - \Pi^{\mathsf{T}}_{\mathbf{Y}^{(2)}} \right\| + \left\| \Pi^{\mathsf{T}}_{\mathbf{X}^{(1)}} - \Pi^{\mathsf{T}}_{\mathbf{Y}^{(1)}} \right\|.$$

Let $A \in \mathcal{B}(\mathfrak{H})$; then

$$\|(\Pi_{\mathbf{X}} - \Pi_{\mathbf{Y}})(A)\|$$
$$= \left\| \sum_{j=1}^{k} \pi(x_j^*) A \,\pi(x_j) - \pi(y_j^*) A \,\pi(y_j) \right\|$$
$$\le \left\| \sum_{j=1}^{k} (\pi(x_j^*) - \pi(y_j^*)) \, A \,\pi(x_j) \right\| + \left\| \sum_{j=1}^{k} \pi(y_j^*) \, A \,(\pi(x_j) - \pi(y_j)) \right\|$$
$$\le 2\|A\| \sum_{j=1}^{k} \|x_j - y_j\|,$$

and therefore

$$\|\Pi^{\mathsf{T}}_{\mathbf{X}} - \Pi^{\mathsf{T}}_{\mathbf{Y}}\| = \|\Pi_{\mathbf{X}} - \Pi_{\mathbf{Y}}\| \le 4 \sum_{j=1}^{k} \|x_j - y_j\|.$$

We can now estimate the difference of the entropies

$$\left| H[\omega, \mathbf{X}^{(2)} \circ \mathbf{X}^{(1)}] - H[\omega, \mathbf{Y}^{(2)} \circ \mathbf{Y}^{(1)}] \right|$$
$$= \left| S\left(\Pi^{\mathsf{T}}_{\mathbf{X}^{(2)}} \circ \Pi^{\mathsf{T}}_{\mathbf{X}^{(1)}}(|\Omega\rangle\langle\Omega|) \right) - S\left(\Pi^{\mathsf{T}}_{\mathbf{Y}^{(2)}} \circ \Pi^{\mathsf{T}}_{\mathbf{Y}^{(1)}}(|\Omega\rangle\langle\Omega|) \right) \right|$$
$$\le \left\| \Pi^{\mathsf{T}}_{\mathbf{X}^{(2)}} \circ \Pi^{\mathsf{T}}_{\mathbf{X}^{(1)}} - \Pi^{\mathsf{T}}_{\mathbf{Y}^{(2)}} \circ \Pi^{\mathsf{T}}_{\mathbf{Y}^{(1)}} \right\| \log(2k^2)$$
$$+ \eta\left(\left\| \Pi^{\mathsf{T}}_{\mathbf{X}^{(2)}} \circ \Pi^{\mathsf{T}}_{\mathbf{X}^{(1)}} - \Pi^{\mathsf{T}}_{\mathbf{Y}^{(2)}} \circ \Pi^{\mathsf{T}}_{\mathbf{Y}^{(1)}} \right\| \right)$$
$$\le 4 \log(2k^2) \sum_{r=1}^{2} \epsilon_r + \eta\left(2k \sum_{r=1}^{2} \epsilon_r \right).$$

$\square$

We now consider the entropy of dynamical systems that are composed of independent subsystems.

Lemma 10.15 *Let $(\mathfrak{A}^{(1)}, \Theta^{(1)}, \omega^{(1)})$ and $(\mathfrak{A}^{(2)}, \Theta^{(2)}, \omega^{(2)})$ be two quantum dynamical subsystems of a third dynamical system $(\mathfrak{A}, \Theta, \omega)$ with distinguished subalgebras $\mathfrak{A}_0^{(1)}$ and $\mathfrak{A}_0^{(2)}$ such that the $*$-algebra generated by $\mathfrak{A}_0^{(1)}$ and $\mathfrak{A}_0^{(2)}$ is contained in $\mathfrak{A}_0$. Suppose, furthermore, that ω is product in the sense that for $n \in I\!N$ and $x_1, \dots, x_n \in \mathfrak{A}^{(1)}$, $y_1, \dots, y_n \in \mathfrak{A}^{(2)}$*

$$\omega(x_1 y_1 \dots x_n y_n) = \omega^{(1)}(x_1 \dots x_n) \, \omega^{(2)}(y_1 \dots y_n).$$

Then, if at least for one of the systems 1 or 2, $h[\omega^{(i)}, \Theta^{(i)}, \mathfrak{A}^{(i)}]$ exists as a limit

$$h[\omega, \Theta, \mathfrak{A}_0] \geq h[\omega^{(1)}, \Theta^{(1)}, \mathfrak{A}_0^{(1)}] + h[\omega^{(2)}, \Theta^{(2)}, \mathfrak{A}_0^{(2)}].$$

Proof Suppose that the condition holds for the first system. For $\epsilon > 0$ we can find a partition $\mathbf{X}$ in $\mathfrak{A}_0^{(1)}$ such that

$$\lim_{n \to \infty} \frac{1}{n} S(\rho[\mathbf{X}^n]) \geq h[\omega^{(1)}, \Theta^{(1)}, \mathfrak{A}_0^{(1)}] - \epsilon.$$

There is also a partition $\mathbf{Y}$ in $\mathfrak{A}_0^{(2)}$ such that

$$\limsup_{n \to \infty} \frac{1}{n} S(\rho[\mathbf{Y}^n]) \geq h[\omega^{(2)}, \Theta^{(2)}, \mathfrak{A}_0^{(2)}] - \epsilon.$$

Considering now the partition $\mathbf{X} \circ \mathbf{Y}$ in $\mathfrak{A}_0$, we find

$$\begin{aligned}
h[\omega, \Theta, \mathfrak{A}_0] &\geq h[\omega, \Theta, \mathbf{X} \circ \mathbf{Y}] \\
&\geq \limsup_{n \to \infty} \frac{1}{n} \big(S(\rho[\mathbf{X}^n]) + S(\rho[\mathbf{Y}^n]) \big) \\
&\geq \lim_{n \to \infty} \frac{1}{n} S(\rho[\mathbf{X}^n]) + \limsup_{n \to \infty} \frac{1}{n} S(\rho[\mathbf{Y}^n]) \\
&\geq h[\omega^{(1)}, \Theta^{(1)}, \mathfrak{A}_0^{(1)}] + h[\omega^{(2)}, \Theta^{(2)}, \mathfrak{A}_0^{(2)}] - 2\epsilon.
\end{aligned}$$

$\square$

We cannot prove that the dynamical entropy scales linearly with respect to the time,

$$h[\omega, \Theta^q, \mathfrak{A}_0] = q \, h[\omega, \Theta, \mathfrak{A}_0], \quad q = 2, 3, \dots$$

but we mention a partial result.

Lemma 10.16 *Let $q \in I\!N_0$; then*

$$q \, h[\omega, \Theta, \mathfrak{A}_0] \leq h[\omega, \Theta^q, \mathfrak{A}_0]. \tag{10.18}$$

Proof Consider a partition $\mathbf{X}$ in $\mathfrak{A}_0$. Then, denoting, as in (10.11),

$$\mathbf{X}^q = \Theta^{q-1}(\mathbf{X}) \circ \cdots \circ \Theta(\mathbf{X}) \circ \mathbf{X},$$

we find

$$\limsup_{n \to \infty} \frac{1}{n} \mathsf{H}\left[\omega, \Theta^{q(n-1)}(\mathbf{X}^q) \circ \cdots \circ \Theta^q(\mathbf{X}^q) \circ \mathbf{X}\right]$$
$$= \limsup_{n \to \infty} \frac{q}{qn} S\left(\rho[\mathbf{X}^{qn}]\right)$$
$$= q \limsup_{j \to \infty} \frac{1}{j} S\left(\rho[\mathbf{X}^j]\right)$$
$$= q\, \mathsf{h}[\omega, \Theta, \mathbf{X}].$$

The last-but-one equality follows from the triangle inequality for the von Neumann entropy. Indeed, if $j, n \in \mathbb{N}_0$ are such that $(q-1)n \le j < qn$ and if $\rho_{\{1,2,\dots,j\}}$ denotes the reduced density matrix of a state ω on a spin half-chain at the sites $\{1, 2, \dots, j\}$, then

$$\left|S\left(\rho_{\{1,2,\dots,j\}}\right) - S\left(\rho_{\{1,2,\dots,(q-1)n\}}\right)\right| \le S\left(\rho_{\{(n-1)q,(n-1)q+1,\dots,j\}}\right)$$
$$\le q \log |\mathbf{X}|.$$

Hence,

$$q\, \mathsf{h}[\omega, \Theta, \mathfrak{A}_0] = q \sup_{\mathbf{X} \in \mathcal{P}(\mathfrak{A}_0)} \mathsf{h}[\omega, \Theta, \mathbf{X}] = \sup_{\mathbf{X} \in \mathcal{P}(\mathfrak{A}_0)} \mathsf{h}[\omega, \Theta^q, \mathbf{X}^q]$$
$$\le \sup_{\mathbf{X} \in \mathcal{P}(\mathfrak{A}_0)} \mathsf{h}[\omega, \Theta^q, \mathbf{X}] = \mathsf{h}[\omega, \Theta^q, \mathfrak{A}_0].$$

$\square$

As an immediate consequence of Lemma 10.16, we see that a trivial dynamical system $(\mathfrak{A}, \mathrm{id}, \omega)$ has either zero or infinite entropy. In Section 10.4 we shall actually produce examples with infinite entropy. The reason for this unexpected behaviour is that entropy can be produced both by the dynamics and by the measurement process. It is, therefore, essential to limit measurements to these that do not generate entropy by themselves at a constant rate.

10.4   Examples

The aim of this section is to compute the dynamical entropy that was introduced in Section 10.2 for a few dynamical systems that may be viewed as simplified models of infinite quantum dynamical systems. Two more advanced examples of infinite dynamical systems will be considered in Chapter 13.

10.4.1   *The quantum shift*

The dynamical triple describing the shift on a quantum spin chain with the $d \times d$ matrices as single-site observables is $(\mathfrak{A}, \gamma, \omega)$ with $\mathfrak{A}$ the quasi-local algebra

$(\mathcal{M}_d)^{\mathbb{Z}}$, γ the right-shift and ω a translation-invariant state. We compute in this subsection the dynamical entropy for partitions in strictly local observables $\mathfrak{A}_{\mathrm{loc}}$ and show that the dynamical entropy is given in terms of the mean von Neumann entropy of ω.

Theorem 10.17 *The dynamical entropy of the shift on a quantum spin chain with $d \times d$ matrices as single-site observables is given by*

$$\mathsf{h}[\omega, \gamma, \mathfrak{A}_{\mathrm{loc}}] = \sigma(\omega) + \log d,$$

where $\sigma(\omega)$ is the mean entropy of ω.

Proof The proof consists of two parts. In the first part we obtain an upper bound for the dynamical entropy of a given operational partition which does not depend on the partition. Next we show that this upper bound is attained by producing an explicit partition. This is the general scheme for all dynamical entropy computations that we present.

Choose a partition $\mathbf{X} = (x_1, x_2, \ldots, x_k)$ of size k in local elements. By translation invariance of the state ω we can, without loss of generality, assume that the x_i belong to $\mathfrak{A}_{\{1,2,\ldots,\ell\}}$ for some $\ell = 1, 2, \ldots$. Then, the dynamics shifts these elements to the right and after m time steps, the refined operational partition

$$\mathbf{X}^m = \gamma^{m-1}(\mathbf{X}) \circ \ldots \circ \gamma(\mathbf{X}) \circ \mathbf{X}$$

belongs to $\mathfrak{A}_{\{1,2,\ldots,m+\ell\}}$. As the dimension of this algebra grows exponentially as d^m, we can use the upper bound of Lemma 10.11(a),

$$\mathsf{H}[\omega, \mathbf{X}^m] = S(\rho[\mathbf{X}^m]) \leq S(\rho_{\{1,2,\ldots,m+\ell\}}) + \log d^{m+\ell-1}.$$

Dividing both sites by m and taking the limit $m \to \infty$, this inequality becomes

$$\mathsf{h}[\omega, \gamma, \mathbf{X}] \leq \sigma(\omega) + \log d.$$

In order to prove that the upper bound is reached, we consider the explicit partition

$$\mathbf{X} := \left(\frac{1}{\sqrt{d}} e_{ij} \,\middle|\, i, j = 1, 2, \ldots, d \right),$$

where the e_{ij} are the standard matrix units for $\mathcal{M}_d$. We compute the correlation matrix $\rho[\mathbf{X}]$ for this partition:

$$\rho[\mathbf{X}]_{i_1 i_2, j_1 j_2} = \frac{1}{d} \operatorname{Tr} \rho_{\{1\}} \left(e_{j_1 j_2} \right)^* e_{i_1 i_2}$$

$$= \frac{1}{d} \delta_{i_1 j_1} \operatorname{Tr} \rho_{\{1\}} e_{j_2 i_2}$$

$$= \frac{1}{d} \delta_{i_1 j_1} \rho_{\{1\}}{}_{i_2 j_2};$$

therefore,

$$\rho[\mathbf{X}] = \frac{1}{d} \mathbf{I}_d \otimes \rho_{\{1\}} \quad \text{and} \quad \mathsf{H}[\omega, \mathbf{X}] = S(\rho_{\{1\}}) + \log d.$$

We now observe that the refined operational partition $\mathbf{X}^m$ generated by letting m time steps of the dynamics act on $\mathbf{X}$ is nothing but the set of suitably normalized canonical matrix units $e_{\mathbf{ij}}$ of $\mathcal{M}_d^{\otimes m}$, where $\mathbf{i}$ and $\mathbf{j}$ are multi-indices of length m. So

$$\rho[\mathbf{X}^m] = \frac{1}{d^m} \mathbf{I}_{d^m} \otimes \rho_{\{1,2,\dots,m\}}$$

and

$$\mathsf{H}[\omega, \mathbf{X}^m] = S(\rho_{\{1,2,\dots,m\}}) + m \log d.$$

Dividing by m and taking the limit $m \to \infty$ finishes the proof. $\qquad\square$

One can actually allow for more partitions than these in local elements by using the continuity property of Lemma 10.14. The estimate that was obtained in that lemma is not tight enough to allow for general partitions in quasi-local elements and one, therefore, has to consider partitions in a subalgebra $\mathfrak{A}_0$ of quasi-local elements with sufficiently rapidly decreasing tails.

10.4.2 The free shift

This model is probably not so relevant from a physical point of view. It is, however, a model that corresponds to an extremely chaotic quantum dynamics, without any asymptotic Abelianness in time.

Consider the universal C*-algebra $\mathfrak{A}$ generated by $\mathbf{I}$ and elements $\{e_i \mid i \in \mathbb{Z}\}$ which satisfy the relations

$$e_i^* = e_i \quad \text{and} \quad e_i^2 = \mathbf{I}. \tag{10.19}$$

The algebra $\mathfrak{A}_0$ spanned by simple words is norm-dense in $\mathfrak{A}$. A simple word w of length n is a monomial $e_{i_1} e_{i_2} \cdots e_{i_n}$ that cannot be simplified by means of the relations (10.19), i.e. $i_1 \neq i_2, i_2 \neq i_3, \dots, i_{n-1} \neq i_n$. The identity $\mathbf{I}$ is considered as the empty word.

The *free shift* Θ is the automorphism determined by

$$\Theta(e_i) = e_{i+1}, \quad i \in \mathbb{Z}.$$

Clearly, $\mathfrak{A}_0$ is globally invariant under Θ. There is a unique tracial state τ on $\mathfrak{A}$ determined by

$$\tau(w) := \delta_{w,\emptyset}.$$

It is obvious that the state τ is invariant under Θ and we consider the dynamical system $(\mathfrak{A}, \Theta, \tau)$ together with the distinguished subalgebra $\mathfrak{A}_0$.

Theorem 10.18 *With the notation of above, the dynamical entropy* $h[\tau, \Theta, \mathfrak{A}_0]$
of the free shift is infinite.

Proof Consider for $k = 1, 2, \ldots$ the partition

$$\mathbf{X} = \frac{1}{\sqrt{k}} (e_1, e_1 e_2 e_1, \cdots, (e_1 e_2)^{k-1} e_1).$$

The shift transforms it into

$$\Theta(\mathbf{X}) = \frac{1}{\sqrt{k}} (e_2, e_2 e_3 e_2, \cdots, (e_2 e_3)^{k-1} e_2). \tag{10.20}$$

A general element of the partition $\mathbf{X}^m := \Theta^{m-1}(\mathbf{X}) \circ \cdots \circ \Theta(\mathbf{X}) \circ \mathbf{X}$ is of the
form

$$\frac{1}{k^{m/2}} [e_m \cdots e_m] [e_{m-1} \cdots e_{m-1}] \cdots [e_1 \cdots e_1]. \tag{10.21}$$

Products of the adjoint of an element of the form (10.21) with another element
of the same type yield only a multiple of the identity when both elements are
equal. Therefore, the matrix elements for the m-step density matrix $\rho[\mathbf{X}^m]$ are
given by

$$\rho[\mathbf{X}^m]_{\mathbf{i},\mathbf{j}} = \tau([e_1 \ldots e_1]_{j_1} \ldots [e_m \ldots e_m]_{j_m} [e_m \ldots e_m]_{i_m} \ldots [e_1 \ldots e_1]_{i_1})$$
$$= \frac{1}{k^m} \delta_{\mathbf{i},\mathbf{j}},$$

with $\mathbf{i} := (i_1, \ldots, i_m)$ and $\mathbf{j} := (j_1, \ldots, j_m)$. Hence $\rho[\mathbf{X}^m]$ is the normalized trace
on $\bigotimes^m \mathcal{M}_k$. The entropy density of this state is $\log k$. Since the size k of the
starting partition $\mathbf{X}$ was arbitrary, we have

$$h[\tau, \Theta, \mathfrak{A}_0] = \infty.$$

$\square$

10.4.3 *Infinite entropy*

The following example shows that the entropy can jump to infinity if one allows
too large a class of partitions. We consider a trivial dynamical system as far as
its evolution is concerned. The only contribution to the entropy comes from the
non-commutativity of the algebra.

Let $\mathcal{O}_2$ be the *Cuntz algebra* with two generators. It is the universal C*-
algebra generated by an identity element and by elements S_1 and S_2 that satisfy
the relations

$$S_1^* S_1 = S_2^* S_2 = \mathbf{1} \quad \text{and} \quad S_1 S_1^* + S_2 S_2^* = \mathbf{1}.$$

Theorem 10.19 *Let $\mathfrak{A}_0$ be any $*$-subalgebra of $\mathcal{O}_2$ that contains S_1 and S_2; then*

$$h[\omega, \mathrm{id}, \mathfrak{A}_0] = \infty.$$

Proof Because S_1 and S_2 are isometries, $S_1 S_1^*$ and $S_2 S_2^*$ are projectors, and as $S_1 S_1^* + S_2 S_2^* = \mathbf{1}$, we must have that $S_1 S_1^* S_2 S_2^* = 0$. Multiplying this on the left by S_1^* and on the right by S_2, we see that

$$S_1^* S_2 = S_2^* S_1 = 0.$$

Consider now the partition

$$\mathbf{X} := \frac{1}{\sqrt{2}} (S_1, S_2).$$

From the property (10.20), we obtain that for any two n-tuples $\mathbf{i} = (i_1, i_2, \dots, i_n)$ and $\mathbf{j} = (j_1, j_2, \dots, j_n)$ with $i_k, j_k \in \{1, 2\}$

$$S_{i_1}^* S_{i_2}^* \cdots S_{i_n}^* S_{j_n} \cdots S_{j_1} = \delta_{\mathbf{i}, \mathbf{j}}.$$

Therefore, $\rho[\mathbf{X}^n]$ is the normalized trace on $\bigotimes^n \mathcal{M}_2$ and so

$$h[\omega, \mathrm{id}, \mathfrak{A}_0] \geq \log 2.$$

This implies, by the remark following Lemma 10.16, that

$$h[\omega, \mathrm{id}, \mathfrak{A}_0] = \infty.$$

$\square$

The example of Theorem 10.18 extends to a much wider range of algebras than the Cuntz algebras. In fact, we shall always obtain an infinite entropy as soon as we are able to form partitions in Cuntz-like generators, irrespective of either the dynamics or the reference state. Such a situation occurs, for example, for von Neumann type dynamical systems with a von Neumann algebra $\mathfrak{M}$ that contains a properly infinite factor and with $\mathfrak{A}_0 = \mathfrak{M}$. This is typically the case for GNS-representations of temperature states.

10.4.4   *Powers–Price shifts*

We shall compute the dynamical entropy of the Powers–Price shifts $(\mathfrak{A}_c, \Theta, \tau)$, defined in Example 7.8. The subalgebra $\mathfrak{A}_0$ is the $*$-algebra generated by the e_j, and $\mathfrak{A}_{[k,\ell]}$ denotes the algebra generated by $\{e_j \mid k \leq j \leq \ell\}$.

Theorem 10.20 *The dynamical entropy of the Powers–Price shift is given by*

$$h[\tau, \Theta, \mathfrak{A}_0] = \log 2.$$

Proof An optimal lower bound for the dynamical entropy is found by choosing the partition

$$\mathbf{X} = \{(\mathbf{1} + e_1)/2, (\mathbf{1} - e_1)/2\} = \{p_{11}, p_{12}\}$$

in two projectors. Applying j times the dynamics to these projectors, we obtain $p_{j1} = (\mathbf{1} + e_j)/2$ and a similar expression for p_{j2}. In order to compute $\rho[\mathbf{X}^m]$, we have to reorder the projectors in $\mathbf{X}^m$ using the commutation relations (7.12). It turns out that

$$\rho[\mathbf{X}^m] = 2^{-m}\, \mathbf{1}_{2^m},$$

where $\mathbf{1}_{2^m}$ is the identity matrix of dimension 2^m. Therefore,

$$\mathsf{h}[\tau, \Theta, \mathfrak{A}_0] \geq \log 2. \tag{10.22}$$

The argument for the upper bound bears some resemblance to that in the proof of Theorem 10.17: if the elements of a partition $\mathbf{X}$ in $\mathfrak{A}_0$ belong to $\mathfrak{A}_{[k,\ell]}$, then, after m time steps, the elements of the evolved and refined partition $\mathbf{X}^m$ belong to $\mathfrak{A}_{[k,\ell+m]}$. We show that the entropy of a partition living in $\mathfrak{A}_{[1,r]}$ is bounded from above by $r \log 2$. This then provides, using shift invariance, an upper bound for the dynamical entropy that coincides with the lower bound (10.22). The argument is based on dimensional considerations.

Consider a general partition $\mathbf{Y}$ in $\mathfrak{A}_{[1,r]}$. The algebra $\mathfrak{A}_{[1,r]}$, which has as a complex vector space dimension 2^r, can be decomposed into a direct sum of square matrix algebras (see Example 5.2):

$$\mathfrak{A}_{[1,r]} = \bigoplus_{\nu} \mathcal{M}_{n_\nu} \quad \text{with} \quad \sum_{\nu} n_\nu^2 = 2^r.$$

This decomposition can also be carried out for every element in the partition and for the restriction $\tau_{[1,r]}$ of the trace to $\mathfrak{A}_{[1,r]}$:

$$y_j = \bigoplus_{\nu} y_j^\nu \quad \text{and} \quad \tau_{[1,r]} = \bigoplus_{\nu} \delta_\nu\, \tau^\nu.$$

The normalization coefficients δ_ν are strictly positive and $\sum_\nu \delta_\nu = 1$. But this implies that the correlation matrix $\rho[\mathbf{Y}]$ decomposes according to

$$\rho[\mathbf{Y}] = \sum_{\nu} \delta_\nu\, \rho[\mathbf{Y}]^\nu \quad \text{with} \quad \rho[\mathbf{Y}]_{ij}^\nu = \tau\big((y_j^\nu)^* y_i^\nu\big).$$

We then combine the inequalities (see Lemma 10.11 and Theorem 9.14)

$$S(\rho^\nu) \leq 2 \log n_\nu$$

and

$$S(\rho[\mathbf{Y}]) \leq \sum_\nu 2\delta_\nu \log n_\nu + \sum_\nu \eta(\delta_\nu)$$

to get

$$S(\rho[\mathbf{Y}]) \leq \sum_\nu \delta_\nu \log \frac{n_\nu^2}{\delta_\nu} \leq \log \sum_\nu n_\nu^2 = r \log 2.$$

$\square$

10.5   Notes

The approach to dynamical entropy presented in this book is based on operational partitions of unity. It was initiated in (Lindblad 1979, 1993) and developed by the authors and coworkers. The construction of a symbolic dynamics for a quantum dynamical system presented in Section 10.2 is very reminiscent of the construction of finitely correlated states on quantum spin chains in (Fannes *et al.* 1992).

The first result showing the existence of the quantum dynamical entropy as a supremum over a meaningful class of partitions was presented in (Alicki and Fannes 1994) for the shift on a quantum spin chain. The dynamical entropies of the free shift and the Powers–Price shifts were studied in (Alicki and Narnhofer 1995), while the model based on the Cuntz algebra appeared in (Choda and Takehana 1998; Tuyls 1997).

There exist many attempts to define a quantum analogue of Kolmogorov–Sinai dynamical entropy both for finite and infinite quantum systems; see, for example, the paper (Słomczyński and Życzkowski 1994) and the references therein. Nevertheless, at present, mainly three types of approaches can be rigorously applied to a number of models.

Besides the approach of this book, there are the CNT and Voiculescu entropies. After the introduction of a dynamical entropy in (Connes and Størmer 1975) for type II$_1$ factors, a general notion was developed in (Connes *et al.* 1987). In (Sauvageot and Thouvenot 1992) it was shown that the same notion can be obtained by coupling the quantum system to a classical one and then observing the classical system. For a review on this theory and its many applications we refer to the books by Benatti (1993), Ohya and Petz (1993), Connes (1994), and to the references therein. An entropy based on dimension was given in (Voiculescu 1992). Other rigorous results can be found in (Hudetz 1998; Choda 2000). An attempt to unify these ideas is found in (Accardi *et al.* 1996).

The CNT-entropy has been computed for most of the models that we considered, leading often to different answers. For the shift on a quantum spin chain, with an additional clustering condition, it is equal to the von Neumann mean entropy, without the extra $\log d$ term (Connes *et al.* 1987), for the free shift it is zero (Størmer 1992). For some Powers–Price shifts it is bounded from above by $\log 2$, but it vanishes for almost all choices of bit maps (Narnhofer and Thirring 1995).

Quantum information theory is a potential domain of application for the construction of an entropy on the basis of operational partitions. A first application in this context can be found in (Alicki 1997).

11

CLASSICAL DYNAMICAL ENTROPY

In this chapter, we approach classical dynamical systems from an algebraic point of view: we move from the standard description on phase space to one in terms of complex functions on phase space. So we pass from the level of phase space points to that of observables. Replacing the Abelian algebra by a general C*-algebra is the first step in the introduction of quantum groups, quantum stochastic processes or non-commutative differential geometry. The geometrical structure of the phase space is now encoded in the algebra of observables. The dynamical flow on phase space translates into a group of automorphisms and an invariant phase space measure defines a state. Doing so, we obtain a scheme that is quite similar to that of quantum dynamics, the main difference lies of course in the lack of commutativity of the observables of a truly quantum mechanical system.

This chapter contains the results of (Alicki *et al.* 1996*a*), except for the unpublished Lemma 11.8. Another application of the quantum formalism to classical dynamical systems and the proof of Lemma 11.7 were presented in (Fannes 1998).

11.1 The Kolmogorov–Sinai invariant

In this section, we consider classical dynamical systems from a statistical point of view. A description in terms of measure-preserving transformations is adequate for this purpose. Consider a measurable space $\mathfrak{X}$ with σ-algebra $\mathfrak{S}$ and probability measure μ and assume that $\mathfrak{S}$ is complete, i.e. any subset B of a null-set A automatically belongs to $\mathfrak{S}$ and has zero measure. The dynamics is given in terms of an automorphism of $\mathfrak{X}$, i.e. a measure-preserving transformation T of $\mathfrak{X}$ with measure-preserving inverse:

$$\mu(A) = \mu(T(A)) = \mu(T^{-1}(A)), \quad A \in \mathfrak{S}.$$

On the level of observables T translates into a measure-preserving automorphism:

$$\Theta(f)(x) := f(T(x)), \quad x \in \mathfrak{X}, \ f \in \mathcal{L}^\infty(\mathfrak{X}, \mu).$$

The triple $(\mathfrak{X}, T, \mu)$ is called a *measurable classical dynamical system*, although the space $\mathfrak{X}$ carries often more structure such as that of a smooth manifold; see Section 3.3.

The *Kolmogorov–Sinai invariant* $\mathsf{h}^{\mathrm{KS}}[\mu, T]$ of $(\mathfrak{X}, T, \mu)$ is constructed as follows: if $\mathbf{C} = (C_1, C_2, \dots, C_k)$ is a partition of $\mathfrak{X}$ in disjoint, measurable subsets, the entropy $\mathsf{H}^{\mathrm{KS}}[\mu, \mathbf{C}]$ of $\mathbf{C}$ is computed as

$$\mathsf{H}^{\mathrm{KS}}[\mu, \mathbf{C}] := \sum_{j=1}^{k} \eta(\mu(C_j)).$$

If $\mathbf{C}$ and $\mathbf{D}$ are two partitions of $\mathfrak{X}$, then $\mathbf{C} \vee \mathbf{D}$ is the partition of $\mathfrak{X}$ generated by $\mathbf{C}$ and $\mathbf{D}$. It consists of the non-empty intersections of sets C_i and D_j. Choosing a fixed partition $\mathbf{C}$, it can be shown that

$$\mathsf{h}^{\mathrm{KS}}[\mu, T, \mathbf{C}] := \lim_{m \to \infty} \frac{1}{m} \, \mathsf{H}^{\mathrm{KS}} \left[\mu, T^{-m+1}(\mathbf{C}) \vee \cdots T^{-1}(\mathbf{C}) \vee \mathbf{C}\right]$$

exists. The Kolmogorov–Sinai entropy $\mathsf{h}^{\mathrm{KS}}[\mu, T]$ is defined as

$$\mathsf{h}^{\mathrm{KS}}[\mu, T] := \sup_{\mathbf{C}} \, \mathsf{h}^{\mathrm{KS}}[\mu, T, \mathbf{C}].$$

We now turn to the algebraic description of a classical dynamical system in order to apply the techniques that were introduced in the previous chapters for quantum dynamical systems. In the von Neumann description of a classical system, the algebra of classical observables consists of the essentially bounded complex measurable functions $\mathcal{L}^{\infty}(\mathfrak{X}, \mu)$ on the phase space $\mathfrak{X}$. These functions act on the Hilbert space of square integrable functions on $\mathfrak{X}$ as multiplication operators:

$$\varphi \mapsto (M_f \, \varphi)(x) := f(x) \, \varphi(x), \quad \varphi \in \mathcal{L}^2(\mathfrak{X}, \mu).$$

In order to simplify the notation, we identify M_f with f. This is allowed because $M_f = 0$ iff $f = 0$ μ-a.e. The constant function $\mathbf{1}$ is a normalized vector in $\mathcal{L}^2(\mathfrak{X}, \mu)$ as μ is a probability measure and it reproduces the measure μ as a vector state:

$$\mu(M_f) := \int_{\mathfrak{X}} \mu(dx) \, f(x) = \langle \mathbf{1}, f\mathbf{1} \rangle.$$

Moreover, the Koopman formalism (5.24) implements the dynamics in terms of a unitary operator on $\mathcal{L}^2(\mathfrak{X}, \mu)$. For the case of continuous functions on a compact phase space, the Koopman construction provides an explicit realization of Theorem 5.22.

In the case of a classical probability space $(\mathfrak{X}, \mu)$, composing operational partitions amounts to refining measurable partitions. To a partition $\mathbf{C} = (C_1, C_2, \dots, C_k)$ of the phase space $\mathfrak{X}$ into disjoint measurable subsets, we can naturally associate the partition of unity $(I_{C_1}, I_{C_2}, \dots, I_{C_k})$, where I_A is the indicator function of the set A. Intersections of subsets of X correspond to products of indicator functions, which means that the notion of refinement in the classical sense coincides with that of composing operational partitions.

An operational partition $\mathbf{F}$ in $\mathcal{L}^{\infty}(\mathfrak{X}, \mu)$ is a k-tuple $(f_1, f_2, \dots, f_k)$ of μ-measurable functions such that

$$\sum_{j=1}^{k} |f_j|^2 = \mathbf{1}. \tag{11.1}$$

The size k of $\mathbf{F}$ is denoted by $|\mathbf{F}|$. We compute the matrix elements of the density matrix $\rho[\mathbf{F}]$:

$$\rho[\mathbf{F}]_{ij} := \mu(\overline{f_j}\, f_i) = \int_{\mathfrak{X}} \mu(dx)\, \overline{f_j(x)}\, f_i(x).$$

The $k \times k$ correlation matrix $\rho[\mathbf{F}]$ has a special structure: it is a continuous mixture of one-dimensional projectors. In order to see this, it is useful to write it in the following way: let $(e_1, e_2, \ldots, e_k)$ be the standard orthonormal basis for $\mathbb{C}^k$. For each $x \in \mathfrak{X}$, the vector

$$\boldsymbol{\Psi}_{\mathbf{F}}(x) := \sum_{j=1}^{k} f_j(x)\, e_j$$

is normalized because of condition (11.1):

$$\left\| \sum_{j=1}^{k} f_j(x)\, e_j \right\|^2 = \sum_{j=1}^{k} |f_j(x)|^2 = 1$$

Therefore,

$$x \mapsto P[\mathbf{F}](x) := |\boldsymbol{\Psi}_{\mathbf{F}}(x)\rangle\langle\boldsymbol{\Psi}_{\mathbf{F}}(x)|$$

is a measurable function from $\mathfrak{X}$ to the projectors of dimension 1 in the $k \times k$ matrices $\mathcal{M}_k$. The correlation matrix $\rho[\mathbf{F}]$ is a mixture of such projectors:

$$\rho[\mathbf{F}] = \int_{\mathfrak{X}} \mu(dx)\, P[\mathbf{F}](x). \tag{11.2}$$

The composition of several operational partitions translates into a tensor structure for the corresponding density matrix. Consider partitions $\mathbf{F} = (f_1, f_2, \ldots, f_k)$ and $\mathbf{G} = (g_1, g_2, \ldots, g_\ell)$ of sizes k and ℓ and denote by $\{e_1, e_2, \ldots, e_k\}$ and $\{e'_1, e'_2, \ldots, e'_\ell\}$ the standard orthonormal bases of $\mathbb{C}^k$ and $\mathbb{C}^\ell$. We write out the vector-valued function $\boldsymbol{\Psi}_{\mathbf{F} \circ \mathbf{G}}$ in the standard basis $\{e_i \otimes e'_j \mid i, j\}$ of $\mathbb{C}^k \otimes \mathbb{C}^\ell$:

$$\boldsymbol{\Psi}_{\mathbf{F} \circ \mathbf{G}}(x) = \sum_{i=1}^{k} \sum_{j=1}^{\ell} f_i(x)\, g_j(x)\, e_i \otimes e'_j$$

$$= \sum_{i=1}^{k} f_i(x)\, e_i \otimes \sum_{j=1}^{\ell} g_j(x)\, e'_j$$

$$= \boldsymbol{\Psi}_{\mathbf{F}}(x) \otimes \boldsymbol{\Psi}_{\mathbf{G}}(x).$$

But this means that $P[\mathbf{F} \circ \mathbf{G}] = P[\mathbf{F}] \otimes P[\mathbf{G}]$. Iterating this argument yields that for n partitions $\mathbf{F}_i$

$$\rho[\mathbf{F}_1 \circ \mathbf{F}_2 \circ \cdots \circ \mathbf{F}_n] = \int_{\mathfrak{X}} d\mu\, P[\mathbf{F}_1] \otimes P[\mathbf{F}_n] \otimes \cdots \otimes P[\mathbf{F}_n]. \tag{11.3}$$

The commutativity of our classical algebra has some useful consequences for the correlation matrices $\rho[\mathbf{F}]$ that generally fail for truly non-commutative dynamical systems. For three partitions $\mathbf{F}_1$, $\mathbf{F}_2$ and $\mathbf{F}_3$ the following properties hold:

(a) The correlation matrix $\rho[\mathbf{F}_1 \circ \mathbf{F}_2]$ is unitarily equivalent to $\rho[\mathbf{F}_2 \circ \mathbf{F}_1]$.

(b) $\rho[\mathbf{F}_1 \circ \mathbf{F}_3]$ is unitarily equivalent to the partial trace of $\rho[\mathbf{F}_1 \circ \mathbf{F}_2 \circ \mathbf{F}_3]$ over the middle space.

Lemma 11.1 *For any three partitions $\mathbf{F}_1$, $\mathbf{F}_2$, and $\mathbf{F}_3$ of an Abelian algebra, the following properties hold:*

(a) $0 \leq \mathsf{H}[\mu, \mathbf{F}_1] \leq \log |\mathbf{F}_1|$

(b) $\mathsf{H}[\mu, \mathbf{F}_1 \circ \mathbf{F}_2 \circ \mathbf{F}_3] = \mathsf{H}[\mu, \mathbf{F}_2 \circ \mathbf{F}_1 \circ \mathbf{F}_3]$

(c) $\mathsf{H}[\mu, \mathbf{F}_1] \leq \mathsf{H}[\mu, \mathbf{F}_1 \circ \mathbf{F}_2]$

(d) $\mathsf{H}[\mu, \mathbf{F}_1 \circ \mathbf{F}_2] \leq \mathsf{H}[\mu, \mathbf{F}_1] + \mathsf{H}[\mu, \mathbf{F}_2]$

Proof The bound (a) follows from (9.12), (b) is an immediate consequence of the commutativity and (d) follows from subadditivity (9.15).

We first show that (c) holds for the case of an atomic probability measure $\mu = (\mu_1, \mu_2, \dots, \mu_n)$ supported in n points. The result then extends by continuity to general μ. Consider the density matrix

$$\bigoplus_{j=1}^{n} \mu_j \, P[\mathbf{F}_1](j) \otimes P[\mathbf{F}_2](j)$$

on $\mathbb{C}^n \otimes \mathbb{C}^{|\mathbf{F}_1|} \otimes \mathbb{C}^{|\mathbf{F}_2|}$. By strong subadditivity (see (9.16))

$$S\left(\bigoplus_{j=1}^{n} \mu_j \, P[\mathbf{F}_1](j) \otimes P[\mathbf{F}_2](j)\right) + S\left(\sum_{j=1}^{n} \mu_j \, P[\mathbf{F}_1](j)\right)$$

$$\leq S\left(\bigoplus_{j=1}^{n} \mu_j \, P[\mathbf{F}_1](j)\right) + S\left(\sum_{j=1}^{n} \mu_j \, P[\mathbf{F}_1](j) \otimes P[\mathbf{F}_2](j)\right).$$

Using the explicit expressions (11.2) and (11.3), we obtain

$$S\left(\bigoplus_{j=1}^{n} \mu_j \, P[\mathbf{F}_1](j) \otimes P[\mathbf{F}_2](j)\right) = S\left(\bigoplus_{j=1}^{n} \mu_j \, P[\mathbf{F}_1](j)\right)$$

$$= \sum_{j=1}^{n} \eta(\mu_j).$$

It follows that

$$S\left(\int_{\mathfrak{X}} d\mu \, P[\mathbf{F}_1]\right) \leq S\left(\int_{\mathfrak{X}} d\mu \, P[\mathbf{F}_1] \otimes P[\mathbf{F}_2]\right).$$

As the $P[\mathbf{F}]$ are one-dimensional projectors in a Hilbert space of fixed, finite dimension, we may approximate any measure by atomic measures and use the continuity of the entropy to get (c). $\qquad \square$

The next result shows that both the Kolmogorov–Sinai construction and the notion of dynamical entropy in terms of operational partitions coincide for classical dynamical systems in the widest possible setting.

Theorem 11.2 *Let $(\mathfrak{X}, T, \mu)$ be a classical dynamical system with algebraic version $(\mathcal{L}^\infty(\mathfrak{X}, \mu), \Theta, \mu)$, then the dynamical entropy $\mathsf{h}[\mu, \Theta, \mathcal{L}^\infty]$, computed with respect to the full algebra $\mathcal{L}^\infty(\mathfrak{X}, \mu)$, is equal to the Kolmogorov–Sinai invariant $\mathsf{h}^{KS}[\mu, T]$.*

Proof The inequality $\mathsf{h}^{KS}[\mu, T] \leq \mathsf{h}[\mu, \Theta, \mathcal{L}^\infty]$ follows from the observation that each partition $\mathbf{C} = (C_1, C_2, \ldots, C_\ell)$ of $\mathfrak{X}$ in disjoint measurable sets C_j corresponds to an operational partition $\mathbf{C} = (I_{C_1}, I_{C_2}, \ldots, I_{C_\ell})$. Both entropies, $\mathsf{h}^{KS}[\mu, T, \mathbf{C}]$ in the sense of subsets and $\mathsf{h}[\mu, \Theta, \mathbf{C}]$ in the sense of operational partitions, coincide in this case. As $\mathsf{h}[\mu, \Theta, \mathcal{L}^\infty]$ is obtained by taking the supremum over all finite operational partitions, the inequality follows.

We show that $\mathsf{h}^{KS}[\mu, T]$ is an upper bound for the entropy $\mathsf{h}[\mu, \Theta, \mathbf{F}]$ generated by a partition $\mathbf{F}$ of arbitrary size k. Let $\mathbf{C} = (C_1, \ldots, C_\ell)$ be a measurable partition of $\mathfrak{X}$ into disjoint sets. We denote by $\mathbf{C}^{(m)}$ and $\mathbf{F}^{(m)}$ the partitions which are generated from $\mathbf{C}$ and $\mathbf{F}$ by applying $m-1$ time steps of the dynamics:

$$\mathbf{C}^{(m)} := T^{-m+1}(\mathbf{C}) \vee \cdots \vee T^{-1}(\mathbf{C}) \vee \mathbf{C},$$
$$\mathbf{F}^{(m)} := \Theta^{m-1}(\mathbf{F}) \circ \cdots \circ \Theta(\mathbf{F}) \circ \mathbf{F}.$$

We denote the elements of the partition $\mathbf{C}^{(m)}$ by $C_{\mathbf{j}}$ with $\mathbf{j} = (j_1, j_2, \ldots, j_m)$ and

$$C_{\mathbf{j}} = C_{j_1} \cap T^{-1}(C_{j_2}) \cap \cdots \cap T^{-m+1}(C_{j_m}).$$

Using the representation (11.3) of $\rho[\mathbf{F}^{(m)}]$, we may write

$$\rho[\mathbf{F}^{(m)}] = \int_{\mathfrak{X}} \mu(dx) \, P[\mathbf{F}^{(m)}](x)$$

$$= \sum_{\mathbf{j}} \int_{C_{\mathbf{j}}} \mu(dx) \, P[\mathbf{F}^{(m)}](x)$$

$$= \sum_{\mathbf{j}} \int_{C_{\mathbf{j}}} \mu(dx) \, P[\Theta^{m-1}(\mathbf{F})](x) \otimes \cdots \otimes P[\mathbf{F}](x)$$

$$= \sum_{\mathbf{j}} \mu(C_{\mathbf{j}}) \frac{1}{\mu(C_{\mathbf{j}})} \int_{C_{\mathbf{j}}} \mu(dx) \, P[\Theta^{m-1}(\mathbf{F})](x) \otimes \cdots \otimes P[\mathbf{F}](x).$$

In this last expression, we have written $\rho[\mathbf{F}^{(m)}]$ as a convex combination of ℓ^m density matrices on $\bigotimes^m \mathbb{C}^k$. Therefore, we get by (9.14) an upper bound for $\mathsf{H}[\mu, \mathbf{F}^{(m)}]$:

$$\mathsf{H}[\mu, \mathbf{F}^{(m)}] = S\left(\rho[\mathbf{F}^{(m)}]\right)$$

$$\leq \sum_{\mathbf{j}} \eta(\mu(C_{\mathbf{j}})) + \sum_{\mathbf{j}} \mu(C_{\mathbf{j}}) \, S\left(\frac{1}{\mu(C_{\mathbf{j}})} \int_{C_{\mathbf{j}}} \mu(dx) \, P[\mathbf{F}^{(m)}](x)\right).$$

Finally, we use the subadditivity of the entropy to estimate the entropies in this sum by the sum of the entropies of the restrictions of the density matrices to each factor in $\bigotimes^m \mathbb{C}^k$:

$$H[\mu, \mathbf{F}^{(m)}] \leq \sum_j \eta(\mu(C_j)) + \sum_j \mu(C_j) \sum_{i=0}^{m-1} S\left(\frac{1}{\mu(C_j)} \int_{C_j} \mu(dx)\, P[\Theta^i(\mathbf{F})](x)\right).$$

For a suitable choice of $\mathbf{C}$, the $P[\Theta^i(\mathbf{F})]$ remain on each C_j sufficiently close to a one-dimensional projector in a fixed, finite-dimensional space of dimension at most k. As the entropy S is a continuous function on the $k \times k$ density matrices which vanishes on the pure states, we can estimate

$$S\left(\frac{1}{\mu(C_j)} \int_{C_j} \mu(dx)\, P[\Theta^i(\mathbf{F})](x)\right)$$

from above by a small number which is independent of j and m. The proof is finished by dividing by m and taking the limit $m \to \infty$. □

Actually, because of the monotonicity property of Lemma 11.1(c) one can show that for classical dynamical systems the dynamical entropy exists as a limit and not as a lim sup.

11.2 H-density

In specific examples, one would hardly take the supremum in the definition of the entropy over all possible measurable partitions of unity but rather only over a limited class. In particular, we would like to replace the sharp partitions that arise by slicing the phase space in disjoint sets by smoother ones. The aim of this section is to state a condition that guarantees that the Kolmogorov–Sinai entropy is indeed reached by considering this restricted supremum. In order to reach this goal, we introduce an entropic distance between partitions and show that, for computing the dynamical entropy, it is sufficient to consider only partitions that can approximate the sharp ones arbitrarily well.

Definition 11.3 Let $\mathbf{F}$ and $\mathbf{G}$ be two operational partitions of unity of a classical probability space $\mathcal{L}^\infty(\mathfrak{X}, \mu)$. The entropic distance $\Delta[\mathbf{F} \mid \mathbf{G}]$ of $\mathbf{F}$ with respect to $\mathbf{G}$ is

$$\Delta[\mathbf{F} \mid \mathbf{G}] := H[\mu, \mathbf{F} \circ \mathbf{G}] - H[\mu, \mathbf{G}].$$

Lemma 11.4 The entropic distance satisfies the following properties:

(a) $\Delta[\mathbf{F} \mid \mathbf{G}] \geq 0$

(b) $\Delta[\mathbf{F}_1 \circ \mathbf{F}_2 \mid \mathbf{G}] \leq \Delta[\mathbf{F}_1 \mid \mathbf{G}] + \Delta[\mathbf{F}_2 \mid \mathbf{G}]$

(c) $\Delta[\mathbf{F} \mid \mathbf{G}_1 \circ \mathbf{G}_2] \leq \Delta[\mathbf{F} \mid \mathbf{G}_1]$.

Proof Statement (a) is actually the same as statement (c) of Lemma 11.1.

The proof of (b) uses the strong subadditivity of the entropy and remark (a) preceding Lemma 11.1:

$$\Delta[\mathbf{F}_1 \circ \mathbf{F}_2 \,|\, \mathbf{G}] = H[\mu, \mathbf{F}_1 \circ \mathbf{F}_2 \circ \mathbf{G}] - H[\mu, \mathbf{G}]$$
$$\leq H[\mu, \mathbf{F}_1 \circ \mathbf{G}] + H[\mu, \mathbf{F}_2 \circ \mathbf{G}] - 2H[\mu, \mathbf{G}]$$
$$= \Delta[\mathbf{F}_1 \,|\, \mathbf{G}] + \Delta[\mathbf{F}_2 \,|\, \mathbf{G}].$$

Finally, (c) follows from (b) and the same remark as above:

$$\Delta[\mathbf{F}_1 \,|\, \mathbf{G}_1 \circ \mathbf{G}_2] = H[\mu, \mathbf{F} \circ \mathbf{G}_1 \circ \mathbf{G}_2] - H[\mu, \mathbf{G}_1 \circ \mathbf{G}_2]$$
$$= \Delta[\mathbf{F} \circ \mathbf{G}_2 \,|\, \mathbf{G}_1] + H[\mu, \mathbf{G}_1] - \Delta[\mathbf{G}_2 \,|\, \mathbf{G}_1] - H[\mu, \mathbf{G}_1]$$
$$\leq \Delta[\mathbf{F} \,|\, \mathbf{G}_1] + \Delta[\mathbf{G}_2 \,|\, \mathbf{G}_1] - \Delta[\mathbf{G}_2 \,|\, \mathbf{G}_1]$$
$$= \Delta[\mathbf{F} \,|\, \mathbf{G}_1].$$

$$\square$$

Definition 11.5 *A unital subalgebra $\mathfrak{B}$ of $\mathcal{L}^\infty(\mathfrak{X}, \mu)$ is H-dense in $\mathcal{L}^\infty(\mathbf{X}, \mu)$ if for any finite partition $\mathbf{C} = (C_1, C_2, \dots, C_k)$ of $\mathfrak{X}$ in measurable sets and for any $\epsilon > 0$ there exists in $\mathfrak{B}$ an operational partition $\mathbf{G} = (g_1, g_2, \dots, g_\ell)$ of unity such that*

$$\Delta[\mathbf{C} \,|\, \mathbf{G}] \leq \epsilon. \tag{11.4}$$

Theorem 11.6 *Let $(\mathfrak{X}, T, \mu)$ be a measurable classical dynamical system with algebraic version $(\mathcal{L}^\infty(\mathfrak{X}, \mu), \Theta, \mu)$. If a unital subalgebra $\mathfrak{B}$ of $\mathcal{L}^\infty(\mathfrak{X}, \mu)$ is globally invariant under Θ and H-dense, then $h(\mu, \Theta, \mathfrak{B}) = h^{\mathrm{KS}}[\mu, T]$.*

Proof Obviously, due to Theorem 11.2,

$$h[\mu, \Theta, \mathfrak{B}] \leq h[\mu, \Theta, \mathcal{L}^\infty] = h^{\mathrm{KS}}[\mu, T].$$

As $\mathfrak{B}$ is H-dense, we can find for each partition $\mathbf{C}$ of $\mathfrak{X}$ in measurable sets and each $\epsilon > 0$ an operational partition $\mathbf{G}$ in $\mathfrak{B}$ such that

$$\Delta[\mathbf{C} \,|\, \mathbf{G}] \leq \epsilon.$$

We first use the property (c) of Lemma 11.1, which states that refining a partition increases the entropy:

$$H[\mu, \Theta^{m-1}(\mathbf{C}) \circ \cdots \circ \Theta(\mathbf{C}) \circ \mathbf{C}]$$
$$\leq H[\mu, \Theta^{m-1}(\mathbf{G}) \circ \cdots \circ \Theta(\mathbf{G}) \circ \mathbf{G} \circ \Theta^{m-1}(\mathbf{C}) \circ \cdots \circ \Theta(\mathbf{C}) \circ \mathbf{C}]. \tag{11.5}$$

Next, we write, using the definition of entropic distance,

$$\mathsf{H}[\mu, \Theta^{m-1}(\mathbf{G}) \circ \cdots \circ \Theta(\mathbf{G}) \circ \mathbf{G} \circ \Theta^{m-1}(\mathbf{C}) \circ \cdots \circ \Theta(\mathbf{C}) \circ \mathbf{C}]$$
$$= \Delta[\Theta^{m-1}(\mathbf{G}) \circ \cdots \circ \Theta(\mathbf{G}) \circ \mathbf{G} \mid \Theta^{m-1}(\mathbf{C}) \circ \cdots \circ \Theta(\mathbf{C}) \circ \mathbf{C}]$$
$$+ \mathsf{H}[\mu, \Theta^{m-1}(\mathbf{G}) \circ \cdots \circ \Theta(\mathbf{G}) \circ \mathbf{G}]. \tag{11.6}$$

Suppose that we have obtained the upper bound

$$\Delta[\Theta^{m-1}(\mathbf{G}) \circ \cdots \circ \Theta(\mathbf{G}) \circ \mathbf{G} \mid \Theta^{m-1}(\mathbf{C}) \circ \cdots$$
$$\circ \Theta(\mathbf{C}) \circ \mathbf{C}] \leq m \, \Delta[\mathbf{C} \mid \mathbf{G}] \tag{11.7}$$

for the first term in the sum in (11.6). Combining (11.5)–(11.7) and dividing by m, we get

$$\frac{1}{m} \mathsf{H}[\mu, \Theta^{m-1}(\mathbf{C}) \circ \cdots \circ \Theta(\mathbf{C}) \circ \mathbf{C}]$$
$$\leq \frac{1}{m} \mathsf{H}[\mu, \Theta^{m-1}(\mathbf{G}) \circ \cdots \circ \Theta(\mathbf{G}) \circ \mathbf{G}] + \Delta[\mathbf{C} \mid \mathbf{G}].$$

Letting m tend to infinity and using H-density, the theorem follows. It remains to prove (11.7):

$$\Delta[\Theta^{m-1}(\mathbf{G}) \circ \cdots \circ \Theta(\mathbf{G}) \circ \mathbf{G} \mid \Theta^{m-1}(\mathbf{C}) \circ \cdots \circ \Theta(\mathbf{C}) \circ \mathbf{C}]$$
$$\leq \sum_{j=0}^{m-1} \Delta[\Theta^j(\mathbf{C}) \mid \Theta^{m-1}(\mathbf{G}) \circ \cdots \circ \Theta(\mathbf{G}) \circ \mathbf{G}]$$
$$\leq \sum_{j=0}^{m-1} \Delta[\Theta^j(\mathbf{C}) \mid \Theta^j(\mathbf{G})]$$
$$= m \, \Delta[\mathbf{C} \mid \mathbf{G}].$$

The inequalities are an application of Lemma 11.1 and the remarks preceding the lemma. □

There are several natural instances of H-dense algebras, according to the regularity properties of the dynamical system $(\mathfrak{X}, T, \mu)$. We list here a few typical cases.

(a) The algebra of measurable simple functions is H-dense in $\mathcal{L}^\infty(\mathfrak{X}, \mu)$.

(b) The algebra $\mathcal{C}(\mathfrak{X})$ of continuous complex functions on $\mathfrak{X}$ is H-dense for a dynamical system belonging to a compact Hausdorff space $\mathfrak{X}$ with continuous dynamical map T and regular T-invariant reference measure μ.

(c) $\mathcal{C}^\infty(\mathfrak{X})$ is H-dense for a dynamical system belonging to a smooth compact manifold $\mathfrak{X}$ with smooth dynamical map T and regular T-invariant reference measure μ.

In the usual computation of dynamical entropy, one has to study the proliferation of intersections of subsets of the phase space under the dynamics. This becomes quite complicated unless there exists a sufficiently large class of subsets of $\mathfrak{X}$ which transform nicely under the dynamics, e.g. Markov partitions. Passing from the geometrical picture to a more analytic one, allows us to replace sharp partitions by softer ones given in terms of continuous functions. The price to be paid is that we must now consider non-diagonal matrices. A special situation occurs when one is able to find a Fourier version of the dynamics in the sense that one considers partitions in terms of complex functions of constant modulus. Such partitions $\mathcal{P}_u(\mathfrak{A}_0)$ and their corresponding entropies have been introduced at the end of Section 10.1. A suitable choice of partitions might then again lead to essentially diagonal density matrices if the often wildly oscillating phases of the products of different elements in the partition average to zero.

Lemma 11.7 *For a classical dynamical system* $(\mathfrak{X}, T, \mu)$, *we have*

$$h_u[\mu, \Theta, \mathfrak{A}_0] = h^{KS}[\mu, T]$$

where $\mathfrak{A}_0$ *is the algebra of continuous complex functions in the case of a compact Hausdorff space* $\mathfrak{X}$ *with continuous dynamics* T *and regular reference measure* μ. *If* $\mathfrak{X}$ *is a smooth manifold and* T *a diffeomorphism of* $\mathfrak{X}$, *then we may choose* $\mathfrak{A}_0 = C^\infty(\mathfrak{X})$.

Proof Consider a partition $\mathbf{C} = (C_1, C_2, \ldots, C_k)$ of $\mathfrak{X}$ in disjoint measurable sets and associate with it the partition $\mathbf{F} = (f_1/\sqrt{k}, f_2/\sqrt{k}, \ldots, f_k/\sqrt{k})$ with f_j equal to $\exp(2\pi i j\ell/k)$ on C_ℓ,

$$f_j := \sum_{\ell=1}^{k} \exp(2\pi i j\ell/k)\, I_{C_\ell}.$$

A simple calculation shows that the matrices with $j\ell$ entries $\mu(C_j)\, \delta_{j\,\ell}$ and $\mu(\overline{f_\ell}\, f_j)$ are unitarily equivalent. This unitary equivalence is maintained under the time evolution. But this means that the entropy of the partition $\mathbf{C}$ is equal to that of the partition $\mathbf{F}$. By Theorem 11.6, it is sufficient to show that the entropic distance $\Delta[\mathbf{C}\,|\,\mathbf{F}]$ between any partition $\mathbf{C} = (C_1, C_2, \ldots, C_k)$ of $\mathfrak{X}$ and a partition in continuous functions of constant absolute value can be made arbitrarily small. Smoothening the partition $\mathbf{F}$ into a partition in continuous functions of constant absolute value, the result follows. The case of a smooth manifold can be handled along the same lines.   □

Lemma 11.8 *Assume that a unital subalgebra* $\mathfrak{B} \subset \mathcal{L}^\infty(\mathfrak{X}, \mu)$ *contains a unimodular complete orthonormal system* $\{f_j \mid j = 1, 2, \ldots\}$, *i.e.*

$$|f_j(x)| = 1 \quad and \quad \int_{\mathfrak{X}} d\mu\, \overline{f_j}\, f_k = \delta_{jk};$$

then $\mathfrak{B}$ *is* H-*dense.*

Proof For any partition $\mathbf{C} = (C_1, C_2, ..., C_k)$ of the phase space $\mathfrak{X}$, which we identify as always with the partition of unity consisting of the corresponding indicator functions and, for any $\epsilon > 0$, we shall find a partition of the form

$$\mathbf{F}_N = \left(\frac{1}{\sqrt{N}} f_1, \frac{1}{\sqrt{N}} f_2, \dots , \frac{1}{\sqrt{N}} f_N \right)$$

such that the entropic distance $\Delta[\mathbf{C} \,|\, \mathbf{F}_N] < \epsilon$. Due to Theorem 10.4, the entropy $H[\mu, \mathbf{C} \circ \mathbf{F}_N]$ is equal to the von Neumann entropy of the density matrix ρ represented by the integral kernel

$$\rho(x, y) = \frac{1}{N} \sum_{n=1}^{k} \sum_{j=1}^{N} I_{C_n}(x)\, f_j(x)\, \overline{f_j(y)}\, I_{C_n}(y)$$

on $\mathcal{L}^2(\mathfrak{X}, \mu)$. Therefore, ρ can be written as

$$\rho = \frac{1}{N} \sum_{n=1}^{k} Q_n\, P_N\, Q_n \tag{11.8}$$

where

$$(Q_n \psi)(x) := I_{C_n}(x)\, \psi(x) \quad \text{and} \quad P_N := \sum_{j=1}^{N} |f_j\rangle\langle f_j|$$

are projectors acting on $\mathcal{L}^2(\mathfrak{X}, \mu)$. Again as a consequence of Theorem 10.4, the entropy $H[\mu, \mathbf{F}_N]$ is equal to the entropy of $(1/N) \sum_{j=1}^{N} |f_j\rangle\langle f_j|$ and hence equal to $\log N$.

Next, we need the expression for the trace of the operator $Q_n P_N Q_n$. For a bounded operator A, the spectra of AA^* and A^*A are equal up to multiplicities of zero. This is a special case of the general result (5.5). Therefore, the traces of AA^* and A^*A are equal too:

$$\mathrm{Tr}(Q_n P_N Q_n) = \mathrm{Tr}(P_N Q_n P_N)$$

$$= \sum_{j=1}^{N} \int_{\mathfrak{X}} \mu(dx)\, |f_j(x)|^2\, I_{C_n}(x) = N\mu(C_n). \tag{11.9}$$

The density matrix (11.8) is a sum of trace-class operators with orthogonal ranges and we can apply Theorem 9.10. This, together with (11.9), yields

$$\Delta[\mathbf{C} \,|\, \mathbf{F}_N] = \mathsf{H}[\mu, \mathbf{C} \circ \mathbf{F}_N] - \mathsf{H}[\mu, \mathbf{F}_N]$$

$$= \sum_{n=1}^{k} \mu(C_n)\, S\Big(\big(N\mu(C_n)\big)^{-1} Q_n P_N Q_n\Big)$$

$$+ \sum_{n=1}^{k} \eta\big(\mu(C_n)\big) - \log N$$

$$= \frac{1}{N} \sum_{n=1}^{k} \operatorname{Tr} \eta(Q_n P_N Q_n).$$

Using once more the equality of the traces of $Q_n P_N Q_n$ and $P_N Q_n P_N$, the theorem is proved if we show that

$$\lim_{N \to \infty} \frac{1}{N} \operatorname{Tr} \eta(P_N Q_n P_N) = 0$$

for all $n = 1, 2, \ldots, k$.

For any $\epsilon > 0$ there exists a $C(\epsilon) > 0$ such that

$$0 \le \eta(x) \le \epsilon + C(\epsilon) x(1 - x), \quad 0 \le x \le 1.$$

As $0 \le P_N Q_n P_N \le \mathbf{1}$ and as the dimension of the range of $P_N Q_n P_N$ is equal to N, the estimate

$$\frac{1}{N} \operatorname{Tr} \eta(P_N Q_n P_N) \le \epsilon + C(\epsilon) \frac{1}{N} \Big(\operatorname{Tr}(P_N Q_n P_N) - \operatorname{Tr}(P_N Q_n P_N)^2 \Big)$$

holds. We now compute explicitly

$$\lim_{N \to \infty} \frac{1}{N} \Big(\operatorname{Tr}(P_N Q_n P_N) - \operatorname{Tr}(P_N Q_n P_N)^2 \Big)$$

$$= \mu(C_n) - \lim_{N \to \infty} \frac{1}{N} \sum_{i,j=1}^{N} \langle f_j, Q_n f_i \rangle \langle f_i, Q_n f_j \rangle = 0,$$

which implies that $\Delta[\mathbf{C} \,|\, \mathbf{F}_N] < \epsilon$. $\square$

Example 11.9 *(Entropy of expanding maps)*   The expanding circle maps

$$T_k : z \mapsto z^k, \quad |z| = 1, \quad k \in \{2, 3, \ldots\}$$

of Example 7.14, with the Lebesgue measure μ as reference measure, are non-invertible. They do not therefore exactly fit with the kind of dynamical systems that we consider. It is, however, completely straightforward to generalize the whole setup so as to include such maps. The algebra $\mathfrak{B}$ generated by the unimodular orthonormal system

$$\{z \mapsto z^j \mid j \in \mathbb{Z}\}$$

is H-dense by Lemma 11.8 and globally invariant under the dynamics. Therefore,

$$h_u[\mu, \Theta, \mathfrak{B}] = h^{KS}[\mu, T_k]$$

by Theorem 11.6. A straightforward dimension counting , similar to that in Section 10.4, yields $\log k$ for the entropy, which is also the Lyapunov exponent of the map T_k.

Example 11.10 *(Entropy of Arnold cat maps)*   For the dynamical systems of Example 7.15, we can choose for $\mathfrak{B}$ the algebra generated by the Fourier basis

$$\mathbf{x} \mapsto \exp i\mathbf{n} \cdot \mathbf{x}, \quad \mathbf{x} \in \mathbb{T}^2, \quad \mathbf{n} \in \mathbb{Z}^2,$$

which is H-dense by Lemma 11.8. Again,

$$h_u[\mu, \Theta, \mathfrak{B}] = h^{KS}[\mu, T] = \log \lambda^+,$$

with λ^+ the largest eigenvalue of the Arnold matrix (see Example 7.15), which also coincides with the positive Lyapunov exponent. The actual computation of $h_u[\mu, \Theta, \mathfrak{B}]$ is performed in Chapter 13.

12

FINITE QUANTUM SYSTEMS

Non-trivial ergodic properties like mixing or exponential instability in the sense of positive Lyapunov exponents or non-zero dynamical entropy studied in Chapters 7, 10 and 11 require a continuous spectrum of the corresponding evolution maps. In the classical case, one deals with a unitary Koopman operator U_T as in Example 5.24, while for quantum systems the relevant unitary operator is the one implementing the time evolution in the GNS-representation; see Theorem 5.22. For classical Hamiltonian systems with a compact effective phase space, i.e. compact energy shells in the full phase space, exponential instability is a generic phenomenon for dimensions $d \geq 2$. On the contrary, continuous dynamical spectrum for quantum systems of physical origin is a property of infinite systems described in the thermodynamical limit.

Finite quantum systems, with finite-dimensional Hilbert spaces, which correspond to classical models with compact effective phase spaces possess discrete Hamiltonian, in fact Liouvillian, spectra; see Example 5.23. As a consequence, all correlation functions are almost periodic in time; there is no non-trivial ergodic behaviour possible and the dynamical entropy vanishes (Theorem 10.12). This is by no means the breakdown of the correspondence principle, but rather the fact that one cannot exchange the long-time and classical, or thermodynamical, limits when dealing with ergodic properties.

Strictly speaking, a precise definition of chaos is only possible after performing the classical, or thermodynamical, limit. Finite quantum systems may, however, display *transient phenomena at various time scales* which are related to the ergodic properties of the limiting systems. For example, one studies static properties such as the level statistics of the Hamiltonian in relation with the random behaviour of the limiting system.

After a very brief introduction to some problems in quantum chaos, we present in a second section the model of the quantum kicked top. The remaining sections are devoted to two new topics: the use of Gram matrices to study the dynamics at the Heisenberg time scale and a characterization of measurements that extract the maximal information from the dynamics. Several numerical simulations are presented.

12.1 Quantum chaos

The aim of this section is to give a very brief sketch of the main ideas in the vast research domain of quantum evolutions of systems with classically chaotic limits. We only touch some basic themes and refer to the literature for further details.

Many results in quantum chaos are obtained from extensive numerical experiments and have been summarized in phenomenological laws and formulas. These laws have also been put to the experimental test: microwave billiards, ultra-cooled atoms in magnetic traps, atoms in pulsated laser fields, ... The basic setting is a single unitary operator U acting on a finite-dimensional Hilbert space. This quantum map can be obtained from a periodically perturbed system as in Section 3.1.4 or from a quantization of a Poincaré return map.

12.1.1   *Time scales*

The dynamical behaviour of finite quantum systems with very large dimension N depends on the time scale one is considering. Essentially two breaking times are important: a short one t_C of order $\log N$, which could be called *classical* and a much longer one, t_H, of order N: the *Heisenberg time*. For very short times, $t \leq t_C$, well-localized wave packets remain well-localized and their motion is governed by the classical evolution. After some time, this picture breaks down. We can estimate the breaking time by requiring that the total entropy produced by the dynamics reaches its maximal value $2 \log N$ (see (10.17)):

$$t_C \ll \frac{2 \log N}{h^{KS}} \sim \log N,$$

where h^{KS} is the Kolmogorov–Sinai invariant of the classical dynamical system. The Heisenberg time scale is determined by

$$t_H = \frac{\hbar}{\delta E},$$

where δE is the average energy level separation, or separation of quasi-energies for Floquet unitaries. The quasi-energies are the eigenvalues of the Floquet operator. For a high-dimensional spin, typically $t_H = N$; see Example 3.7. For $t \ll t_H$, the almost periodic behaviour of correlation functions due to the discrete nature of the spectrum is not apparent and numerical simulations show statistical relaxation and diffusive behaviour. *Ergodic states* with Wigner functions close to the classical microcanonical distribution are relevant. There is, on the other hand, the possibility of *quantum localization*, i.e. the formation of non-ergodic localized states even for a classically ergodic motion.

12.1.2   *Spectral statistics*

The statistical distribution of energy or quasi-energy level spacings s for the Hamiltonian or Floquet operator of a finite quantum system and its relation to the qualitative behaviour of the corresponding classical system is one of the major themes in quantum chaos. The following statements represent the standard wisdom supported by numerical experiments and some theoretical arguments:

(a) For a generic classically integrable system, the eigenvalue gaps are Poisson
distributed according to

$$\rho(s) = \frac{1}{\delta E} e^{-s/\delta E}.$$

(b) For a generic classically chaotic system, one observes *level repulsion*, described
by the Wigner formula

$$\rho(s) = A s^\beta e^{-\beta s^2}.$$

The *repulsion parameter* β can take the values 1, 2 or 4, depending on the
symmetry of the system. These distributions appear also as limiting distributions of eigenlevel spacings of large matrices in the standard Gaussian orthogonal ensemble (GOE), Gaussian unitary ensemble (GUE) and Gaussian
symplectic ensemble (GSE), respectively.

12.1.3 Semi-classical limits

Because quantum mechanics turns in the classical limit into classical mechanics,
the features of classical dynamics must be encoded in the spectra and structure
of eigenfunctions of the quantum Hamiltonian, at least in the limit of large
quantum numbers. A striking example is found in the Gutzwiller trace formula
which establishes a correspondence between the quantum mechanical spectrum
and the periodic orbits of the classical system, or in the localization of the zeros
of eigenfunctions with large quantum numbers.

12.2 The kicked top

12.2.1 The model

We recall the model of the kicked top in Example 3.3. It is a generic model of a
finite-dimensional quantum system whose classical limit exhibits various degrees
of chaotic behaviour depending on the values of the control parameters of the
model. It is a kicked system which is described by a Hamiltonian periodic in
time. There exist two equivalent versions:

$$\begin{aligned}
H^{(1)}(t) &= aJ^y + b\sum_{n=-\infty}^{\infty}(J^z)^2\delta(t-n), \\
H^{(2)}(t) &= b(J^z)^2 + a\sum_{n=-\infty}^{\infty}J^y\delta(t-n),
\end{aligned} \tag{12.1}$$

where $a, b \in \mathbb{R}$ are the control parameters of the model and $\mathbf{J} = (J^x, J^y, J^z)$ are
generators of a representation of SU(2), i.e. $[J^x, J^y] = iJ^z$, etc. The corresponding Floquet operators (3.17) for a period $T = 1$ are explicitly given by

$$U^{(1)} = e^{-ib(J^z)^2} e^{-iaJ^y} \quad \text{and} \quad U^{(2)} = e^{-iaJ^y} e^{-ib(J^z)^2},$$

where $U^{(1)}$ and $U^{(2)}$ are unitarily equivalent, as their spectra coincide due
to (5.5).

12.2.2   The classical limit

We realize the system by the irreducible representation of SU(2) of dimension $2j+1$ as in Example 3.7 and choose the standard parameterization $b = k/2j$ and $a = p$. The classical limit corresponds to $j \to \infty$ and we shall use spin coherent states to obtain the limit.

Consider the Heisenberg evolution equations for the observables $\mathbf{N} := \mathbf{J}/j$ governed by the Hamiltonians $p\,J^y$ and $k\,(J^z)^2/2j$. For the first Hamiltonian, we obtain the linear operator-valued equation

$$\frac{d\mathbf{N}}{dt} = p\,\mathbf{e}_y \times \mathbf{N},$$

where $\mathbf{e}_y$ is the unit vector in $\mathbb{R}^3$ pointing in the direction of the positive y-axis. Taking the average with respect to the coherent states, we obtain a *closed* classical equation of motion on the phase space S^2, the unit sphere in $\mathbb{R}^3$, which describes precession around the y-axis:

$$\frac{d\mathbf{n}}{dt} = p\,\mathbf{e}_y \times \mathbf{n}.$$

The second Hamiltonian $k\,(J^z)^2/2j$ yields the following non-linear Heisenberg equations of motion:

$$\frac{d\mathbf{N}}{dt} = \frac{k}{2}\left\{(\mathbf{e}_z \cdot \mathbf{N}),(\mathbf{e}_z \times \mathbf{N})\right\}. \tag{12.2}$$

Again computing the averages of both sides of (12.2) in a coherent state, we obtain in the limit $j \to \infty$ a non-linear equation of motion on S^2:

$$\frac{d\mathbf{n}}{dt} = k\,(\mathbf{e}_z \cdot \mathbf{n})\,(\mathbf{e}_z \times \mathbf{n}). \tag{12.3}$$

The solution of the eqns (12.2) and (12.3) yields the dynamics of the classical kicked top. The classical dynamical map $T : \mathbf{n} \to \mathbf{n}'$ can be written as the composition of two maps: a rotation around the y-axis by the angle p and a subsequent torsion around the z-axis by an angle proportional to n_z. The explicit expressions read

$$\begin{aligned}
n'_x &= (n_x \cos p + n_z \sin p)\cos(kn_z \cos p - kn_x \sin p) \\
&\quad + n_y \sin(kn_z \cos p - kn_x \sin p), \\
n'_y &= (n_x \cos p + n_z \sin p)\sin(kn_z \cos p - kn_x \sin p) \\
&\quad - n_y \cos(kn_z \cos p - kn_x \sin p), \\
n'_z &= -n_x \sin p + n_z \cos p.
\end{aligned} \tag{12.4}$$

12.2.3   Kicked mean-field Heisenberg model

The dynamical system discussed below is an idealized model of a possible experimental realization of the periodically kicked top. We consider a collection of N spin-$\frac{1}{2}$ particles with anisotropic ferromagnetic Heisenberg interaction of mean-field type:

$$H_N^0 = -\frac{I}{2N}\left((S^x[N])^2 + (S^y[N])^2 + (1+\alpha)(S^z[N])^2\right), \quad I > 0,$$

where, for example,

$$S^x[N] := \sum_{k=1}^{N} S_k^x.$$

Here S_k^x, S_k^y and S_k^z are spin-$\frac{1}{2}$ operators at the site k which can be represented by Pauli matrices (see Example 4.6), and $\alpha \in [-1, \infty[$ is an anisotropy parameter. The system of spins is perturbed by short magnetic pulses oriented along the y-axis and periodic in time. The full time-dependent Hamiltonian reads

$$H_N(t) = H_N^0 + b\, S^y[N] \sum_{n=-\infty}^{\infty} \delta(t - n), \qquad (12.5)$$

which is essentially equivalent to the Hamiltonian $H^{(2)}$ of (12.1). The corresponding Floquet operator has the form

$$U[N] = \exp\left(-i\tau H_N^0\right) \exp\left(-ibS^y[N]\right). \qquad (12.6)$$

The Hamiltonian (12.5) and the Floquet operator (12.6) are invariant with respect to permutations of the spins and therefore highly reducible. This allows us to perform the mean-field limit; see Example 3.6. We introduce the three fundamental mean-field observables $\mathbf{J}_N := \mathbf{S}[N]/N$. The classical states are product states $\rho_N := \bigotimes_N \rho$, which are uniquely determined by the mean values of $\mathbf{J}_N$ and can therefore be parameterized by three-dimensional vectors $\mathbf{s} := \mathrm{Tr}\,\rho\mathbf{S}$, $\mathbf{s}^2 \leq 1$. The mean-field limit $N \to \infty$ is now quite similar to the classical limit of the kicked top and yields the discrete classical map (12.4) with

$$\mathbf{n} = \frac{\mathbf{s}}{|\mathbf{s}|}, \qquad k = -I\alpha|\mathbf{s}|, \quad \text{and} \quad p = b.$$

12.2.4 *Chaotic properties*

The phase portrait of the classical kicked top is quite complicated. Depending on the values of the control parameters k and p, the behaviour varies from an almost integrable system to fully developed chaos. The borderline between these regimes seems to be roughly determined by the value of pk. In the quasi-regular region, periodic orbits are surrounded by large domains of regular motion on curves of dimension 1. These domains are surrounded by tiny chaotic layers. For large values of the control parameters the stability regions around the periodic orbits virtually disappear and we can observe global chaos.

The quantum kicked top fits quite well the Berry–Tabor conjecture: the statistics of the (quasi-)energies distribution is Poisson when the corresponding classical dynamics is regular and follows that of one of the canonical random matrix ensembles in the chaotic case. Depending on the symmetries of the dynamics, e.g. time-reversal, the statistics is that of the Gaussian orthogonal, unitary or symplectic ensemble. A striking difference between the regular and the chaotic case is that the nearest-neighbour levels repel each other in the chaotic case.

12.3   Gram matrices

As all time-dependent correlation functions of a finite-dimensional quantum dynamical system are almost periodic, one may look for the return probability of an initial state by considering

$$n \in \mathbb{Z} \mapsto |\langle \varphi, U^n \varphi \rangle|^2.$$

Instead of considering only such return probabilities, we consider the full information on transition probabilities between vectors generated by the dynamics as it acts on an initial condition. We are especially looking for the dependence of these probabilities on the dimension N of the quantum system when $N \to \infty$. The choice of initial condition is also important: we consider coherent states or rather classically well-localized initial states. Eigenstates of the Hamiltonian tend to define approximately ergodic states on the classical limit and they give rise to trivial transition probabilities.

Consider a sequence

$$\varphi = \left(\varphi_1, \varphi_2, \dots, \varphi_K \right) \tag{12.7}$$

of normalized vectors in a Hilbert space $\mathfrak{H}$. The Gram matrix $G(\varphi)$ of the sequence φ is the $K \times K$ matrix

$$G(\varphi) = \begin{pmatrix} \langle \varphi_1, \varphi_1 \rangle & \langle \varphi_1, \varphi_2 \rangle & \cdots & \langle \varphi_1, \varphi_K \rangle \\ \langle \varphi_2, \varphi_1 \rangle & \langle \varphi_2, \varphi_2 \rangle & \cdots & \langle \varphi_2, \varphi_K \rangle \\ \vdots & \vdots & \ddots & \vdots \\ \langle \varphi_K, \varphi_1 \rangle & \langle \varphi_K, \varphi_2 \rangle & \cdots & \langle \varphi_K, \varphi_K \rangle \end{pmatrix}.$$

$G(\varphi)$ is positive definite and its rank equals the dimension of the space spanned by the φ_j. In particular, φ is linearly independent iff $\det(G(\varphi)) \neq 0$. The spectrum of $G(\varphi)$ is independent of the order of the vectors in φ and on multiplying the φ_j with a complex number of modulus 1.

Let us, for a moment, consider a classical word $\mathbf{i} = (i_1, i_2, \dots, i_K)$, where the letters i_j are chosen from a given alphabet $\{1, 2, \dots\}$. We can identify $\mathbf{i}$ with a sequence $\mathbf{e} := (e_{i_1}, e_{i_2}, \dots, e_{i_K})$ of vectors, where $\{e_1, e_2, \dots\}$ is an orthonormal basis of a Hilbert space $\mathfrak{H}$. Grouping e_{i_ℓ} with equal index j, the Gram matrix is block-diagonal with blocks $E(j)$ of the type

$$\begin{pmatrix} 1\,1 \ldots\, 1 \\ \vdots\; \vdots\; \ddots\; \vdots \\ 1\,1 \ldots\, 1 \end{pmatrix}.$$

The dimension of $E(j)$ is precisely the multiplicity $m(j)$ of j in $\mathbf{i}$. As the spectrum of $E(j)$ consists of the non-degenerate eigenvalue $m(j)$ and the $m(j) - 1$ degenerated eigenvalue 0, we find that $\mathrm{Sp}(G(\mathbf{e}))$ determines precisely the amount of different letters appearing in $\mathbf{i}$ with their multiplicities, i.e. the relative frequencies of the different letters in $\mathbf{i}$. The spectrum of the Gram matrix of a very regular sequence $\mathbf{i}$ consists of a few large natural numbers and a highly degenerated 0 while for sequences with many different indices the spectrum is concentrated on small natural numbers appearing with high multiplicities. Similar interpretation remains valid for the general non-commutative case: a Gram matrix $G(\varphi)$ with spectrum concentrated around small natural numbers points at a vector wandering wildly through the Hilbert space of the system and is therefore a sign of chaotic behaviour. More regular motion, such as precession or slow diffusion, signals its presence by large eigenvalues and a high occurrence of eigenvalues close to 0. In contrast to the classical case, however, eigenvalues different from 0 are generically non-degenerate. The multiplicity of the eigenvalue 0 is equal to the number of linear dependencies in φ.

The following general procedure may be followed. Consider a sequence of finite-dimensional quantum systems belonging to $\mathbb{C}^N$ and which tend, as $N \to \infty$, to a classical dynamical system as sketched in Section 3.2. Let x be a point in the classical phase space $\mathfrak{X}$ and $\varphi_x^N \in \mathbb{C}^N$ a coherent state centred around x. We follow the evolution of the initial state φ_x^N under the dynamics $U^N(t)$ up to a time suitably scaled with respect to N. We assume, for the sake of simplicity, discrete time steps of length t_0 and put $U^N := U^N(t_0)$. In this way we obtain a sequence φ^N of vectors

$$\varphi^N = \left(\varphi_x^N,\, U^N\varphi_x^N,\, (U^N)^2\varphi_x^N,\, \ldots,\, (U^N)^{K-1}\varphi_x^N\right) \tag{12.8}$$

with $K = \tau N^\alpha$ for a suitable exponent α. τ can be viewed as a rescaled time. Let $\{\lambda_j^N \mid j = 1, 2, \ldots, K\}$ denote the eigenvalues of $G(\varphi^N)$. We associate with the Gram matrix $G(\varphi^N)$ its *empirical eigenvalue distribution*, which is the probability measure

$$\mu_{x,N}(d\lambda) := \frac{1}{K}\sum_{j=1}^{K}\delta(\lambda_j^N - \lambda)\,d\lambda.$$

The aim is then to study the $*$-weak limit of the $\mu_{x,N}$ as $N \to \infty$ in function of the time scale α, the initial condition x and possible control parameters of the classical dynamical system.

Example 12.1 *(Integrable system)*   An integrable classical system with 1 degree of freedom is described in action-angle variables (J, ω) by the phase space $\mathbb{R} \times \mathbb{T}$. Its Hamiltonian is given by $H = \nu_0 J$ and it generates the flow

$$(J, \omega) \mapsto (J, \omega - \nu_0 t).$$

We consider here the dynamics

$$(J, \omega) \mapsto (J, \omega - \omega_0)$$

in discrete time with $\omega_0 = \nu_0 t_0$ an irrational multiple of 2π and t_0 the elementary time step.

   This system can be quantized as follows. For $\hbar > 0$, the basic observables act on $\mathfrak{H} := \mathcal{L}^2(\mathbb{T}, d\theta)$ as

$$(F \varphi)(\theta) := e^{i\theta} \varphi(\theta),$$
$$(P \varphi)(\theta) := -i\hbar \varphi'(\theta), \quad \varphi \text{ absolutely continuous and } \varphi' \in \mathfrak{H}.$$

F is unitary, P is self-adjoint and they satisfy the commutation relations

$$[P, F] = \hbar F \quad \text{on } \mathrm{Dom}(P).$$

Let ψ be a Schwartz function on $\mathbb{R}$, normalized in $\mathcal{L}^2$-sense. The states $\psi_x^\hbar$, approximately given by rescaling ψ,

$$\psi_x^\hbar(\theta) :\approx \frac{1}{\hbar^{1/4}} e^{iJ\theta/\hbar} \psi\left(\frac{\theta - \omega}{\sqrt{\hbar}}\right).$$

are well-localized around the point $x = (J, \omega)$. In the classical limit $\hbar \to 0$, F tends to the classical phase observable $\exp(i\omega)$ and P to the action J. Finally, the quantum evolution is given by the unitary operator

$$(U_0 \varphi)(\theta) := \varphi(\theta - \omega_0 \mod 2\pi).$$

   When $\hbar \to 0$, the support of $\psi_x^\hbar$ shrinks as $\sqrt{\hbar}$. Therefore, the overlap between $\psi_x^\hbar$ and $(U_0)^j \psi_x^\hbar$ is always extremely small except when $j\omega_0$ is an integer multiple of 2π, up to a correction term of the order $\sqrt{\hbar}$. For time spans of the order $\tau/\hbar$ this condition is fulfilled $\hbar^{-1/2}$ times. In an idealized situation where the overlap between $\psi_x^\hbar$ and its evolutes is either 1 or 0, the Gram matrix is an orthogonal projector multiplied by a factor $\hbar^{-1/2}$. Its empirical eigenvalue distribution tends *-weakly to a Dirac measure at 0. The actual behaviour of the eigenvalue distribution is shown in Fig. 12.1. Here $\hbar = 0.001$, $\tau = 0.5$, $\omega_0 = 1$ and ψ is the standard Gaussian. The histogram shows the distribution of the eigenvalues of a Gram matrix of dimension $\tau/\hbar = 500$.

FIG. 12.1. Empirical eigenvalue distribution of Gram matrix, integrable
case.

Example 12.2 *(Random vectors)* This model is not a true quantum dynamical system as the sequence of vectors in (12.7) is randomly chosen and not
generated by a deterministic evolution such as in (12.8). More precisely, consider
random sequences of normalized vectors

$$\varphi^N = (\varphi_1, \varphi_2, \ldots, \varphi_K)$$

with $\varphi_j \in \mathbb{C}^N$, $K = \tau N$ and all vectors chosen independently and uniformly
distributed on the unit sphere of $\mathbb{C}^N$. The corresponding Gram matrix $G(\varphi^N)$
and its associated empirical eigenvalue distribution measure μ_N are now random
objects.

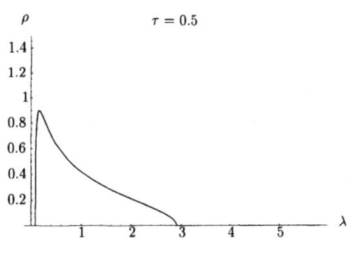

FIG. 12.2. $\mu_{0.5}$.

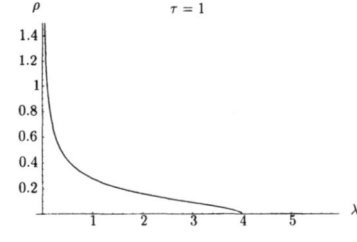

FIG. 12.3. μ_1.

It turns out that in the limit $N \to \infty$ the measures μ_N converge $*$-weakly
to a deterministic probability measure: the *Marchenko–Pastur* measure μ_τ with
density τ.

$$\mu_\tau(d\lambda) = \rho_\tau(\lambda) \, d\lambda = \frac{\sqrt{4\tau\lambda - (\lambda + \tau - 1)^2}}{2\pi\tau\lambda} \, d\lambda, \quad 0 < \tau \leq 1$$

$$\mu_\tau(d\lambda) = \frac{\tau - 1}{\tau}\delta(\lambda) \, d\lambda + \rho_\tau(\lambda) \, d\lambda, \quad 1 < \tau.$$

Figures 12.2 and 12.3 show the densities of the Marchenko–Pastur measure for
$\tau = 0.5$ and $\tau = 1$.

Example 12.3 *(Quantum kicked top)* The classical kicked top, introduced in Section 12.2 is a dynamical system with a complex phase space portrait controled by the parameters k and p. In Figs 12.4–12.7, we show histograms of the spectra of Gram matrices for the quantized top of dimension 201 for the values $\tau = 0.5$ and $\tau = 1$. The initial vector is a coherent state centred around the (111)-direction. The figures show a situation with essentially regular classical limit ($k = 1$, $p = 1.5$) and one with developed chaos ($k = 6.5$, $p = 1.5$). In the chaotic cases, the Marchenko–Pastur densities for $\tau = 0.5$ and $\tau = 1$ are superimposed on the plots.

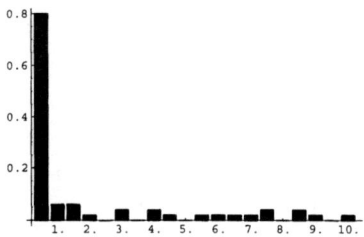

FIG. 12.4. $\tau = 0.5$, $k = 1$, $p = 1.5$.

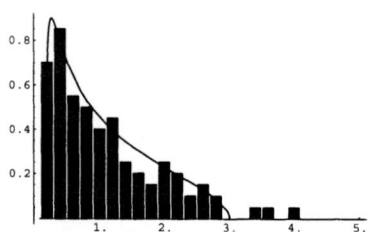

FIG. 12.5. $\tau = 0.5$, $k = 6.5$, $p = 1.5$.

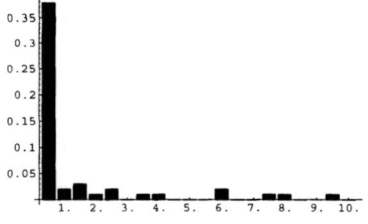

FIG. 12.6. $\tau = 1$, $k = 1$, $p = 1.5$.

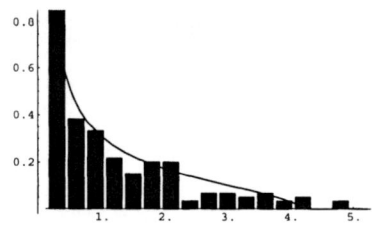

FIG. 12.7. $\tau = 1$, $k = 6.5$, $p = 1.5$.

From the Examples 12.1–12.3 the following picture emerges: the nature of the limiting eigenvalue distribution of a Gram matrix of a sequence of vectors of length proportional to the Heisenberg time of the system seems to be correlated with the behaviour of the corresponding classical limit. Integrable classical systems or systems with a regular behaviour give rise to a degenerate eigenvalue distribution centred at 0. If the classical system has good ergodic properties, then the limiting eigenvalue distribution seems to be absolutely continuous for times less than the Heisenberg time, has a compact support and a distribution that looks similar to the Marchenko–Pastur distribution.

12.4 Entropy production

In this section, we deal with finite-dimensional quantum dynamical systems $(\mathfrak{A}_N, \Theta_N, \omega_N)$ with $\mathfrak{A}_N = \mathcal{M}_N$, Θ_N determined by a unitary U_N in the usual way: $\Theta_N(A) = U_N^* A U_N$ and ω_N is the normalized trace τ_N on $\mathcal{M}_N$. We apply to such a system the scheme presented in Chapter 10 to compute dynamical entropies in terms of operational partitions. More specifically, we consider families of size k partitions $\mathbf{X}_N \subset \mathcal{M}_N$ which tend to a classical operational partition as $N \to \infty$ and their associated entropies. Although there is the estimate

$$S(\rho[\mathbf{X}_N^n]) \leq 2 \log N,$$

uniform in time n (see Lemma 10.11), it may still be interesting to consider the behaviour of this entropy for various time scales. For numerical computations, one would fix a value of N and then consider the time dependence $n \mapsto S(\rho[\mathbf{X}_N^n])$. We may then compare the results for various N. For fixed N and for a size-k partition $\mathbf{X}$, the correlation matrix $\rho[\mathbf{X}_N^n]$ has dimension k^n which grows too fast with n. It is useful, for numerical computations, to turn to the GNS-representation of τ_N, the dimension of which is limited to N^2. In the sequel we shall often drop the index N for notational convenience.

We realize the GNS-representation of τ in the spirit of Example 5.23 and Theorem 10.6 in order to obtain a doubled dynamical system. Let $\mathfrak{H}$ denote the space $\mathbb{C}^N$. The GNS-space, which is the space of matrices with the Hilbert–Schmidt inner product (2.4), can be identified with $\mathfrak{H} \otimes J\mathfrak{H}$ through

$$|\varphi\rangle\langle\psi| \iff \varphi \otimes J\psi,$$

where $\psi \mapsto J\psi$ is a complex conjugation which preserves the inner product. The precise choice of conjugation turns out to be irrelevant for the final result. A $N \times N$ matrix X acts on $\mathfrak{H} \otimes J\mathfrak{H}$ as $X \otimes \mathbf{1}$. The cyclic vector Ω is the maximally entangled vector

$$\Omega = \frac{1}{\sqrt{N}} \sum_{j=1}^{N} e_j \otimes J e_j,$$

where $\{e_1, e_2, \ldots, e_N\}$ is an arbitrary orthonormal basis in $\mathfrak{H}$, and Ω is independent of the choice of basis. Indeed, let $\{f_1, f_2, \ldots, f_N\}$ be another orthonormal base and put

$$\Psi = \frac{1}{\sqrt{N}} \sum_{j=1}^{N} f_j \otimes J f_j;$$

then

$$\langle \Omega, \Psi \rangle = \frac{1}{N} \sum_{j,\ell=1}^{N} \langle e_j \otimes J\, e_j, f_\ell \otimes J\, f_\ell \rangle$$

$$= \frac{1}{N} \sum_{j,\ell=1}^{N} \langle e_j, f_\ell \rangle \langle J\, e_j, J\, f_\ell \rangle$$

$$= \frac{1}{N} \sum_{j,\ell=1}^{N} \langle e_j, f_\ell \rangle \langle f_\ell, e_j \rangle = 1.$$

Hence, by Schwarz's inequality (2.5), $\Omega = \Psi$. The time evolution is implemented by the unitary $U \otimes JU\,J$ as in Theorem 5.22:

$$U\,X\,U^* \otimes \mathbf{1} = (U \otimes JU\,J)(X \otimes \mathbf{1})(U \otimes JU\,J)^*,$$
$$(U \otimes JU\,J)\,\Omega = \Omega.$$

We now compute the entropy of an evolved partition in GNS-space as in Section 10.2. Using the invariance of the trace under unitary transformations, the entropy $S(\rho[\mathbf{X}_N^n])$ equals the entropy of the density matrix:

$$\sigma[\mathbf{X}, n] := \frac{1}{N} \sum_{j,\ell=1}^{N} \left(\Pi_{U\mathbf{X}}^{\mathsf{T}} \right)^n \left(|e_j\rangle\langle e_\ell| \right) \otimes |f_j\rangle\langle f_\ell|. \tag{12.9}$$

The different symbols appearing in (12.9) have the following meaning:

$$\Pi_{\mathbf{Y}}^{\mathsf{T}}(\sigma) := \sum_{j=1}^{k} Y_j \, \sigma \, Y_j^*,$$

where $\mathbf{Y} = (Y_1, Y_2, \dots, Y_k)$ is a partition and σ a matrix. Further,

$$U\,\mathbf{X} := (U\,X_1, U\,X_2, \dots, U\,X_k),$$

and $\{e_1, e_2, \dots, e_N\}$ and $\{f_1, f_2, \dots, f_N\}$ are arbitrary orthonormal bases in $\mathfrak{H}$, which should be chosen conveniently for computational purposes.

In an optimal situation, the partitions $\mathbf{X}_N$ are such that they tend in the limit $N \to \infty$ to a classical partition that closely reproduces the Kolmogorov–Sinai entropy. The following theorem states a number of equivalent conditions which guarantee that for finite N repeated measurements extract the maximal information from the dynamical system.

Theorem 12.4 *In the above scheme, let* $\mathbf{X}$ *be a bistochastic partition; then the following are equivalent:*

(a) $\lim_{n\to\infty} S(\rho[\mathbf{X}^n]) = 2\log N$,

(b) $\lim_{n\to\infty} \sigma[\mathbf{X}, n] = \frac{1}{N^2}\,\mathbf{1}$,

(c) $\lim_{n\to\infty} \left(\Pi_{U\mathbf{X}} \right)^n (A) = \frac{1}{N}(\operatorname{Tr} A)\,\mathbf{1}, \quad A \in \mathcal{M}_N,$

(d) $\lim_{n\to\infty} \left(\Pi_{U\mathbf{X}}^{\mathsf{T}}\right)^n (A) = \frac{1}{N}\left(\operatorname{Tr} A\right) \mathbf{1}, \quad A \in \mathcal{M}_N.$

Moreover, if the commutant of $\mathbf{X} \circ \mathbf{X}^*$ *or* $\mathbf{X}^* \circ \mathbf{X}$ *is trivial, i.e.*

$$(\mathbf{X} \circ \mathbf{X}^*)' := \{Y \mid [Y, X_j X_k^*] = 0 \; \forall j, k\} = \mathbb{C}\mathbf{1}, \tag{12.10}$$

then (a)–(d) *hold for an arbitrary unitary matrix* U.

Proof The proof of the first part of the theorem is very simple.

(a) $\iff$ (b) is obvious,
(b) $\iff$ (d) follows directly from the expression (12.9) of $\sigma[\mathbf{X}, n]$ and
(c) $\iff$ (d) is obvious too.

In the proof (12.10) $\Rightarrow$ (d), we use the short-hand notation

$$\mathbf{Y} \equiv U\mathbf{X}, \quad \Phi \equiv \Pi_{U\mathbf{X}}, \text{ and } \Phi^* \equiv \Phi^{\mathsf{T}},$$

where the adjoint $\Phi \to \Phi^*$ is meant with respect to $\mathcal{M}_N$ considered as an N^2-dimensional Hilbert space equipped with the scalar product and associated norm

$$\langle A, B \rangle := \operatorname{Tr} A^* B, \quad \|A\|_2 := \sqrt{\langle A, A \rangle}.$$

We define a sesquilinear positive form $d(\cdot, \cdot)$ on $\mathcal{M}_N$:

$$d(A, B) := \sum_{i,j=1}^{k} \operatorname{Tr}\left([Y_i Y_j^*, A]^* [Y_i Y_j^*, B]\right) = \langle A, (\operatorname{id} - \Phi^* \Phi)(B) \rangle$$

with the obvious property

$$\{B \in \mathcal{M}_N \mid d(B, B) = 0\} = \{\mathbf{Y} \circ \mathbf{Y}^*\}'.$$

Assume that the commutant of $\mathbf{Y} \circ \mathbf{Y}^*$ is trivial, then $d(\cdot, \cdot)$ is non-degenerate on the $(N^2 - 1)$-dimensional subspace:

$$\mathcal{Q}_N := \{B \in \mathcal{M}_N \mid \operatorname{Tr} A = 0\} \quad \text{and} \quad \mathcal{M}_N = \mathcal{Q}_N \oplus \mathbb{C}\mathbf{1}.$$

As all norms on finite-dimensional spaces are equivalent, there exists a constant $\gamma > 0$ such that

$$d(B, B) \geq \gamma \|B\|_2^2 \quad \text{for all } B \in \mathcal{Q}_N. \tag{12.11}$$

A straightforward computation yields, for any $A \in \mathcal{M}_N$,

$$\|\Phi^k(A)\|_2^2 - \|\Phi^{k+1}(A)\|_2^2 = d(\Phi^k(A), \Phi^k(A)). \tag{12.12}$$

Taking into account (12.11) and (12.12), we obtain, for any $B \in \mathcal{Q}_N$, the inequality

$$\|B\|_2^2 \geq \gamma \sum_{k=0}^{n} \|\Phi^k(B)\|_2^2, \quad \text{for any } n \in \mathbb{N},$$

and therefore,

$$\lim_{n\to\infty} \Phi^n(B) = 0.$$

Exchanging $\mathbf{Y}$ with $\mathbf{Y}^*$, we obtain the analogous dual relation. $\square$

Example 12.5 As an illustration of the result of above, we consider the kicked top of Example 3.3, with the partition

$$\mathbf{X} := \mathbf{X}^* = \frac{1}{\sqrt{j(j+1)}}\mathbf{J}.$$

From the commutation relations (3.18), it follows that the commutant of $\mathbf{J} \circ \mathbf{J}$ coincides with the commutant of $\mathbf{J}$. The later is trivial because $\mathbf{J}$ generates an irreducible representation of the Lie algebra SU(2). Therefore we can, by Theorem 12.4, always extract the maximal information using this partition for an arbitrary dynamics but the rate of entropy change for different time scales depends on the detailed chaotic properties of U.

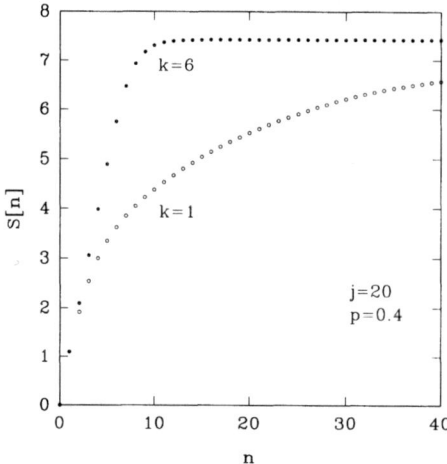

FIG. 12.8. Entropy of the quantum kicked top.

Figure 12.8 shows the entropy $S(n) = S(\rho[\mathbf{X}^n])$ as a function of n for a spin-20 representation, i.e. a Hilbert space of dimension 41. The dotted line corresponds to an essentially regular dynamics with the values $p = 1.25$ and $k = 1$, while the dashed line represents a chaotic evolution with $p = 1.25$ and $k = 6$. In both cases the entropy tends asymptotically to its maximal value $2\log 41 \approx 7.43$. In the regular case the maximal entropy is slowly attained while in the chaotic regime it increases linearly for very short times, up to $n = 5$, with a slope equal to the Lyapunov exponent of the classical system, approximately 1.

12.5 Notes

There is a vast body of literature on quantum chaos. One can, for example, find a lot of information in the books by Gutzwiller (1990), Haake (1991), Casati and Chirikov (1995), and Stöckmann (1999), and in the references therein. Many results are numerical or partially based on heuristic arguments. Introducing randomness, one can obtain a number of rigorous results. Basic references for random matrix theory are the works by Mehta (1991) and Hiai and Petz (2000). Many references can also be found in the review paper (Pastur 2000).

The classical and quantum kicked tops were studied in the paper (Haake et al. 1987), the book by Haake (1991) and the paper (Życzkowski 1993). For a treatment of the classical regime, see, for example, the book by Robert (1987) and the papers (Combescure and Robert 1997; De Bièvre 1998).

Gram matrices, as a tool to study chaos in quantum systems, have been introduced in (De Cock et al. 1999). Theorem 12.4 is unpublished. For numerical evidence on entropy production, we refer to the papers (Alicki et al. 1996b) and (Słomczyński and Życzkowski 1994).

13

MODEL SYSTEMS

In this chapter we consider some more involved entropy computations. Instead of presenting the arguments in full detail, we rather try to sketch the line of argumentation and refer the interested reader to the literature for further technical details. The chapter deals with three topics: the quantum cat maps, Ruelle's inequality and quasi-free fermion evolutions. The quantum cat maps, also called non-commutative toral automorphisms, are examples of quantum dynamical systems that allow a differentiable structure. Quantization makes this kind of systems rather rigid and atypical. The section on Ruelle's inequality aims at establishing a connection between non-commutative Riemannian structures and statistical descriptions of some dynamical systems. Finally, the section on quasi-free fermionic evolutions deals with a more physical situation where the evolution is governed by a Hamiltonian and where one has to consider a spectral density of dynamical entropy.

13.1 Entropy of the quantum cat map

We return to the dynamical system of Example 7.9. It will be convenient to introduce

$$W(f) := \sum_{\mathbf{n} \in \mathbb{Z}^2} f(\mathbf{n}) W(\mathbf{n}), \quad f \in \ell^1(\mathbb{Z}^2).$$

This is a slight abuse of notation as we identify $\mathbf{m} \in \mathbb{Z}^2$ with its indicator function. The algebraic relations of the $W(\mathbf{m})$ translate into

$$W(f)^* = W(f^\dagger) \quad \text{and} \quad W(f)W(g) = W(f *_\theta g)$$

with

$$f^\dagger(\mathbf{n}) := \overline{f(-\mathbf{n})},$$
$$(f *_\theta g)(\mathbf{n}) = \sum_{\mathbf{m}} e^{i\theta\sigma(\mathbf{m},\mathbf{n})} f(\mathbf{m}) g(\mathbf{n}-\mathbf{m}).$$

There exists a natural $*$-subalgebra $\mathfrak{A}_\theta^\infty$ of *smooth functions on the non-commutative torus* given by

$$\mathfrak{A}_\theta^\infty = \{W(f) \mid f \in \mathcal{S}(\mathbb{Z}^2)\}. \tag{13.1}$$

Here $\mathcal{S}(\mathbb{Z}^2)$ is the space of sequences of rapid decrease: $f \in \mathcal{S}(\mathbb{Z}^2)$ iff

$$\lim_{\|\mathbf{n}\|\to\infty} \|\mathbf{n}\|^k \, |f(\mathbf{n})| = 0, \quad k \in \mathbb{N}.$$

In the non-deformed case $\mathfrak{A}_0^\infty$ is isomorphic to the space of infinitely differentiable functions on $\mathbb{T}^2$.

We consider the automorphic dynamics Θ_T on $\mathfrak{A}_\theta$ given in terms of a matrix $T \in SL(2,\mathbb{Z})$, i.e. matrices with integer entries and determinant 1, by

$$\Theta_T\big(W(\mathbf{n})\big) := W(T^\mathsf{T}\mathbf{n}), \tag{13.2}$$

where T^T is the transpose of T. Θ_T leaves $\mathfrak{A}_\theta^\infty$ globally invariant and is therefore a diffeomorphism of the non-commutative torus.

There exists a tracial state τ on $\mathfrak{A}_\theta$, explicitly given by

$$\tau\big(W(\mathbf{n})\big) := \delta_{\mathbf{n},0} \quad \text{and} \quad \tau\big(W(f)^*W(g)\big) = \langle f,g \rangle. \tag{13.3}$$

For irrational multiples θ of π, the trace is unique. The state τ is invariant under Θ_T. The following result holds:

Theorem 13.1 *Let λ^+ be the largest eigenvalue of the matrix $T \in SL(2,\mathbb{Z})$ and assume that $\mathrm{Tr}\, T > 2$. For any $0 \le \theta < 2\pi$, we have that*

$$\mathsf{h}_\mathsf{u}[\tau,\Theta_T,\mathfrak{A}_\theta^\infty] = \mathsf{h}[\tau,\Theta_T,\mathfrak{A}_\theta^\infty] = \log\lambda^+.$$

Proof We sketch the proof leaving aside some technical complications. We prove the result for partitions in elements of finite support. This can be extended to general partitions in $\mathfrak{A}_\theta^\infty$ using the continuity property (10.14). We therefore introduce the algebra $\mathfrak{D}_\theta$ of strictly local elements on the non-commutative torus

$$\mathfrak{D}_\theta = \{W(f)|f \in \ell^0(\mathbb{Z}^2)\},$$

$\ell^0(\mathbb{Z}^2)$ being the space of complex sequences indexed by $\mathbb{Z}^2$ with only a finite number of elements different from 0. The *support* $\mathrm{Supp}\big(W(f)\big)$ of an element $W(f) \in \mathfrak{D}_\theta$ is the support of f and the support $\mathrm{Supp}(\mathbf{X})$ of a partition $\mathbf{X} = \big(W(f_1),W(f_2),\dots,W(f_k)\big)$ in $\mathfrak{D}_\theta$ is

$$\mathrm{Supp}(\mathbf{X}) := \bigcup_{j=1}^k \mathrm{Supp}(f_j).$$

Next, we obtain upper and lower bounds for $\mathsf{h}[\tau,\Theta_T,\mathfrak{D}_\theta]$. Consider a partition $\mathbf{X} = \big(W(f_1),W(f_2),\dots,W(f_k)\big)$ in $\mathfrak{D}_\theta$. In the GNS-representation $(\mathfrak{H},\pi,\Omega)$ of τ, we may compute the entropy of $\rho[\mathbf{X}]$ in terms of the density matrix $\hat\rho[\mathbf{X}] = \Pi_\mathbf{X}^\mathsf{T}(|\Omega\rangle\langle\Omega|)$ as in Theorem 10.4. This density matrix is obviously supported by the subspace of $\mathfrak{H}$ generated by the vectors $\pi(W(\eta))\,\Omega$ with η in Supp($\mathbf{X}$). As this space is at most $\#(\mathrm{Supp}(\mathbf{X}))$-dimensional,

$$S(\rho[\mathbf{X}]) \le \log \#(\mathrm{Supp}(\mathbf{X})). \tag{13.4}$$

Let $\mathbf{u}^+$ and $\mathbf{u}^-$ be the eigenvectors of T^{T} which belong to the eigenvalues λ^+ and $\lambda^- = 1/\lambda^+$. For any $\mathbf{X} \subset \mathfrak{D}_\theta$, there exist $a, b > 0$ with Supp($\mathbf{X}$) $\subset R(a, b)$, where

$$R(a, b) := \{\mathbf{n} \in \mathbb{Z}^2 \mid \mathbf{n} = \alpha \mathbf{u}^+ + \beta \mathbf{u}^-, \, |\alpha| < a, \, |\beta| < b\}.$$

Then

$$\mathrm{Supp}\left(\mathbf{X}^m\right) = \mathrm{Supp}\left(\Theta_T^{m-1}(\mathbf{X}) \circ \ldots \circ \Theta_T(\mathbf{X}) \circ \mathbf{X}\right)$$
$$\subset R(a\frac{\lambda^{+m}-1}{\lambda^+ - 1}, b\frac{1}{1-\lambda^-}).$$

If we apply inequality (13.4) to the evolved partition $\mathbf{X}^m$ as in (10.11), divide by m and take the limit $m \to \infty$ we find

$$\mathsf{h}[\tau, \Theta_T, \mathbf{X}] \le \log \lambda^+.$$

Finally, taking the supremum over all partitions in $\mathfrak{D}_\theta$,

$$\mathsf{h}[\tau, \Theta_T, \mathfrak{D}_\theta] \le \log \lambda^+. \tag{13.5}$$

We now obtain a lower bound for the dynamical entropy by using unitary partitions of the type

$$\mathbf{X} = \left(\frac{W(\mathbf{n}_1)}{\sqrt{k}}, \frac{W(\mathbf{n}_2)}{\sqrt{k}}, \cdots, \frac{W(\mathbf{n}_k)}{\sqrt{k}}\right)$$

with $\mathbf{n}_j \in \mathbb{Z}^2$ and $k = [(\lambda^+)^m]$, the integer part of $(\lambda^+)^m$. Under the mth power of the dynamics an element $W(\mathbf{n})$ transforms into $W((T^{\mathsf{T}})^m\,\mathbf{n})$ and the labels of such elements approach the line $\mathbb{R}\mathbf{u}^+$ exponentially fast. Choosing the first k vectors $\mathbf{n}_j$ in $\mathbb{Z}^2$ closest to the half-line $\mathbb{R}^+\mathbf{u}^+$, we see that Θ^m is close to the classical expanding maps

$$z \mapsto z^{[(\lambda^+)^m]}, \quad |z| = 1$$

of Example 11.9 and that the correlation matrix asymptotically coincides with that of the optimal partition of the classical expanding map. It has therefore entropy $\log[(\lambda^+)^m]$. The lower bound now follows from Lemma 10.16.   □

13.2 Ruelle's inequality

In the classical ergodic theory of smooth dynamical systems, one of the fundamental results is Ruelle's inequality which establishes the relation

$$h^{KS}[T, \mu] \leq \sum_{\chi_j > 0} \chi_j, \tag{13.6}$$

between the Kolmogorov–Sinai entropy h^{KS} and the Lyapunov exponents chi_j for any ergodic diffeomorphism T on a compact Riemannian manifold with an invariant Borel probability measure μ; see Section 7.3. Moreover, under certain regularity assumptions, the equality holds, this is *Pesin's theorem.*

Recall that for a smooth and ergodic classical dynamical system, the Lyapunov exponents measure the rate of exponential separation of initial conditions under the dynamics. Therefore, we consider, on a d-dimensional compact manifold $\mathfrak{X}$, a Riemannian metric determined by the tensor $g_{rs}(x)$. Using Einstein's summation convention, the infinitesimal distance $\delta\ell$ between points is given by $(\delta\ell)^2 = g_{rs}(x)\delta x^r \delta x^s$. In a local coordinate system, the dynamical map T is given by a set of equations

$$T(x)^k = t^k(x^1, x^2, \dots, x^d)$$

and the distance between initial points changes after n time steps into

$$(\delta\ell)^2(n) = \partial_r t^p_{(n)}(x)\, g_{pq}(T^n(x))\, \partial_s t^q_{(n)}(x)\delta x^r \delta x^s$$

where we use the notation

$$T^n(x)^k = t^k_{(n)}(x^1, x^2, \dots, x^d)$$

for the powers of the dynamical map T. Depending on the direction of the vector δx^r, the distance $\delta\ell(n)$ increases or decreases according to the exponential law

$$\delta\ell(n) \sim \exp \chi_p n,$$

where $\{\chi_p \mid p = 1, 2, \dots, d\}$ is a set of Lyapunov exponents which are independent of the metric g_{rs} and constant almost everywhere by the assumed ergodicity of the system.

The same Lyapunov exponents appear in the asymptotic expression for the quadratic form on $C^\infty(\mathfrak{X}) \subset \mathcal{L}^2(\mathfrak{X}, \mu)$ given in local coordinates by

$$(f, g)_\xi := \int_{\mathfrak{X}} \mu(dx)\, \xi^k \partial_k \overline{f(x)}\, \xi^l \partial_l g(x).$$

Namely, taking into account the identity

$$\partial_k f(T^n(x)) = \partial_k t^p_{(n)}(x)\, (\partial_p f)(T^n(x)),$$

we have, for $\xi \in \mathbb{R}^d$ and a non-trivial $f \in C^\infty(\mathfrak{X})$,

$$\log(\Theta^n(f), \Theta^n(f))_\xi \sim 2\chi_p n + o(n) \tag{13.7}$$

as $n \to \infty$. We pick up the different Lyapunov exponents by varying ξ and f. Here Θ is the dynamical map at the level of phase space functions: $\Theta(f)(x) := f(T(x))$.

In this section, we show that Ruelle's inequality holds for a class of non-commutative smooth dynamical systems with a tracial invariant state. The entropy of Kolmogorov–Sinai will be generalized to the bistochastic dynamical entropy (see Definition (10.10)), while the positive Lyapunov exponents, which in classical case determine the rates of divergence of trajectories in different directions, will be defined in terms of non-commutative Laplacians or their Dirichlet forms, slightly generalizing the idea expressed in (13.7). We show that this result covers the case of classical ergodic diffeomorphisms and of the quantum cat map, but one can hope to extend its range of applicability. The setting seems to be more general than the horocyclic actions discussed in Section 7.3.2.

13.2.1 *Non-commutative Riemannian structures*

As already mentioned, in the classical theory we need a metric to define Lyapunov exponents as the rates of divergence of trajectories. The final results, however, are independent of the choice of this metric and what really matters is the change of a metric under the action of a dynamical map. Hence we need in fact a whole family of Riemannian metrics on a given compact manifold.

A quantum version of the Riemannian spin space was proposed by A. Connes. The fundamental object in his approach is an unbounded self-adjoint operator which is a non-commutative analogue of the *Dirac operator*. Here, the metric is given in terms of a *Laplacian* or its *Dirichlet form*. One may expect a deeper relation to Connes' approach as well as to other non-commutative structures which are classically linked to Riemannian metrics. We think, in particular, of diffusion processes which are governed on the quantum level by non-commutative symmetric completely positive Markov semigroups and their generators.

We consider a C*-algebra $\mathfrak{A}$ with a normalized tracial state τ and a distinguished subalgebra of smooth elements $\mathfrak{A}_0$. We denote by $(\mathfrak{H}_\tau, \Omega, \pi)$ the GNS-triple of τ. Assume that such a system can be equipped with a *non-commutative Riemannian structure* in the following sense: there exists a positive sesquilinear form $\Delta(\cdot, \cdot)$ on $\mathbb{C}^d \otimes \mathfrak{A}_0$ that satisfies the following conditions:

- For any $\mathbf{u} \in \mathbb{C}^d$ and any two bistochastic partitions $\mathbf{X}, \mathbf{Y} \in \mathcal{P}_{\mathrm{b}}(\mathfrak{A}_0)$

$$\left(\sum_{j,k} \Delta(\mathbf{u} \otimes x_j y_k, \mathbf{u} \otimes x_j y_k) \right)^{1/2}$$
$$\leq \left(\sum_j \Delta(\mathbf{u} \otimes x_j, \mathbf{u} \otimes x_j) \right)^{1/2} + \left(\sum_k \Delta(\mathbf{u} \otimes y_k, \mathbf{u} \otimes y_k) \right)^{1/2}. \quad (13.8)$$

- The sesquilinear form has a closed extension $\overline{\Delta}(\cdot, \cdot)$ on a dense domain in $\mathbb{C}^d \otimes \mathfrak{H}_\tau$ such that

$$\overline{\Delta}(\mathbf{u} \otimes \pi(x)\Omega, \mathbf{v} \otimes \pi(y)\Omega) = \Delta(\mathbf{u} \otimes x, \mathbf{v} \otimes y)$$

for $\mathbf{u}, \mathbf{v} \in \mathbb{C}^d$ and all $x, y \in \mathfrak{A}_0$.

- For any strictly positive $d \times d$ matrix g with spectral decomposition

$$g = \sum_{s=1}^{d} g_s |\mathbf{u}_s\rangle\langle\mathbf{u}_s|, \tag{13.9}$$

we define a positive closed sesquilinear form with domain $\mathrm{Dom}(\overline{\Delta_g}) \supset \mathfrak{H}_0 :=$ $\pi(\mathfrak{A}_0)\Omega$ by

$$\overline{\Delta_g}(\Psi, \Phi) := \sum_{s=1}^{d} g_s \overline{\Delta}(\mathbf{u}_s \otimes \Psi, \mathbf{u}_s \otimes \Phi), \quad \Psi, \Phi \in \mathfrak{H}_0. \tag{13.10}$$

- For any $g > 0$, $\overline{\Delta}_g$ uniquely determines an unbounded self-adjoint positive operator—a Laplacian—with dense domain in $\mathfrak{H}_\tau$. We denote this operator by the same symbol $\overline{\Delta}_g$. We assume, moreover, that $\overline{\Delta}_g$ possesses a discrete spectrum and that $\exp(-t\overline{\Delta}_g)$ is trace-class for all $t > 0$. Finally, for any sequence of positive matrices $g^{(n)}$ (13.9) with fixed eigenvectors and vanishing eigenvalues as $n \to \infty$, we assume the following asymptotic behaviour for the spectrum of the associated Laplacians:

$$\lim_{n\to\infty} \frac{\log\left(\mathrm{Tr}\exp -\overline{\Delta}_{g^{(n)}}\right)}{\log(\det g^{(n)})} = -\frac{1}{2}. \tag{13.11}$$

The structure $(\mathfrak{A}, \mathfrak{A}_0, \tau, \Delta)$ is called a *non-commutative d-dimensional compact Riemannian manifold* and the different Laplacians $\overline{\Delta}_g$ determine different but equivalent *non-commutative Riemannian metrics*. The Conditions (13.8)–(13.11) are motivated by the properties of the classical Laplace operators on compact differentiable manifold and of their corresponding Dirichlet forms written in terms of derivations.

Condition (13.11) is motivated by Weyl's theorem on the asymptotic behaviour of the spectrum of a self-adjoint Laplacian on a compact Riemannian space; see Section 3.4. Taking the logarithm of both sides of (3.28), we see how $\log(\mathrm{Tr}\exp -\overline{\Delta})$ depends on the overall variation of $\log\det(g)$. This is exactly the meaning of condition (13.11).

While conditions (13.9) and (13.10) are rather technical, condition (13.8) encodes the properties of Dirichlet forms given in terms of derivations. To clarify this point we consider the following example. Define the sesquilinear form Δ as

$$\Delta(\mathbf{u} \otimes x, \mathbf{v} \otimes y) := \sum_{s=1}^{d} \langle\mathbf{u}, e_s\rangle \, \tau(\nabla_s(x^*)\nabla_s(y)) \, \langle e_s, \mathbf{v}\rangle, \tag{13.12}$$

where $\{e_s \mid s = 1, 2, ..., d\}$ is an orthonormal basis in $\mathbb{C}^d$ and the ∇_s, $s = 1, 2, ..., d$ are derivations on the joint domain $\mathfrak{A}_0$, i.e.

$$\nabla_s(x^*) = \nabla_s(x)^* \quad \text{and} \quad \nabla_s(xy) = \nabla_s(x)y + x\nabla_s(y). \tag{13.13}$$

A simple direct computation, using the properties (13.13) of the derivation, the tracial property, and the bistochasticity, shows that this example satisfies condition (13.8).

13.2.2 *Non-commutative Lyapunov exponents*

Consider a smooth dynamical system on a non-commutative Riemannian manifold described by $(\mathfrak{A}, \Theta, \tau, \mathfrak{A}_0, \Delta)$. For any $\mathbf{u} \in \mathbb{R}^d$, we define a possible infinite *Lyapunov exponent* $\chi(\mathbf{u})$,

$$
\begin{aligned}
\chi(\mathbf{u}, x) &:= \limsup_{n \to \infty} \tfrac{1}{2n} \log \Delta\big(\mathbf{u} \otimes \Theta^n(x), \mathbf{u} \otimes \Theta^n(x)\big), \\
\chi(\mathbf{u}) &:= \sup_{x \in \mathfrak{A}_0} \chi(\mathbf{u}, x).
\end{aligned}
\tag{13.14}
$$

The following Lemma is a quantum analogue of Theorem 7.16. Its proof is also quite similar.

Lemma 13.2 *The function* $\mathbb{R}^d \ni \mathbf{u} \mapsto \chi(\mathbf{u}) \in \mathbb{R} \cup \{\pm\infty\}$ *has the following properties:*

(a) $\chi(\alpha\mathbf{u}) = \chi(\mathbf{u})$ *for* $\alpha \in \mathbb{R} \setminus \{0\}$,
(b) $\chi(\mathbf{u} + \mathbf{v}) \leq \max\{\chi(\mathbf{u}), \chi(\mathbf{v})\}$,
(c) *if* $\chi(\mathbf{u}) \neq \chi(\mathbf{v})$, *then* $\chi(\mathbf{u} + \mathbf{v}) = \max\{\chi(\mathbf{u}), \chi(\mathbf{v})\}$.

It follows from Lemma 13.2 that for $\chi \in \mathbb{R}$ the set

$$
E_\chi := \{\mathbf{u} \in \mathbb{R}^d \mid \chi(\mathbf{u}) \leq \chi\}
$$

is a linear subspace of $\mathbb{R}^d$ and that $E_\chi \subset E_{\chi'}$ whenever $\chi \leq \chi'$. We have therefore a finite, possibly degenerate, spectrum of Lyapunov exponents

$$
-\infty \leq \chi_1 \leq \chi_2 \leq \cdots \leq \chi_d \leq \infty
$$

and corresponding orthogonal vectors $\{e_1, e_2, \ldots, e_d\}$ which span $\mathbb{R}^d$ and

$$
\chi(e_r) = \chi_r, \quad r = 1, 2, \ldots, d.
$$

In the forthcoming case of a dynamical system $(\mathfrak{A}, \Theta, \tau, \mathfrak{A}_0, \Delta)$, we shall assume that the $*$-subalgebra $\mathfrak{A}_0$ is generated by a finite set of unitary elements. In such a situation, the Lyapunov exponents can be computed using the following lemma.

Lemma 13.3 *Let* $\mathfrak{A}_0$ *be generated by unitaries* $\{u_1, u_2, \ldots, u_m\}$. *Then, for any* $\mathbf{u} \in \mathbb{R}^d$

$$
\chi(\mathbf{u}) = \max\{\chi(\mathbf{u}, u_r) \mid r = 1, 2, \ldots, m\}.
$$

Proof Obviously, $\chi(\mathbf{u}) \geq \max\{\chi(\mathbf{u}, u_r) \mid r = 1, 2, \ldots, m\}$. By assumption, an arbitrary element x in $\mathfrak{A}_0$ can be written as a finite linear combination of products of unitaries u_j. Using the triangle inequality for the seminorm $\sqrt{\Delta(\cdot, \cdot)}$ and (13.8), we have

$$\left(\Delta(\mathbf{u} \otimes \Theta^n(x), \mathbf{u} \otimes \Theta^n(x))\right)^{1/2}$$

$$\leq \sum |\alpha_{j_1, j_2, \dots, j_N}| \sum_{k=1}^{N} \left(\Delta(\mathbf{u} \otimes \Theta^n(u_{j_k}), \mathbf{u} \otimes \Theta^n(u_{j_k}))\right)^{1/2}.$$

As the summation involves only a finite number of terms, taking the logarithm of both sides, dividing by $2n$ and taking the limit yields the reversed inequality.
□

13.2.3 Ruelle's inequality

Theorem 13.4 is a generalization to the quantum domain of the classical inequality (13.6). We prove it for the bistochastic dynamical entropy h_b assuming that the quantum dynamical system possesses a Riemannian structure as described in Section 13.2.1.

Theorem 13.4 *Let $(\mathfrak{A}, \Theta, \tau, \mathfrak{A}_0, \Delta)$ be a quantum dynamical system on a non-commutative Riemannian manifold as described in Section 13.2.1. Then the following inequality holds:*

$$h_b[\tau, \Theta, \mathfrak{A}_0] \leq \sum_{r=1}^{d} \chi_r^+,$$

where the χ_j are the Lyapunov exponents of the system and $\chi_r^+ = \chi_r$ for $\chi_r > 0$ or 0 otherwise.

Proof From the inequality $\mathrm{Tr}(\rho \log \rho - \rho \log \sigma) \geq 0$ that holds for any two density matrices ρ and σ (see Theorem 9.17(a)), it follows that for any density matrix ρ on $\mathfrak{H}_\tau$ and any $g > 0$

$$S(\rho) \leq \mathrm{Tr}\, \rho \, \overline{\Delta_g} + \log\left(\mathrm{Tr} \exp -\overline{\Delta_g}\right). \tag{13.15}$$

Take a bistochastic partition $\mathbf{X} = \{x_1, x_2, \dots, x_k\} \subset \mathfrak{A}_0$ and the density matrix $\hat{\rho}[\mathbf{X}^n]$ corresponding to the evolved partition $\mathbf{X}^n$ as in (10.13) and (10.11):

$$\hat{\rho}[\mathbf{X}^n] = \sum_{j_1, \dots, j_n = 1}^{k} |\pi(\Theta^{n-1}(x_{j_n})) \cdots \pi(x_{j_1}) \Omega\rangle\langle\pi(\Theta^{n-1}(x_{j_n})) \cdots \pi(x_{j_1}) \Omega|.$$

$$\tag{13.16}$$

Choosing for $\epsilon > 0$, $d \times d$ positive matrices $g^{(n)}$ of the form

$$g^{(n)} := \sum_{r=1}^{d} e^{-2(\chi_r^+ + \epsilon)n} |e_r\rangle\langle e_r|$$

and using (13.16), (13.9) and (13.10), we obtain

$$\operatorname{Tr}\hat{\rho}[\mathbf{X}^n]\,\overline{\Delta_{g^{(n)}}} = \sum_{r=1}^{d} e^{-2(\chi_r^+ + \epsilon)n}$$

$$\times \sum_{j_1,\dots,j_n=1}^{k} \Delta(e_r \otimes \pi(\Theta^{n-1}(x_{j_n})) \cdots \pi(x_{j_1}),$$

$$e_r \otimes \pi(\Theta^{n-1}(x_{j_n})) \cdots \pi(x_{j_1})). \tag{13.17}$$

Due to the definition of the Lyapunov exponent in (13.14), there exists a constant $C > 0$ such that

$$e^{-(\chi_r^+ + \epsilon)n}\left(\sum_{j_1,\dots,j_n=1}^{k} \Delta(e_r \otimes \pi(\Theta^{n-1}(x_{j_n})) \cdots \pi(x_{j_1}),\right.$$

$$\left. e_r \otimes \pi(\Theta^{n-1}(x_{j_n})) \cdots \pi(x_{j_1})) \right)^{1/2}$$

$$\leq e^{-(\chi_r^+ + \epsilon)n} \times \sum_{l=0}^{n-1}\sum_{j=1}^{k}\left(\Delta(e_r \otimes \pi(\Theta^l(x_j)), e_r \otimes \pi(\Theta^l(x_j))) \right)^{1/2}$$

$$\leq C. \tag{13.18}$$

Therefore, $\operatorname{Tr}\hat{\rho}[\mathbf{X}^n]\,\overline{\Delta_{g^{(n)}}}$ is bounded and from (13.15), (13.17) and (13.18) it follows that

$$\limsup_{n\to\infty}\frac{1}{n}S(\hat{\rho}[\mathbf{X}^n]) \leq \sum_{r=1}^{d}(\chi_r^+ + \epsilon),$$

which completes the proof.   □

13.2.4   Examples

We illustrate the general result by two examples of dynamical systems: classical ergodic diffeomorphisms and quantum cat maps.

Example 13.5 *(Classical systems)*   We consider an ergodic diffeomorphism $T : \mathfrak{X} \mapsto \mathfrak{X}$ on a d-dimensional smooth compact manifold $\mathfrak{X}$ equipped with a T-invariant probability measure μ. As a commutative C*-algebra we take the algebra $\mathcal{C}(\mathfrak{X})$ and $\mathcal{C}^\infty(\mathfrak{X})$ as the algebra of smooth elements. By ∇_s, $s = 1, 2, \dots, d$, we denote derivations on $\mathcal{C}^\infty(\mathfrak{X})$ given in terms of pointwise linearly independent vector fields. In local coordinates, the derivations act as

$$\nabla_s := a_s^k(x)\,\partial_k, \quad s = 1, 2, \dots, d.$$

The sesquilinear form Δ is given by an expression of the type (13.12) and the operators $\overline{\Delta_g}$ are self-adjoint Laplacians acting on the GNS-Hilbert space $\mathcal{L}^2(\mathfrak{X}, \mu)$.

Applying Weyl's theorem we obtain the required asymptotic estimate (13.11). For ergodic systems, our definition of Lyapunov exponents coincides with the usual classical definition discussed in Section 7.3.1 and hence Theorem 13.4 reproduces the usual classical Ruelle inequality (13.6).

Example 13.6 *(Quantum cat maps)* The algebra of smooth elements is $\mathfrak{A}_\theta^\infty$; see (13.1). There are two canonical derivations acting on $\mathfrak{A}_\theta^\infty$ which are the generators of the canonical groups in (5.19):

$$\nabla_1(W(\mathbf{n})) := in_1\, W(\mathbf{n}) \quad \text{and} \quad \nabla_2(W(\mathbf{n})) := in_2\, W(\mathbf{n}),$$

with $\mathbf{n} = (n_1, n_2) \in \mathbb{Z}^2$. These derivations define a sesquilinear form Δ according to formula (13.12). An obvious computation involving (5.25) shows that the vectors

$$\phi(\mathbf{n}) := \pi\big(W(\mathbf{n})\big)\,\Omega, \quad \mathbf{n} \in \mathbb{Z}^2$$

form an orthonormal basis in the GNS-Hilbert space $\mathfrak{H}_\tau$ of τ and diagonalize any Laplacian $\overline{\Delta_g}$, i.e.

$$\langle \phi(\mathbf{n}), \overline{\Delta_g}\, \phi(\mathbf{m}) \rangle = \delta_{\mathbf{nm}}\langle \mathbf{n}, g\,\mathbf{n}\rangle.$$

As the spectrum of $\overline{\Delta_g}$ is the same as the spectrum of the corresponding Laplacian on a commutative torus, Weyl's formula (13.11) holds. The matrix T^T has a pair of eigenvectors and eigenvalues

$$T^\mathsf{T}\mathbf{u}^\pm := \lambda^\pm\mathbf{u}^\pm.$$

Combining eqns (13.2) and (13.3) and choosing a vector $\mathbf{k}$ such that $\langle \mathbf{k}, g\,\mathbf{u}^+\rangle \neq 0$, we obtain

$$\langle \pi\big(\Theta^n(W(\mathbf{k}))\big)\,\Omega, \overline{\Delta_g}\,\pi\big(\Theta^n(W(\mathbf{k}))\big)\,\Omega\rangle = \langle T^n\mathbf{k}, g\,T^n\mathbf{k}\rangle \sim (\lambda^+)^n.$$

Applying Lemma 11.4, we find a single positive Lyapunov exponent

$$\chi^+ = \log\lambda^+.$$

The direct computation of Section 13.1 shows that the dynamical entropy is also equal to χ^+.

13.3 Quasi-free Fermion dynamics

The quantum dynamical system studied in this section differs in several aspects from the simple models discussed in Chapter 10. First of all, as argued in Chapter 6, quasi-free fermionic models provide a reasonable approximation for a number of real physical systems. The natural setup for the dynamics is not a single dynamical map but a Hamiltonian of the system with its corresponding one-parameter group of unitaries. Moreover, if the spectrum of the one-particle Hamiltonian is unbounded from above or infinitely degenerated, then external perturbations, e.g. the measurement process, can produce arbitrarily large number of particles in a finite time and hence infinite dynamical entropy. Therefore, the natural notion in this case is rather a *renormalized spectral density of dynamical entropy* which is obtained by computing a dynamical entropy with partitions of unity restricted to different *spectral CAR-subalgebras* and by extracting a vacuum contribution.

13.3.1 *Description of the model*

From Chapter 7 we know that interesting ergodic behaviour of the quasi-free systems is due to the absolutely continuous part of the energy spectrum. Therefore, we restrict ourselves to one-particle Hamiltonians H_1 acting as multiplication operators on the one-particle Hilbert space $\mathfrak{H}$:

$$\mathfrak{H} = \mathcal{L}^2([0, E_{\max}[, \rho(\epsilon)d\epsilon) \otimes \ell^2(\mathfrak{J})$$

and

$$(H_1\psi)(\epsilon, j) = \epsilon\,\psi(\epsilon, j), \quad \epsilon \in [0, E_{\max}[, \; j \in \mathfrak{J}.$$

Here $E_{\max}$ is the, possibly infinite, maximal one-particle energy, $\rho(\epsilon)$ measures the density of states and $\mathfrak{J}$ is a discrete set of additional quantum numbers which, together with ϵ, fully parameterize the quantum states of our particles. This model covers all examples of fermionic particles and quasi-particles propagating in infinite continuous or discrete (lattice) space which do not interact or whose interactions can be approximated by a self-consistent external field. Without any loss of generality, we can assume that the time-invariant quasi-free reference state ω_Q (see (6.25)) is defined by the multiplication operator

$$(Q\psi)(\epsilon, j) = q(\epsilon, j)\,\psi(\epsilon, j), \quad 0 \le q(\epsilon, j) \le 1.$$

Moreover, we assume for simplicity that the function $\epsilon \mapsto q(\epsilon, j)$ is continuous.
  Fixing a time step t, we put

$$U \equiv U_t = \exp\{-itH_1\}, \qquad (U\psi)(\epsilon, j) = e^{-it\epsilon}\psi(\epsilon, j)$$

and from now on we consider the quasi-free dynamical system $(\mathfrak{A}(\mathfrak{H}), \alpha_U, \omega_Q)$ with all ingredients as above. This system is highly reducible in the sense that it contains many invariant subsystems. Take any measurable subset Λ of the spectrum $[0, E_{\max}[\times\mathfrak{J}$ and denote by $\mathfrak{H}_\Lambda$ the subspace $\mathcal{L}^2(\Lambda)$ of $\mathfrak{H}$. The CAR-subalgebra $\mathfrak{A}(\mathfrak{H}_\Lambda)$ is invariant under α_U. We denote by α_Λ and ω_Λ the restrictions of α_U and the state ω_Q to the subalgebra $\mathfrak{A}(\mathfrak{H}_\Lambda)$. Then $(\mathfrak{A}(\mathfrak{H}_\Lambda), \alpha_\Lambda, \omega_\Lambda)$ is a well-defined C*-algebraic dynamical system in the sense of Definition 5.19. For a particular choice of $\Lambda = [0, E[\times F$ where F is a finite subset of $\mathfrak{J}$, we define a smooth subalgebra $\mathfrak{A}_0(\mathfrak{H}_\Lambda)$ as the *-algebra generated by the elements

$$c(\phi_i), \quad \phi_i(\epsilon, j) = f_i(\epsilon)\,\delta_{ij}, \quad f_i \in C_0^\infty([0, E[), \quad i \in F.$$

Here $C_0^\infty([0, E[)$ denotes the infinitely differentiable functions with compact supports in $]0, E[$.

13.3.2 Main result

Using the notation introduced above, we can formulate the main result.

Theorem 13.7 *Let $(\mathfrak{A}(\mathfrak{H}), \alpha_U, \omega_Q)$ with $U \equiv U_t = \exp\{-itH_1\}$ be a fermionic quasi-free dynamical system as described above. This system possesses a* spectral *density h of dynamical entropy given by*

$$h(\epsilon, j) = -\frac{1}{2\pi}\left[q(\epsilon, j)\log q(\epsilon, j) + (1 - q(\epsilon, j))\log(1 - q(\epsilon, j)) - \log 2\right]$$

which for any $\Lambda = [0, E[\times F$ with $F \subset \mathfrak{J}$ satisfies

$$h[\omega_Q, \alpha_{U_t}, \mathfrak{A}_0(\mathfrak{H}_\Lambda)] = t \sum_{j \in F} \int_0^E d\epsilon\, \rho(\epsilon)\, h(\epsilon, j). \tag{13.19}$$

In formula (13.19), the dependence of the dynamical entropy on the time step t is explicitly displayed. The spectral density of the dynamical entropy consists of two terms. For suitable choices of the state ω_Q the first term can produce a finite dynamical entropy even for $E \to E_{\max} = \infty$ and infinite F, while the second always yields infinite dynamical entropy in these cases. As the first term is absent for the vacuum state $Q = 0$, we can call

$$h_{\rm ren}(\epsilon, j) := -\frac{1}{2\pi}\left[q(\epsilon, j)\log q(\epsilon, j) + (1 - q(\epsilon, j))\log(1 - q(\epsilon, j))\right]$$

the *renormalized spectral density of dynamical entropy.* One should notice that the dynamical entropy computed with $h_{\rm ren}(\epsilon, j)$ coincides with the CNT-entropy.

13.3.3 Sketch of the proof

The full proof of Theorem 13.7 is rather involved and contains tedious technical computations, so we present the main steps only. Moreover, we restrict ourselves to the case $\rho(\epsilon) = 1$. The main idea is simple, we have to prove additivity of the dynamical entropy with respect to the spectral decomposition of the invariant subsystems of above into smaller constituents and then reduce the problem to the most elementary system: the quasi-free shift.

Quasi-free shift

Our first step is a computation of the dynamical entropy for the system $(\mathfrak{A}(\mathfrak{H}_\Lambda), \alpha_\Lambda, \omega_\Lambda)$ with the special choice $\Lambda = [0, 2\pi/t[\times \{j\}, \, j \in \mathcal{J}$, i.e.

$$\mathfrak{H}_\Lambda \equiv \mathcal{L}^2[0, 2\pi/t[.$$

This choice of Λ makes our subsystem completely equivalent to a one-dimensional lattice with sites occupied by fermions. The dynamics is equivalent to a shift γ

and the state ω_Λ is a shift-invariant quasi-free state as described in Section 9.1.4 with symbol given by $q(\epsilon, j)$ in terms of the rescaled variable $\theta = \epsilon t$. To compute the entropy of the quasi-free shift we combine some previously obtained results. First, we apply the Jordan–Wigner transformation (6.18) and Example 4.6 to establish the isomorphism of our system with a one-dimensional spin-$\frac{1}{2}$ chain. Then, we apply Theorem 10.17 with the entropy density $\sigma(\omega)$ given by the formula (9.26). Up to a change of variable, we obtain exactly the expression (13.19) for our particular Λ. The only subtle point is that the dynamical entropy in Theorem 10.17 has been computed for a local subalgebra. In our case, it corresponds to the *-subalgebra of $\mathfrak{A}(\mathfrak{H}_\Lambda)$ generated by the $c(\phi_k)$'s with test functions $\varphi_k(\epsilon) \sim exp(ikt\epsilon)$, $k \in \mathbb{Z}$, while we need smooth functions with compact supports in $]0, 2\pi/t[$. Fortunately, applying the continuity property (11.7), one can show that the dynamical entropy remains unchanged.

Powers of the shift
There are obvious generalizations of this result. First of all, one notices that for

$$\Lambda = [0, 2\pi r/t[\times\{j\}, \quad r \in \mathbb{N} \tag{13.20}$$

the corresponding dynamical subsystem $(\mathfrak{A}(\mathfrak{H}_\Lambda), \alpha_\Lambda, \omega_\Lambda)$ is equivalent to the rth power γ^r of the quasi-free shift. The dynamical entropy of γ^r is simply r times the dynamical entropy of γ because the rth power of the shift on a quantum spin chain with $d \times d$ matrices at a single site is equivalent to a shift on a chain with $d^r \times d^r$ matrices. Therefore, after rescaling of variables, we again obtain formula (13.19) with the special choice (13.20) of Λ.

Shift with multiplicities
The next simplest generalization consists in replacing a single point $j \in \mathfrak{J}$ by a finite subset $F \subset \mathfrak{J}$ yielding

$$\Lambda = [0, 2\pi r/t[\times F, \quad r \in \mathbb{N}. \tag{13.21}$$

Again such a system is equivalent to a shift on a chain with an enlarged algebra of single-site observables. This fact, together with the factorization property of quasi-free states (Theorem 6.12) and the additivity of the entropy (10.16) leads to a simple additivity of the dynamical entropy with respect to the points j and hence again to the formula (13.19) with Λ given by (13.21).

Upper bound
Consider a more general

$$\Lambda = [0, 2\pi r/st[\times F, \quad r, s \in \mathbb{N}, \quad 2\pi r/st \le E_{\max}.$$

Obviously, we have now a dynamical system equivalent to a rational power r/s of a shift. In order to reduce the problem to a natural power of a shift, we use inequality (10.18) with q replaced by s and Θ by γ^r. Then, applying the results of above, we obtain

$$h[\omega_Q, \alpha_{U_t}, \mathfrak{A}_0(\mathfrak{H}_\Lambda)] \leq t \sum_{j \in F} \int_0^E d\epsilon\, h(\epsilon, j),$$

first with $E = 2\pi r/st$ and then with any $E \in [0, E_{\max}[$ by continuity of the right-hand side of (13.19) and the obvious monotonicity of the dynamical entropy with respect to Λ .

Lower bound
Fixing a finite set $F \subset \mathfrak{J}$, we can consider the entropy $h[\omega_Q, \alpha_{U_t}, \mathfrak{A}_0(\mathfrak{H}_\Lambda)]$ with $\Lambda = [0, E[\times F$ as a function of $E \in [0, E_{\max}[$, denoted in short by $h[E]$. We already know that $h[E]$ grows monotonically with E and

$$h[E] \leq t \sum_{j \in F} \int_0^E d\epsilon\, h(\epsilon, j). \tag{13.22}$$

Moreover, equality in (13.22) holds for $E = 2\pi r/t$, $r \in \mathbb{N}$. Taking into account the graded tensor product structure of CAR-algebras (see (6.21) and quasi-free states (see Theorem 6.12) and the super-additivity of dynamical entropy with respect to composition of dynamical systems (see Lemma 11.8), we obtain for any $p \in \mathbb{N}$

$$h[E + 2\pi/tp] - h[E] \geq \sum_{j \in F} h[\omega_Q, \alpha_{U_t}, \mathfrak{A}_0(\mathfrak{H}_{([E, E+2\pi/tp[\times\{j\})})].$$

If we can prove that

$$\lim_{p \to \infty} \frac{tp}{2\pi} h[\omega_Q, \alpha_{U_t}, \mathfrak{A}_0(\mathfrak{H}_{([E, E+2\pi/tp[\times\{j\})}))] \geq h(E, j)$$

then, due to (13.22), the proof of our Theorem 13.7 is complete.

As we are interested in the asymptotic behaviour $p \to \infty$, the function $\epsilon \mapsto q(\epsilon, j)$ is essentially constant on the interval $[E, E + 2\pi/tp[$ and equal to $q(E, j)$. Therefore, for a large p, our dynamical subsystem corresponding to $\Lambda = ([E, E + 2\pi/tp[\times\{j\})$ is asymptotically equivalent to the pth root of the quasi-free shift with the quasi-free state defined by $Q = q(E)\mathbf{1}$. Taking the already computed dynamical entropy of the shift and the inequality (10.18), we have

$$h[\omega_{\lambda\mathbf{1}}, \gamma^{1/p}, \mathfrak{A}_0] \leq \frac{1}{p} h[\omega_{\lambda\mathbf{1}}, \gamma, \mathfrak{A}_0] = \frac{1}{p}(\eta(\lambda) + \eta(1 - \lambda) + \log 2)$$

where, as usual, $\eta(\lambda) = -\lambda \log \lambda$, $\lambda \in [0, 1]$.

To complete the arguments and to prove equation (13.19), we have to construct for any sequence $p_m \to \infty$, $m = 1, 2, 3, \ldots$ a sequence of partitions of unity $\mathbf{X}_m$ such that

$$\lim_{m \to \infty} p_m \, \mathsf{h}[\omega_{\lambda\mathbf{1}}, \gamma^{1/p_m}, \mathbf{X}_m] = \eta(\lambda) + \eta(1 - \lambda) + \log 2. \tag{13.23}$$

Optimal partitions for roots of the shift

The construction of the partitions which satisfy (13.23) is the most involved part of the proof. We consider a quasi-free shift in Fourier representation, i.e. with one-particle Hilbert space $\mathfrak{H} = \mathcal{L}^2([0, 2\pi[, d\theta/2\pi)$, reference state $\omega_{\lambda\mathbf{1}}$ and dynamics given by $\gamma^{1/p}$, an automorphism generated by a pth root $S^{1/p}$ of the shift S:

$$(S\varphi)(\theta) = e^{i\theta} \, \varphi(\theta).$$

Take a finite-rank projector P and a number $\nu \in]0, 1[$ to construct the completely positive unity-preserving and bistochastic quasi-free map Γ (see Section 8.5) uniquely determined by

$$\omega_Q \circ \Gamma = \omega_{Q'} \quad \text{with} \quad Q' := (\mathbf{1} - \nu P)Q(\mathbf{1} - \nu P) + (\nu^2/2 - \nu)P$$

for any quasi-free state ω_Q. Because P has a finite rank, the map Γ acts only non-trivially on a finite-dimensional CAR-subalgebra and therefore there exists a finite partition of unity $\mathbf{P}$ corresponding to Γ. In the following, we do not need to know the explicit form of this partition, all computations are performed in terms of the quasi-free dynamical map. The refined partitions $\mathbf{P}^m$ correspond to the dynamical maps $\Gamma^{(m)}$ given by the expression

$$\omega_Q \circ \Gamma^{(m)} = \omega_{Q'^{(m)}} \quad \text{with} \quad Q'^{(m)} = V_m^* Q V_m + \tfrac{1}{2}(\mathbf{1} - V_m^* V_m),$$

where

$$V_n = (S^{1/p} - \nu P S^{1/p})^n S^{-n/p}. \tag{13.24}$$

In order to compute the entropy of the partition $\mathbf{P}^m$, we use the explicit GNS-representation of our dynamical system on the double Fock space $\Gamma_{\text{a.s.}}(\mathfrak{H} \oplus \mathfrak{H})$; see Example 6.13. Direct computation shows that, in the notation of (11.4), we have

$$\mathsf{H}[\omega_{\lambda\mathbf{1}}, \mathbf{P}^m] = S(\rho[\mathbf{P}^m]) = S(\hat{\rho}[\mathbf{P}^m]),$$

where $\hat{\rho}[\mathbf{P}^m]$ is unitary equivalent to the quasi-free density matrix ω_{R_m} on the Fock space $\Gamma_{\text{a.s.}}(\mathfrak{H} \oplus \mathfrak{H})$ determined by the symbol

$$R_m = \begin{pmatrix} \lambda + (1/2 - \lambda)K_m & \sqrt{\lambda(1 - \lambda)(\mathbf{1} - K_m)} \\ \sqrt{\lambda(1 - \lambda)(\mathbf{1} - K_m)} & 1 - \lambda \end{pmatrix}$$

on $\mathfrak{H} \oplus \mathfrak{H}$, where $K_m := \mathbf{I} - V_m V_m^*$, is a positive finite-rank operator with spectral decomposition

$$K_m = \sum_j k_j^{(m)} |\xi_j\rangle \langle \xi_j|.$$

Denoting by $q_\pm(k)$ the two eigenvalues of the 2×2 matrix

$$R(k) = \begin{pmatrix} \lambda + (1/2 - \lambda)k & \sqrt{\lambda(1-\lambda)(1-k)} \\ \sqrt{\lambda(1-\lambda)(1-k)} & 1 - \lambda \end{pmatrix}$$

with $k \in [0, 1]$, we write the entropy as

$$H[\omega_{\lambda \mathbf{1}}, \mathbf{P}^m] = S(\hat{\rho}[\mathbf{P}^m]) = S(\omega_{R_m}) = \sum_j \tilde{\eta}(k_j^{(m)}),$$

where

$$\tilde{\eta}(k) := \eta(q_+(k)) + \eta(1 - q_+(k)) + \eta(q_-(k)) + \eta(1 - q_-(k)).$$

It is easy to show that the function $\tilde{\eta}(k)$ is concave on $[0, 1]$, $\tilde{\eta}(0) = 0$ and $\tilde{\eta}(1) = \eta(\lambda) + \eta(1 - \lambda) + \log 2$. Therefore, for $k \in [0, 1]$,

$$\tilde{\eta}(k) \geq k(\eta(\lambda) + \eta(1 - \lambda) + \log 2).$$

Hence, we obtain a lower bound for the dynamical entropy of the pth root of the shift

$$h[\omega_{\lambda \mathbf{1}}, \gamma^{1/p}, \mathfrak{A}_0] \geq \left(\eta(\lambda) + \eta(1 - \lambda) + \log 2\right) \lim_{m \to \infty} \frac{1}{m} \operatorname{Tr} K_m.$$

Writing the spectral decomposition of P as

$$P = \sum_{\ell = -N}^{N} |\psi_\ell\rangle \langle \psi_\ell|$$

one obtains from (13.24)

$$\operatorname{Tr} K_m = (2\nu - \nu^2) \sum_{\ell = -N}^{N} \sum_{k=0}^{m-1} \|W_k^* \psi_\ell\|^2,$$

where

$$W_k := [S^{-1/p}(\mathbf{I} - \nu P)]^k S^{k/p}.$$

Here $W := s - \lim_{k \to \infty} W_k$ is a kind of discrete-time wave operator which exists by the standard Cook criterion of scattering theory and hence

$$\lim_{m \to \infty} \frac{1}{m} \operatorname{Tr} K_m = (2\nu - \nu^2) \sum_{\ell = -N}^{N} \|W^* \psi_\ell\|^2. \tag{13.25}$$

Again, applying the techniques of standard scattering theory, one can find the equation for $W^* \psi_\ell$, solve it, and plug the solution in eqn (13.25). With the optimal choices $\nu = 2/p$, $\psi_\ell(\theta) = (2\pi)^{-1/2} \exp(i\ell\theta)$, and $N \to \infty$ for $p \to \infty$, we obtain the desired asymptotic behaviour:

$$\lim_{m \to \infty} \frac{1}{m} \operatorname{Tr} K_m = \frac{1}{p} + o(1/p).$$

Strictly speaking, the functions ψ_ℓ should be modified at the endpoints of the interval $[0, 2\pi[$ in order to become functions of $\mathcal{C}_0^\infty$-class but this does not alter the final result due to the continuity property (10.14).

13.4   Notes

The quantum cat maps of Section 13.1 have been widely studied in the context of dynamical entropy. Moreover, they yield different values for the entropy depending on the construction that one considers. The entropy in terms of operational partitions is completely insensitive to the value of the deformation parameter θ and it coincides with the positive Lyapunov exponenteither introduced in terms of the horocyclic actions or in terms of non-commutative Riemannian manifolds (Andries *et al.* 1995). A more general computation than that presented here can be found in (Fannes and Tuyls 1999). The CNT-entropy is quite sensitive to the algebraic structure of the algebra of observables, which amounts in the case of the cat maps to arithmetic properties of θ. It turns out that one finds almost always a vanishing entropy (Narnhofer and Thirring 1995). The positive Lyapunov exponent is an upper bound for one of the entropies introduced in (Voiculescu 1992). For comparisons of entropies in different models, see the paper (Alicki and Narnhofer 1995).

The section on Ruelle's inequality (Ruelle 1978) is a first attempt to arrive at a quantum version of Pesin's theorem (Pesin 1977; Ledrappier 1984) in that it means to link the statistical notion of entropy to non-commutative geometry. A geometrical approach to general non-commutative spaces can be found, e.g. in works of Woronowicz (1989) and Connes (1994). Dirichlet forms on non-commutative algebras have been studied in (Davies and Lindsay 1992). The idea of the proof of the inequality as presented here is similar to that in (Alicki 1998a); see also (Alicki 1998b).

The last section contains largely unpublished material. It is a first example of an entropy computation for a physically more realistic system. A version in discrete time has been considered in Andries (2000). The CNT-entropy for a version of this dynamics in discrete time has been computed in (Størmer and Voiculescu 1990). The main difficulty in our computation lies in the construction of partitions that provide an optimal information gain under the time evolution. This leads technically to the problem of existence of scattering operators, guaranteed by Cook's criterion (Kato 1984), and their explicit construction.

14

EPILOGUE

During the International Congress on Mathematical Physics held in London in the year 2000, J. L. Lebowitz expressed his opinion that one of the great challenges to mathematical physics in the twenty-first century is the theory of heat conductivity and other transport phenomena in macroscopic bodies. Indeed, Fourier's law described by the equation

$$c\frac{\partial}{\partial t}T(\mathbf{r}, t) = \nabla \cdot \big(\kappa \nabla T(\mathbf{r}, t)\big), \qquad (14.1)$$

where $T(\mathbf{r}, t)$ is the local temperature, $c = c(T)$ the specific heat, and $\kappa = \kappa(T)$ the heat conductivity, holds with high precision. *There is, however, no rigorous mathematical derivation of* (14.1) *for any classical or quantum model with a Hamiltonian microscopic evolution*, see (Bonetto *et al.* 2000). The problem is quite fundamental and related to the question: *Can a reversible and deterministic microscopic dynamics fully explain the behaviour of macroscopic matter?*

The answer is not so obvious as we have many examples where it is possible to derive rigorously the transport laws by adding a non-Hamiltonian component, either stochastic (Davies 1978) and Spohn (1991) or deterministic (Gallavotti and Cohen 1996), to the microscopic Hamiltonian equations of motion; see also the paper (Maes 1999). A similar situation occurs in quantum mechanics, where some experts propose various modifications of Schrödinger's equation to deal with the state vector reduction in quantum measurement theory or to explain the absence of superpositions of macroscopically distinguishable states (Penrose 2000). The only way to provide a positive answer for this question is to present a rigorous derivation of the transport equations for a Hamiltonian model which is simple enough to be manageable but still acceptable from the physical point of view.

The main difficulty is that only truly interacting systems—in other words systems with non-linear interactions—could display nontrivial transport phenomena. Quasi-free or linear systems, as those studied in Section 6.2 and 13.3, may serve as good models of heat baths but their ergodic properties are too weak to produce internal dissipation and hence finite heat conductivity. In the classical case, non-linear interactions in finite systems produce exponential instability, positive Lyapunov exponents and positive Kolmogorov–Sinai entropy. For interacting systems in the thermodynamic limit continuous spectra of Lyapunov exponents and densities of dynamical entropy appear (Sinai 1996; Dolgopyat 1997) and they should be related to the transport coefficients. It is quite plausible that for quantum models the techniques developed in this book, like those related to the quantum dynamical entropy, should be helpful in solving the problem of transport phenomena. Let us discuss briefly one of the possible links.

Consider a macroscopic body, initially in thermal equilibrium, which is put locally in contact with a heat source at a much higher temperature. The ensuing time-dependent non-uniform temperature distribution $T(\mathbf{r}, s)$ is a source of entropy increase given by the following standard phenomenological formula:

$$\delta S(t) = \int_0^t ds \int d\mathbf{r} \, \kappa (\nabla T(\mathbf{r}, s))^2 \ .$$

Passing to an approximate microscopic quantum model in discrete time, we study a dynamical system $(\mathfrak{A}, \Theta, \omega)$ with the equilibrium reference state perturbed by a completely positive map $\Pi_\mathbf{X}$ as in (10.13), which corresponds to putting the system in contact with a heat source. Then, due to the inequality in Lemma 10.11(c) and the remark following the same lemma, the asymptotic entropy production per unit time step, which should be identified with the phenomenological entropy production,

$$s = \lim_{t \to \infty} \frac{\delta S(t)}{t}$$

is bounded from above by the dynamical entropy $\mathsf{h}[\omega, \Theta, \mathbf{X}]$; see Definition 10.8.

The natural model for further studies seems to be a system of interacting fermions. It is not difficult to define rigorously perturbations of the quasi-free Fermi gas and one can hope that in the future the techniques presented in Section 13.3 will be extended to the interacting case too.

REFERENCES

Books

Agarwal, G. S. (1971). *Quantum Optics. Quantum Statistical Theories of Spontaneous Emission and their Relation to Other Approaches*, volume 70 of *Springer Tracts in Modern Physics*. Springer, Berlin.

Alicki, R. and Lendi, K. (1987). *Quantum Dynamical Semigroups and Applications*. Springer, Berlin.

Alicki, R., Bożejko, M., and Majewski, W., editors (1998). *Quantum Probability*, volume 43, Warsaw. Banach Center Publications.

Andries, J. (2000). *Correlation Functions and Dynamical Entropy in Relaxing Quantum Systems*. PhD thesis, Katholieke Universiteit Leuven, Belgium.

Arnold, V. I. and Avez, A. (1967). *Problèmes Ergodiques de la Mécanique Classique*. Gauthier-Villars, Paris.

Beltrametti, E. and Cassinelli, G. (1981). *The Logic of Quantum Mechanics*. Addison-Wesley, Reading Mass.

Benatti, F. (1993). *Deterministic Chaos in Infinite Quantum Systems*. Springer, Berlin.

Berline, N., Getzler, E., and Vergne, M. (1992). *Heat Kernels and Dirac Operators*. Springer, Berlin.

Blank, J., Exner, P., and Havlíček, M. (1994). *Hilbert Space Operators in Quantum Physics*. AIP Press, New York.

Bratteli, O. and Robinson, D. W. (1979). *Operator Algebras and Quantum Statistical Mechanics: C*- and W*-Algebras. Symmetry Groups. Decomposition of States*, volume 1. Springer, Berlin.

Bratteli, O. and Robinson, D. W. (1997). *Operator Algebras and Quantum Statistical Mechanics: Equilibrium States. Models in Quantum Statistical Mechanics*, volume 2. Springer, Berlin, second edition.

Busch, P., Grabowski, M., and Lahti, P. J. (1995). *Operational Quantum Physics*. Springer-Verlag, Berlin.

Carmichael, H. J. (1993). *An Open Systems Approach to Quantum Optics*. Springer, Berlin.

Casati, G. and Chirikov, B. (1995). *Quantum Chaos*. Cambridge University Press, Cambridge.

Connes, A. (1994). *Non-commutative Geometry*. Academic Press, New York.

Cornfeld, I. P., Fomin, S. V., and Sinai, Y. G. (1982). *Ergodic Theory*. Springer, Berlin.

Cycon, H. L., Froese, R. G., Kirsch, W., and Simon, B. (1987). *Schrödinger Operators*. Springer, Berlin.

Davidson, K. R. (1991). *C*-Algebras by Example*, volume 6 of *Fields Institute*

Monographs. AMS, Providence.

Davies, E. B. (1976). *Quantum Theory of Open Systems*. Academic Press, London.

Emch, G. G. (1972). *Algebraic Methods in Statistical Mechanics and Quantum Field Theory*. Wiley, New York.

Evans, D. E. and Kawahigashi, Y. (1998). *Quantum Symmetries on Operator Algebras*. Clarendon Press, Oxford.

Folland, G. B. (1989). *Harmonic Analysis in Phase Space*. Princeton University Press, Princeton.

Grenander, U. and Szegö, G. (1958). *Toeplitz Forms and their Applications*. University of California Press, Berkeley.

Gruska, J. (1999). *Quantum Computing*. McGraw Hill, London.

Gutzwiller, M. C. (1990). *Chaos in Classical and Quantum Mechanics*. Springer, New York.

Haake, F. (1973). *Statistical treatment of open systems by generalized master equations*, volume 66 of *Springer Tracts in Modern Physics*. Springer, Berlin.

Haake, F. (1991). *Quantum Signatures of Chaos*. Springer, Berlin.

Hiai, F. and Petz, D. (2000). *The Semicircle Law, Free random variables and Entropy*, volume 77 of *Mathematical Surveys and Monographs*. AMS, Providence.

Holevo, A. S. (1982). *Probabilistic and Statistical Aspects of Quantum Theory*. North-Holland, Amsterdam.

Ingarden, R. S., Kossakowski, A., and Ohya, M. (1997). *Information Dynamics and Open Systems*. Kluwer, Dordrecht.

Isham, C. J. (1989). *Modern Differential Geometry for Physicists*. World Scientific, Singapore.

Kadison, R. V. and Ringrose, J. R. (1986a). *Fundamentals of the Theory of Operator Algebras*, volume 1. Academic Press, New York.

Kadison, R. V. and Ringrose, J. R. (1986b). *Fundamentals of the Theory of Operator Algebras. Advanced Theory*, volume 2. Academic Press, New York.

Kato, T. (1984). *Perturbation Theory for Linear Operators*. Springer, Berlin, second edition.

Katok, A. and Hasselblatt, B. (1996). *Introduction to the Modern Theory of Dynamical Systems*. Cambridge University Press, Cambridge.

Klauder, J. and Skagerstam, B.-S. (1985). *Coherent States, Applications in Physics and Mathematical Physics*. World Scientific, Singapore.

Kraus, K. (1983). *States, Effects, and Operations*, volume 190 of *Lecture Notes in Physics*. Springer, Berlin.

Ludwig, G. (1983). *Foundations of Quantum Mechanics*, volume 1 of *Texts and Monographs in Physics*. Springer, Berlin.

Ludwig, G. (1985). *Foundations of Quantum Mechanics*, volume 2 of *Texts and Monographs in Physics*. Springer, Berlin.

Mackey, G. (1963). *Mathematical Foundations of Quantum Mechanics*. Benjamin, New York.

Mehta, M. L. (1991). *Random Matrices*. Academic Press, London.

Nash, C. and Sen, S. (1983). *Topology and Geometry for Physicists*. Academic Press, London.

Nielsen, M. A. and Chuang, I. L. (2000). *Quantum Computation and Quantum Information*. Cambridge University Press, Cambridge.

Ohya, M. and Petz, D. (1993). *Quantum Entropy and Its Use*. Springer, Berlin.

Parthasarathy, K. R. (1992). *An Introduction to Quantum Stochastic Calculus*. Birkhäuser, Basel.

Paulsen, V. I. (1986). *Completely Bounded Maps and Dilations*, volume 146 of *Pitman research notes in mathematics*. Longman, London.

Perelomov, A. (1986). *Generalized Coherent States and their Applications*. Texts and Monographs in Physics. Springer, Berlin.

Peres, A. (1993). *Quantum Theory: Concepts and Methods*. Kluwer Academic Publishers. Dordrecht.

Petersen, K. (1983). *Ergodic Theory*. Cambridge University Press, Cambridge.

Petz, D. (1990). *An Invitation to the Algebra of Canonical Commutation Relations*, volume 2 of *A: Mathematical Physics*. Leuven University Press, Leuven.

Plymen, R. and Robinson, P. (1994). *Spinors in Hilbert Space*. Cambridge University Press, Cambridge.

Quantum Probability and Applications, Accardi, L. *et al.*, editors, volumes 1055, 1136, 1303, 1396, 1442 of *Lecture Notes in Mathematics*, Springer.

Quantum Probability & Related Topics, Accardi, L., editor, volumes 7–10, World Scientific.

Reed, M. and Simon, B. (1972). *Methods of Modern Mathematical Physics*, volume 1, Academic press, New York.

Richtmyer, R. (1978). *Principles of Advanced Mathematical Physics I*. Springer, Berlin.

Robert, D. (1987). *Autour de l'Approximation Sémi-Classique*. Birkhäuser, Boston.

Schmidt, K. (1990). *Algebraic Ideas in Ergodic Theory*, volume 76 of *Regional Conference Series in Mathematics*. American Mathematical Society, Providence.

Sewell, G. L. (1986). *Quantum Theory of Collective Phenomena*. Clarendon Press, Oxford.

Spohn, H. (1991). *Large Scale Dynamics of Interacting Particles*. Springer, Berlin.

Stöckmann, H.-J. (1999). *Quantum Chaos, An Introduction*. Cambridge University Press, Cambridge.

Streater, R. F. (1995). *Statistical Dynamics*. Imperial College Press, London.

Thirring, W. (1981). *A Course in Mathematical Physics 3. Quantum Mechanics of Atoms and Molecules*. Springer, New York.

Thirring, W. (1983). *A Course in Mathematical Physics 4. Quantum Mechanics of Large Systems*. Springer, New York.

Tuyls, P. (1997). *Towards Quantum Kolmogorov–Sinai Entropy*. PhD thesis,

Katholieke Universiteit Leuven, Belgium.

Varadarajan, V. S. (1968). *Geometry of Quantum Theory*. Van Nostrand, New York.

Voiculescu, D. V., Dykema, K. J., and Nica, A. (1992). *Free Random Variables*, volume 1 of *CRM Monograph Series*. AMS, Providence.

von Neumann, J. (1932). *Mathematische Grundlagen der Quantenmechanik*. Springer, Berlin.

Walters, P. (1982). *An Introduction to Ergodic Theory*. Springer, Berlin.

Yosida, K. (1965). *Functional Analysis*. Springer, Berlin.

Papers

Accardi, L., Lu, Y. G., Alicki, R., and Frigerio, A. (1992). An invitation to the weak coupling and low density limits, volume 6 of *Quantum Probability and Related Topics*, pages 237–66, Singapore. World Scientific.

Accardi, L., Ohya, M., and Watanabe, N. (1996). Note on quantum dynamical entropies. *Reports on Mathematical Physics*, **38**, 457–69.

Alicki, R. (1978). The theory of open systems in application to unstable particles. *Reports on Mathematical Physics*, **14**, 27–42.

Alicki, R. (1997). Quantum ergodic theory and communication channels. *Open Systems and Information Dynamics*, **4**, 53–69.

Alicki, R. (1998a). Quantum geometry of non-commutative Bernoulli shift. In Alicki, R., Bożejko, M., and Majewski, W., editors, *Quantum Probability*, volume 43, pages 25–9. Banach Center Publications.

Alicki, R. (1998b). Quantum mechanical tools in applications to classical dynamical systems. *Physical Review Letters*, **81**, 2040–3.

Alicki, R., Andries, J., Fannes, M., and Tuyls, P. (1996a). An algebraic approach to the Kolmogorov-Sinai entropy. *Reviews in Mathematical Physics*, **8**, 167–84.

Alicki, R. and Fannes, M. (1987). Dilations of quantum dynamical semigroups with classical Brownian motion. *Communications in Mathematical Physics*, **108**, 353–61.

Alicki, R. and Fannes, M. (1994). Defining quantum dynamical entropy. *Letters in Mathematical Physics*, **32**, 75–82.

Alicki, R., Makowiec, D., and Miklaszewski, W. (1996b). Quantum chaos in terms of entropy for the periodically kicked top. *Physical Review Letters*, **77**, 838–41.

Alicki, R. and Narnhofer, H. (1995). Comparison of dynamical entropies of non-commutative shifts. *Letters in Mathematical Physics*, **33**, 241–7.

Andries, J., Benatti, F., Cock, M. D., and Fannes, M. (2000). Multi-time correlations in relaxing quantum dynamical systems. *Reviews in Mathematical Physics*, **12**, 921–44.

Andries, J., Fannes, M., Tuyls, P., and Alicki, R. (1995). The dynamical entropy of the quantum Arnold cat map. *Letters in Mathematical Physics*, **35**, 375–83.

Araki, H. (1976). Relative entropy for states on von Neumann algebras. *Publications of the RIMS Kyoto University*, **11**, 808–33.

Araki, H. (1977). Relative entropy for states on von Neumann algebras II. *Publications of the RIMS Kyoto University*, **13**, 173–92.

Araki, H. and Matsui, T. (1985). Ground states of the XY-model. *Communications in Mathematical Physics*, **101**, 213–45.

Bach, V., Fröhlich, J., and Sigal, I. M. (1998). Quantum electrodynamics of confined nonrelativistic particles. *Advances in Mathematics*, **137**, 299–395.

Balazs, N. L. and Voros, A. (1989). The quantized baker's transformation. *Annals of Physics*, **190**, 1–31.

Balslev, E. and Verbeure, A. (1968). States on Clifford algebras. *Communications in Mathematical Physics*, **7**, 55–76.

Benatti, F. and Fannes, M. (1998). Statistics and quantum chaos. *Journal of Physics A*, **31**, 9123–30.

Benatti, F., Narnhofer, H., and Sewell, G. L. (1991). A non-commutative version of the Arnold cat map. *Letters in Mathematical Physics*, **21**, 157–72.

Biedenharn, L. C. (1990). A q-boson realization of the quantum group $SU_q(2)$, and the theory of q-tensor operators. In Doebner, H.-D. and Hennig, J.-D., editors, *Quantum groups*, volume 370 of *Lecture Notes in Physics*, pages 67–88, Berlin. Springer.

Bonetto, F., Lebowitz, J. L., and Rey-Bellet, L. (2000). Fourier's law: A challenge to theorists. In Fokas, A., Grigoryan, A., Kibble, T., and Zegarlinski, B., editors, *Mathematical Physics 2000*, pages 128–50, London. Imperial College Press.

Bożejko, M. and Speicher, R. (1996). Interpolation between bosonic and fermionic relations given by generalized Brownian motions. *Mathematisches Zeitschrift*, **222**, 135–60.

Chebotarev, A. M. and Fagnola, F. (1998). Sufficient conditions for conservativity of minimal quantum dynamical semigroups. *Journal of Functional Analysis*, **153**, 382–404.

Choda, M. (2000). A C*-dynamical entropy and applications to canonical endomorphisms. *Journal of Functional Analysis*, **173**, 453–80.

Choda, M. and Takehana, H. (1998). A finite partition in a dynamical system with entropy depending on the size. *Preprint Osaka Kyoiku University*.

Choi, M.-D. (1972). Positive linear maps on C*-algebras. *Canadian Journal of Mathematics*, **24**, 520–6.

Choi, M.-D. (1975). Completely positive linear maps on complex matrices. *Linear Algebra and Applications*, **10**, 285–90.

Christensen, E. and Evans, D. E. (1979). Cohomology of operator algebras and quantum dynamical semigroups. *Journal of the London Mathematical Society*, **20**, 358–68.

Combescure, M. and Robert, D. (1997). Semiclassical spreading of quantum wave packets and applications near unstable fixed points of the classical flow. *Asymptotic Analysis*, **14**, 377–404

Connes, A., Narnhofer, H., and Thirring, W. (1987). Dynamical entropy of C*-algebras and von Neumann algebras. *Communications in Mathematical Physics*,

112, 691–719.

Connes, A. and Størmer, E. (1975). Entropy for automorphisms of II_1 von Neumann algebras. *Acta Mathematica*, **134**, 289–306.

Cushen, C. D. and Hudson, R. L. (1971). A quantum mechanical central limit theorem. *Journal of Applied Probability*, **8**, 249–52.

Davies, E. B. (1974). Markovian master equations. *Communications in Mathematical Physics*, **39**, 91–110.

Davies, E. B. (1977). Quantum dynamical semigroups and the neutron diffusion equation. *Reports on Mathematical Physics*, **11**, 169–88.

Davies, E. B. (1978). A model of heat conduction. *Journal of Statistical Physics*, **18**, 161–70.

Davies, E. B. and Lewis, J. T. (1970). An operational approach to quantum probability. *Communications in Mathematical Physics*, **17**, 239–59.

Davies, E. B. and Lindsay, J. M. (1992). Non-commutative symmetric Markov semigroups. *Mathematisches Zeitschrift*, **210**, 379–411.

De Bièvre, S. (1998). Chaos, quantization and the classical limit of the torus. In Ali, S. T., Antoine, J.-P., Gazeau, J.-P., Odzijewicz, A., and Strasburger, A., editors, *Quantization, Coherent States, and Poisson Structures*, pages , Warsaw, Polish Scientific Publishers.

De Cock, M. D., Fannes, M., and Spincemaille, P. (1999). Quantum dynamics and Gramm's matrix. *Europhysics Letters*, **49**, 403–9

Demoen, B., Vanheuverzwijn, P., and Verbeure, A. (1977). Completely positive maps on the CCR-algebra. *Letters in Mathematical Physics*, **2**, 161–6.

Dolgopyat, D. (1997). Entropy of coupled map lattices. *Journal of Statistical Physics*, **86**, 377–89.

Duffield, N. G. and Werner, R. F. (1992). Local dynamics of mean-field quantum systems. *Helvetica Physica Acta*, **65**, 1016–54.

Dümcke, R. (1985). The low density limit for an n-level system interacting with a free Bose or Fermi gas. *Communications in Mathematical Physics*, **97**, 331–411.

Emch, G. G., Narnhofer, H., Thirring, W., and Sewell, G. L. (1994). Anosov actions on non-commutative algebras. *Journal of Mathematical Physics*, **35**, 5582–99.

Evans, D. E. (1979). Completely positive quasi-free maps on the CAR-algebra. *Communications in Mathematical Physics*, **70**, 53–68.

Evans, D. E. (1980). Dissipators for symmetric quasi-free dynamical semigroups on the CAR-algebra. *Journal of Functional Analysis*, **37**, 318–30.

Evans, D. E. and Lewis, J. T. (1986). On a C*-algebra approach to phase transition in the two-dimensional Ising model II. *Communications in Mathematical Physics*, **102**, 521–35.

Fannes, M. (1973*a*). Entropy density of quasi-free states. *Communications in Mathematical Physics*, **31**, 279–90.

Fannes, M. (1973*b*). A continuity property of the entropy density for spin lattice systems. *Communications in Mathematical Physics*, **31**, 291–4.

Fannes, M. (1998). Using density matrices in classical dynamical systems. In Alicki, R., Bożejko, M., and Majewski, W. A., editors, *Quantum Probability*, volume 43, pages 175–82. Banach Center Publications.

Fannes, M., Nachtergaele, B., and Werner, R. F. (1992). Finitely correlated states on quantum spin chains. *Communications in Mathematical Physics*, **144**, 443–90.

Fannes, M. and Rocca, F. (1980). A class of dissipative evolutions with applications in thermodynamics of Fermion systems. *Journal of Mathematical Physics*, **21**, 221–6.

Fannes, M. and Tuyls, P. (1999). A continuity property of quantum dynamical entropy. *Infinite Dimensional Analysis, Probability Theory and Related Topics*, **2**, 511–27.

Fidaleo, F. and Liverani, C. (1999). Ergodic properties for a quantum nonlinear dynamics. *Preprint University of Rome II*.

Gallavotti, G. and Cohen, E. G. D. (1996). Dynamical ensembles and stationary states. *Journal of Statistical Physics*, **80**, 931–70.

Gärding, L. and Wightman, A. S. (1954). Representations of the anticommutation relations. *Proceedings of the National Academy of Sciences*, **40**, 617–21.

Gleason, A. M. (1957). Measures on the closed subspaces of a Hilbert space. *J. Math. Mech.*, **6**, 91–110.

Goderis, D., Verbeure, A., and Vets, P. (1989). Non-commutative central limits. *Probability Theory and Related Fields*, **82**, 527–44.

Gorini, V. and Kossakowski, A. (1976). n-level system in contact with singular reservoir. *Journal of Mathematical Physics*, **17**, 1298–305.

Gorini, V., Kossakowski, A., and Sudarshan, E. C. G. (1976). Completely positive dynamical semigroups of n-level systems. *Journal of Mathematical Physics*, **17**, 821–5.

Haag, R., Kadison, R. V., and Kastler, D. (1973). Asymptotic orbits in a free Fermi gas. *Communications in Mathematical Physics*, **33**, 1–22.

Haake, F., Kuś, M., and Scharf, R. (1987). Classical and quantum chaos for a kicked top. *Zeitschrift für Physik B*, **65**, 381–95.

Hiai, F., Ohya, M., and Tsukuda, M. (1983). Sufficiency and relative entropy in ∗-algebras with applications to quantum systems. *Pacific Journal of Mathematics*, **107**, 117–40.

Holevo, A. S. (1996). On dissipative stochastic equations in a Hilbert space. *Probability Theory and Related Fields*, **104**, 483–500.

Howland, J. S. (1974). Stationary theory for time-dependent Hamiltonians. *Math. Ann.*, **207**, 315–33.

Hudetz, T. (1998). Quantum dynamical entropy revisited. In Alicki, R., Bożejko, M., and Majewski, W. A., editors, *Quantum Probability*, volume 43, pages 241–51. Banach Center Publications.

Hudson, R. L. (1973). A central limit theorem for anticommuting observables. *Journal of Applied Probability*, **10**, 502–9.

Hudson, R. L. and Parthasarathy, K. R. (1984). Quantum Ito's formula and stochastic evolutions. *Communications in Mathematical Physics*, **93**, 301–23.

Jaynes, E. T. (1957). Information theory and statistical mechanics. *Physical Review*, **106**, 620–30.

Kishimoto, A. (1979). On invariant states and the commutant of a group of quasi-free automorphisms of the CAR-algebra. *Reports on Mathematical Physics*, **15**, 21–6.

Kraus, K. (1971). General state changes in quantum theory. *Annals of Physics*, **64**, 311–35.

Lanford, O. E. and Robinson, D. W. (1972). Approach to equilibrium of free quantum systems. *Communications in Mathematical Physics*, **24**, 193–210.

Ledrappier, F. (1984). Quelques propriétés des exposants caracteristiques. In Hennequin, P. L., editor, *Ecole d'été de probabilités de Saint-Flour XII*, volume 1097 of *Lecture Notes in Mathematics*, pages 306–96 Springer.

Lieb, E. H. and Ruskai, M. B. (1973). Proof of the strong subadditivity of quantum mechanical entropy. *Journal of Mathematical Physics*, **14**, 1938–41.

Lindblad, G. (1975). Completely positive maps and entropy inequalities. *Communications in Mathematical Physics*, **40**, 147–51.

Lindblad, G. (1976). On the generators of quantum dynamical semigroups. *Communications in Mathematical Physics*, **48**, 119–30.

Lindblad, G. (1979). Non-Markovian stochastic processes and their entropy. *Communications in Mathematical Physics*, **65**, 281–94.

Lindblad, G. (1993). Determinism and randomness in quantum dynamics. *Journal of Physics A*, **26**, 7193–211.

Ludwig, G. (1974). Measuring and preparing process. In: *Foundations of quantum mechanics and ordered linear spaces*, volume 29 of *Lecture Notes in Physics*, pages 122–61. Springer.

Maassen, H., Guţă, M., and Botvich, D. (1999). Stability of Bose dynamical systems and branching theory. *Preprint University of Nijmegen*.

Maes, C. (1999). The fluctuation theorem as a Gibbs property. *Journal of Statistical Physics*, **95**, 367–92.

Majewski, W. A. and Kuna, M. (1993). On quantum characteristic exponents. *Journal of Mathematical Physics*, **34**, 5007–15.

Narnhofer, H. (1992). Quantized Arnold cat maps can be entropic K-systems. *Journal of Mathematical Physics*, **33**, 1502–10.

Narnhofer, H. and Thirring, W. (1994). Clustering for algebraic K-systems. *Letters in Mathematical Physics*, **30**, 307–16.

Narnhofer, H. and Thirring, W. (1995). C*-Dynamical systems that are asymptotically highly anticommutative. *Letters in Mathematical Physics*, **35**, 145–54.

Oseledec, V. I. (1968). A multiplicative ergodic theorem. Lyapunov characteristic numbers for dynamical systems. *Transactions of the Moscow Mathematical Society*, **19**, 197–221.

Pastur, L. (2000). Random matrices as paradigm. In Fokas, A., Grigoryan, A., Kibble, T., and Zegarlinski, B., editors, *Mathematical Physics 2000*, pages 216–65, London. Imperial College Press.

Penrose, R. (2000). Wave function collapse as a real gravitational effect. In Fokas, A., Grigoryan, A., Kibble, T., and Zegarlinski, B., editors, *Mathematical Physics 2000*, pages 266–82, London. Imperial College Press.

Perelomov, A. M. (1972). Coherent states for arbitrary Lie groups. *Communications in Mathematical Physics*, **26**, 222–36.

Pesin, Ya. B. (1977). Characteristic exponents and smooth ergodic theory. *Russian Mathematical Surveys*, **32**, 55–114.

Powers, R. T. (1970). UHF-algebras and their applications to representations of the anticommutation relations, volume 4 of *Cargèse Lecture Notes in Physics*, pages 137–68, New York. Gordon and Breach.

Powers, R. T. (1988). An index theory for semigroups of *-endomorphisms of $\mathcal{B}(\mathfrak{H})$ and type II$_1$ factors. *Canadian Journal of Mathematics*, **40**, 86–114.

Powers, R. T. and Størmer, E. (1970). Free states of the canonical anti-commutation relations. *Communications in Mathematical Physics*, **16**, 1–33.

Price, G. L. (1987). Shifts on type II$_1$ factors. *Canadian Journal of Mathematics*, **39**, 492–511.

Ruelle, D. (1978). An inequality for the entropy of differentiable maps. *Boletim da Sociedade Brasileira Matemática*, **9**, 83–7.

Sauvageot, J.-L. and Thouvenot, J.-P. (1992). Une nouvelle définition de l'entropie dynamique des systèmes non commutatifs. *Communications in Mathematical Physics*, **145**, 411–23.

Shale, D. and Stinespring, W.-F. (1964). States on the Clifford algebra. *Annals of Mathematics*, **80**, 365–81.

Sinai, Ya. (1996). A remark concerning the thermodynamic limit of the Lyapunov spectrum. *International Journal of Bifurcation and Chaos*, **6**, 1137–42.

Sinha, K. (1994). Quantum dynamical semigroups. volume 70 of *Operator Theory: Advances and Applications*, pages 161–9, Basel. Birkhäuser.

Słomczyński, W. S. and Życzkowski, K. (1994). Quantum chaos: An entropy approach. *Journal of Mathematical Physics*, **35**, 5674–99.

Stinespring, W. F. (1955). Positive functions on C*-algebras. *Proceedings of the American Mathematical Society*, **6**, 211–6.

Størmer, E. (1963). Positive linear maps on operator algebras. *Acta Mathematica*, **110**, 233–78.

Størmer, E. (1992). Entropy of some automorphisms of the II$_1$ factor of the free group in infinite number of generators. *Inventiones Mathematicae*, **110**, 63–73.

Størmer, E. and Voiculescu, D. (1990). Entropy of Bogoliubov automorphisms of the canonical anticommutation relations. *Communications in Mathematical Physics*, **133**, 521–42.

Umegaki, H. (1962). Conditional expectations in an operator algebra IV. *Kodai Mathematical Seminar Reports*, **14**, 59–85.

Vanheuverzwijn, P. (1977). Generators for quasi-free completely positive semigroups. *Annales de l'Institut Henri Poincaré*, **29**, 123–38.

Verbeure, A. and Zagrebnov, V. A. (1993). Quantum fluctuations in many-body problems, volume 70 of *Operator Theory: Advances and Applications*, pages 207–12, Basel. Birkhäuser.

Voiculescu, D. V. (1992). Dynamical approximation entropies and topological entropy in operator algebras. *Communications in Mathematical Physics*, **144**, 443–90.

von Neumann, J. (1931). Die Eindeutigkeit der Schrödingerschen Operatoren. *Mathematische Annalen*, **104**, 570–8.

Wehrl, A. (1978). General properties of entropy. *Reviews on Modern Physics*, **50**, 221–60.

Weidlich, W. and Haake, F. (1965). Coherence properties of the statistical operator in laser model. *Physik*, **185**, 30–47.

Werner, R. F. (1990). Remarks on a quantum state extension problem. *Letters in Mathematical Physics*, **19**, 319–26.

Werner, R. F. (1994). The classical limit of quantum theory. *Preprint Osnabrück*.

Wolfe, J. C. (1975). Free states and automorphisms of the Clifford algebra. *Communications in Mathematical Physics*, **45**, 53–8.

Woronowicz, S. (1989). Differential calculus on compact matrix pseudogroups. *Communications in Mathematical Physics*, **122**, 125–70.

Życzkowski, K. (1993). Parametric dynamics of quantum systems and transitions between ensembles of random matrices. *Acta Physica Polonica*, **24**, 967–1025.

INDEX

Lightning Source UK Ltd.
Milton Keynes UK
10 March 2010

151146UK00001B/148/P